ADVANCED TOPICS
IN SCIENCE AND TECHNOLOGY IN CHINA

ADVANCED TOPICS IN SCIENCE AND TECHNOLOGY IN CHINA

Zhejiang University is one of the leading universities in China. In Advanced Topics in Science and Technology in China, Zhejiang University Press and Springer jointly publish monographs by Chinese scholars and professors, as well as invited authors and editors from abroad who are outstanding experts and scholars in their fields. This series will be of interest to researchers, lecturers, and graduate students alike.

Advanced Topics in Science and Technology in China aims to present the latest and most cutting-edge theories, techniques, and methodologies in various research areas in China. It covers all disciplines in the fields of natural science and technology, including but not limited to, computer science, materials science, life sciences, engineering, environmental sciences, mathematics, and physics.

Guoping Zhang
Chengdao Li

Genetics and Improvement of Barley Malt Quality

With 63 figures

ZHEJIANG UNIVERSITY PRESS
浙江大学出版社

Springer

Editors
Prof. Guoping Zhang
Department of Agronomy
Zhejiang University
268 Kaixuan Road
Hangzhou, 310029, China
E-mail: zhanggp@zju.edu.cn

Prof. Chengdao Li
Department of Agriculture
Government of Western Australia
3 Baron-Hay Court
South Perth, WA6151, Australia
E-mail: cli@agric.wa.gov.au

ISSN 1995-6819 .e-ISSN 1995-6827
Advanced Topics in Science and Technology in China

ISBN 978-7-308-06382-1
Zhejiang University Press, Hangzhou

ISBN 978-3-642-01278-5 e-ISBN 978-3-642-01279-2
Springer Dordrecht Heidelberg London New York

Library of Congress Control Number: 2009930029

Cover design: Frido Steinen-Broo, EStudio Calamar, Spain

Printed on acid-free paper

Springer is a part of Springer Science+Business Media (www.springer.com)

Contributors

Rachel A. Burton — Australian Centre for Plant Functional Genomics, School of Agriculture, Food and Wine, University of Adelaide, Adelaide, Australia

Mehmet Cakir — Western Australian Agricultural Biotechnology Centre, Murdoch University, Perth, Australia

Helen M. Collins — Australian Centre for Plant Functional Genomics, School of Agriculture, Food and Wine, University of Adelaide, Adelaide, Australia

Arnis Druka — Scottish Crop Research Institute (SCRI), Dundee, Scotland, UK

Evan. Evans — Tasmanian Institute of Agricultural Research, University of Tasmania, Tasmania, Australia

Jason K. Eglinton — School of Agriculture, Food and Wine, University of Adelaide, Adelaide, Australia

Geoffrey B. Fincher — Australian Centre for Plant Functional Genomics, School of Agriculture, Food and Wine, University of Adelaide, Adelaide, Australia

Glen P. Fox — Department of Primary Industries & Fisheries, Queensland Grains Research Laboratory, Queensland, Australia

Xue Gong — Huazhong Agricultural University, Wuhan, China

Reg Lance — Department of Agriculture & Food, Government of Western Australia, Perth, Australia

Mike Jones — Murdoch University, Perth, Australia

Chengdao Li — Department of Agriculture & Food, Government of Western Australia, Perth, Australia

Dongfa Sun — Huazhong Agricultural University, Wuhan, China

Junmei Wang — Zhejiang Academy of Agricultural Sciences, Hangzhou, China

Robbie Waugh — Scottish Crop Research Institute (SCRI), Dundee, Scotland, UK

Guoping Zhang — Agronomy Department, Zhejiang University, Hangzhou, China

Meixue Zhou — Tasmanian Institute of Agricultural Research, University of Tasmania, Tasmania, Australia

Preface

Barley is ranked second in the industrial use of grain and is projected to total 27.4 million tonnes in 2008. A significantly high value use of barley is for malting to produce malt as a raw material for the production of beer. Obviously, malting quality is the primary objective of various barley research and breeding programs around the world as there is a premium for malting quality in many markets. Conventional approaches to the improvement of malting quality have been very successful in the last decade and this is especially true for Australia. Recently, releases of new malting barley varieties Baudin, Hamelin and Flagship resulted in a quantum leap in Australian malting barley quality. Knowledge of the physiological processes, the biochemistry of the pathways involved and the genetic control of these processes and their interaction with the environment has made a great contribution to the development of new malting barley varieties with excellent malting quality.

The world malting barley market keeps expanding constantly and it may be expected that the global demand for malt will increase by 27% to reach 19 million tonnes by 2010. At the same time, brewing style tends to be diversified significantly in order to meet different market requirements. As such, there are different quality requirements for different brewing styles. In addition to the traditional malting quality parameters, including malt extract, malt protein, soluble protein, Kolbach Index, free amino nitrogen, diastatic power, apparent attenuation limit, wort viscosity, alpha-amylase, beta glucanase and wort beta-glucan, new requirements for mouth-feel, flavor, haze, foam head retention and antioxidants have increased the complexity of malting quality improvement. It is one of the objectives for us in writing this book to address these challenges based on current knowledge.

The improvement in malting quality appears to have peaked in some countries. Lack of genetic variation in the current germplasm presents a major challenge for further quality improvement. Chapter 2 provides a comprehensive overview of the genus Hordeum, the relationships of the species among the genus, the origin and evolution of cultivated barley, and utilization of barley germplasm in breeding, including some successful examples and the potential for improvement of cultivated barley. The information about the Qinghai-Tibetan Plateau barley should be interesting to some readers. Diastatic power (DP) has long been used as an indicator for starch degrading enzymes and fermentability. Recently, a new DP enzyme analysis method was developed and a better model for prediction of fermentability was proposed.

In Chapter 6, a new malting paradigm was thus outlined where selection of the appropriate barley variety, growing location and malting conditions was described, providing the maltster sufficient latitude to produce malt within an appropriate range of fermentability characteristics to satisfy brewers varying requirements.

The endosperm cell wall can have direct and often detrimental effects on various processes in the brewery and on the quality and shelf life of the final beer. The same wall components can have indirect effects on other quality characteristics, such as malt extract and fermentability. Chapter 7 provides updated information on the chemical composition, the structures of individual constituents, synthesis and metabolism of cell walls. Most of genes/QTLs controlling malting quality have been mapped. Marker-assisted selection (MAS) has evolved from selection of single gene/traits toward multiple traits/loci and the whole genome. Simple and rapid DNA extraction methods as well as high-throughput MAS systems make marker assisted selection more efficient and cost-effective for malting barley breeding, as presented in Chapter 9.

Chapter 8 outlines how the new genomics data and approaches may enhance our understanding of malting quality. The chapter detailes two genomics platforms and how these technologies can potentially be applied to address malting quality traits. Both platforms are excellent examples of international collaboration in barley research. China is the world's largest and fastest growing beer market. It is also the largest malting barley importer. Australia is the largest malting barley supplier for China. This book is a symbol of collaboration between Australian and Chinese barley researchers. We deeply believe such collaboration to be not only of benefit for both countries but also to the whole barley industry around the world. We'd like to thank China's National Natural Science Foundation and the Australian Grain Research and Corporation for their support on the specific projects.

This book is divided into nine chapters, including barley production and consumption, germplasm and utilization, chemical composition, protein and protein components, carbohydrates and sugars, starch degrading enzymes, endosperm cell walls and malting quality, genomics and malting quality improvement, and marker-assisted selection for malting quality. We hope that the information present here will be especially useful to barley breeders, malsters, brewers, biochemists, barley quality specialists, molecular geneticists and biotechnologists. This book may also serve as reference text for postgraduate students and barley researchers. The authors for each chapter are the experts and frontier researchers in the specific areas. We appreciate their great efforts in providing updated information and sharing their specialist knowledge. We would like thank the authors for their outstanding and timely work in producing the chapters.

Guoping Zhang, Hangzhou, China
Chengdao Li, Perth, Australia
Feb. 28, 2009

Contents

1

Barley Production and Consumption

M. X. Zhou
Tasmanian Institute of Agricultural Research
University of Tasmania, P.O. Box 46, Kings Meadows
TAS 74249, Australia
Meixue.Zhou@dpiw.tas.gov.au

1.1 Introduction

Barley (*Hordeum vulgare*) is a very important grain in the world today. It ranks fourth in both quantity produced and in area of cultivation of cereal crops in the world. The annual world harvest of barley in the late century was approximately 140 million tonnes from about 55 million ha. It is very versatile in every way and has well adapted through its evolution. In fact, it is the most adaptable cereals. Much of the world's barley is produced in the regions where cereals such as maize and rice can not grow well. It extends into the arctic or subarctic. Some species approach the subtropical Zone. *Hordeum* species are found in most areas with Mediterranean climate. The genus is also represented in zones with an oceanic as well as a continental climate (Rasmusson 1985). Barley also has a very good resistance to dry heat compared to other small grains. This feature allows it to grow near desert areas such as North Africa.

Barley is a cereal. It belongs to the grass family. There are three types of barley: 1) *Hordeum vulgare*: a six-rowed type of barley that has a spike notched on opposite sides with three spikelets on each notch. At each notch, there is a flower or floret that later develops into a kernel. 2) *Hordeum distichum*: a two-rowed type of barley that has central florets producing kernels and it has lateral florets that are sterile. 3) *Hordeum irregulare*: the least cultivated, with fertile central florets and varying proportions of fertile and sterile lateral florets.

Barley has a very debatable origin. Egypt or China is most likely the centre for the origin of cultivated barley. Barley cultivation is also evident in other parts of the world at later times. For example, north-western Europe is estimated to have been cultivated barley around 3000 BC. Since barley is so old and records of trade were non-existent, the real answer for the origin of barley will probably never be known. The pathways for the domestication of barley also have some points in doubt. The earliest known and most probable ancestor of barley was the two-rowed variety. Most indications point to *spontaneum* as the most probably immediate ancestor of cultivated barley, and all the six-rowed forms are the results of accumulated mutations and hybridization. The ancestor of *spontaneum* was a form with more or less developed lateral spikelets, perhaps male-fertile like *H. murinum* and *H. bulbosum*. Barley cultivation probably originated in the highlands of Ethiopia and in Southeast Asia in prehistoric times. It is believed to extend back to 5000 BC in Egypt, 3500 BC in Mesopotamia, 3000 BC in north-western Europe, and 2000 BC in China. Barley was the chief bread plant of the Hebrews, Greeks, and Romans and most Europeans through the 16th century.

Barley has many economic uses today. Barley is produced primarily as animal feed. For example, over half of the barley is used for livestock feed in the United States. Barley as feed has the same nutritive value as corn. Barley is high in carbohydrates, with moderate amounts of protein, calcium and phosphorus. It also has small amounts of vitamin B. The entire barley kernel is used as feed after it has been steam rolled or gone through a grinding process. By products from the brewing process and malt sprouts are also used in livestock feed. The two-rowed barley is most often used for animal feed because it produces higher weight and superior kernel production. Barley is also used in the production of beer and some wines. About 25% of the cultivated barley in the United States is used for malting, with about 80% used in beer production, 14% used in distilled alcohol production, and 6% used for malt syrup, malted milk, and breakfast foods. A small amount of the barley is used for human food in the form of pearl barley or flour for porridge. Sometimes, barley is grown as a hay crop in some areas. Only the smooth varieties and awnless varieties are used in hay production. Winter barley also can be used for hay if pasteurized before the stems start to elongate. The amount of barley used for ethanol has been increased significantly in the last decade, especially in some EU countries (www.biofeulreview.com).

1.2 World Barley Production

Barley production traditionally has been important in the world. The total area harvested each year is around 50~80 million ha and ranked 4th after wheat (~200 million ha), rice (120~150 million ha) and corn (100~150 million ha) (Fig. 1.1). In the recent two decades, the area has been declining from more than 80 million ha to around 55 million ha. Among the major barley production countries, Russia and other countries from former USSR, United States, India, China are the major countries showing the significant decrease in barley cropping area (Fig. 1.2). The production area in the USSR has decreased from around 30 million ha in the 1980s to around 17 million ha recently with Russia being one of the major countries showing a significant decrease (from more than 16 million ha in the 1980s to around 10 million ha in the 2000s). In the United States, the production area has reduced from around 5 million ha in the 1960s to around 1.4 million ha recently. Barley was replaced by more commercially attractive corn and feed wheat in the feed mix. The Chinese barley production area has reduced from more than 5 million ha in the 1960s to less than 0.8 million ha today. India showed a similar trend to China, from more than 3 million ha to less than 0.8 million ha, but the decrease in the barley production area is mainly due to the significant increase in the wheat production area (from 13 million ha to 28 million ha) while the Chinese wheat production area has been relatively consistent or even reduced slightly. In contrast to the above major barley production countries, Australia showed a significant increase in production area, from 1 million ha in the 1960s to 4 million ha in the 2000s, mainly due to the total increased cropping area.

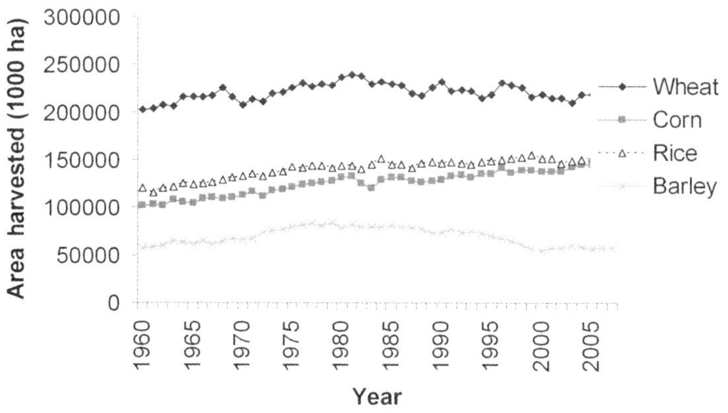

Fig. 1.1. World production of corn, rice, wheat and barley: area harvested

The average yield of the major cereal crops has been continuously increasing since the 1960s (Fig. 1.3). The increasing rate in the average yield for corn was much higher than that for other crops with the average yield of corn being increased by nearly 1.5 times (from 2.0 tonnes/ha in the 1960s to nearly 5 tonnes/ha in the 2000s) in the last 45 years. Even though the average yield of barley had increased from 1.4 tonnes/ha in the early 1960s to 2.4 tonnes/ha in the 2000s, compared to other crops barley showed a lower increasing rate in the average yield, mainly due to the fact that relatively less resource was allocated to barley variety improvement. The EU has a consistently higher yield than other countries. But China is closing the gap with the fastest increasing

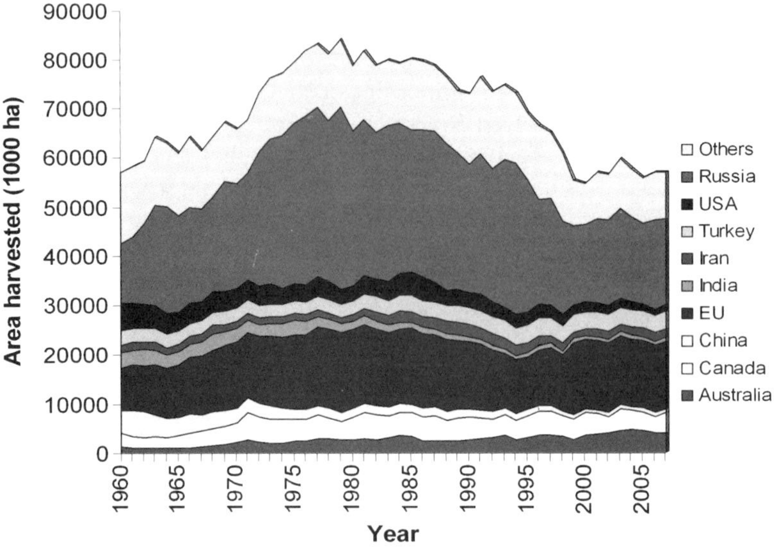

Fig. 1.2. World production of barley (including major barley production countries): area harvested

rate. The average yield of barley in China, increased by nearly 3 times in the last 45 years, from 1 tonnes/ha in the 1960s to 4 tonnes/ha in the 2000s (Fig. 1.4). The slight decrease in world barley yield in the last couple of years was caused by severe drought in some countries such as Australia with its yield dropping from 2.15 tonnes/ha in 2005 to 0.95 tonnes/ha in 2006 (Fig. 1.4). Of the other major barley production countries shown in Fig. 1.2, USA and Canada showed a slightly higher yield and increasing rate than the world average, while USSR, Turkey, Iran and India had similar yields to Australia and all below the world average.

World barley production increased from about 100 million tonnes in the 1960s to more than 170 million tonnes in the mid-1980s and decreased to 140 million tonnes in the 2000s. In contrast, the total production of wheat, rice and corn has been increasing continuously with production in the recent years, being nearly three times that in the early 1960s. For example, corn production was 200 million tonnes in 1960 and it has increased to 768 million tonnes in 2007 (Fig. 1.5). The increase in the total production of barley is mainly due to the increase in yield (Fig. 1.4) since the production area was relatively steady or even decreased in recent years (Figs. 1.1, 1.2). Among the major barley production countries, USSR, USA, India and China all showed a significant decrease in total barley production due to decreased production area (Fig. 1.2). In contrast, the total production in the EU, Turkey, Iran and Australia has been increasing and was quite steady in recent decades. The decreased

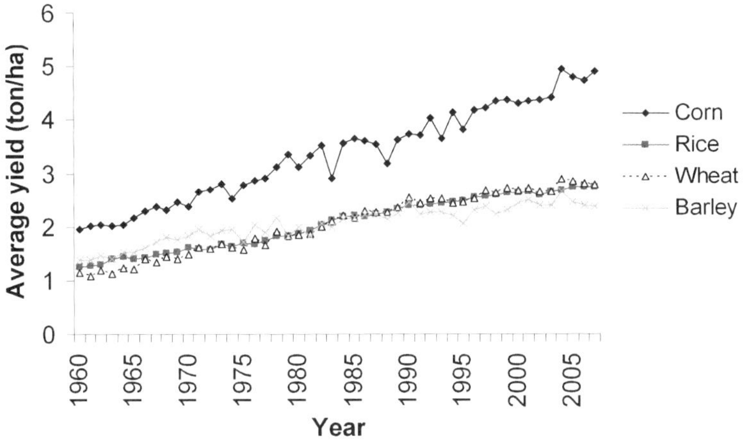

Fig. 1.3. Average yields of corn, rice, wheat and barley

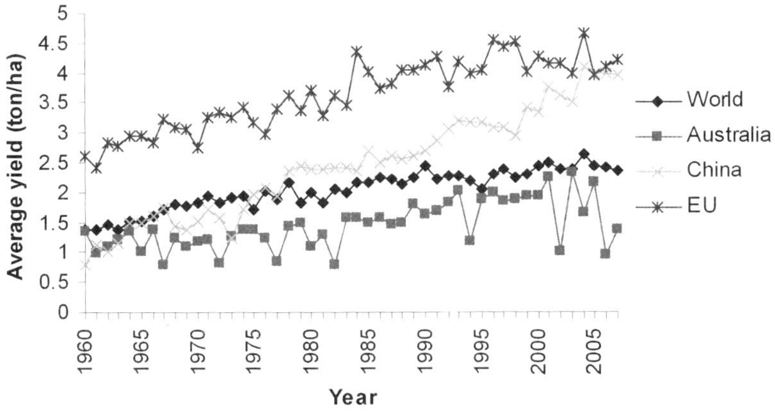

Fig. 1.4. Average yields of barley in Australia, China, EU countries and worldwide

production in Australia in 2002 and the last two years was mainly caused by severe drought (Fig. 1.6). Heat and dryness also contributed to the decrease of barley in the former USSR in 2007, especially in Ukraine which saw its total production reduced from 11.4 millions tonnes to 6.2 million tonnes.

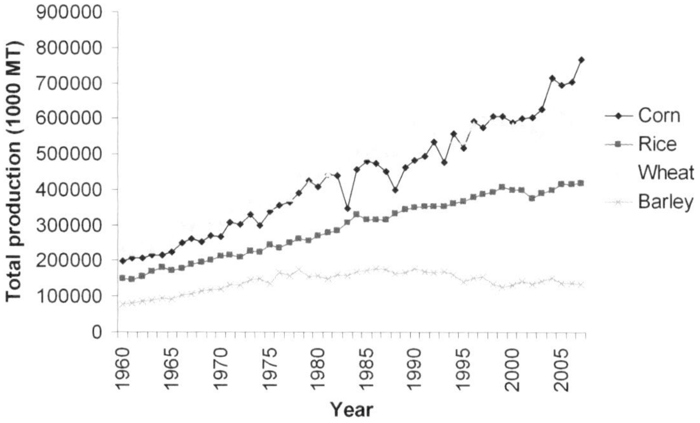

Fig. 1.5. World production of corn, rice, wheat and barley: total production

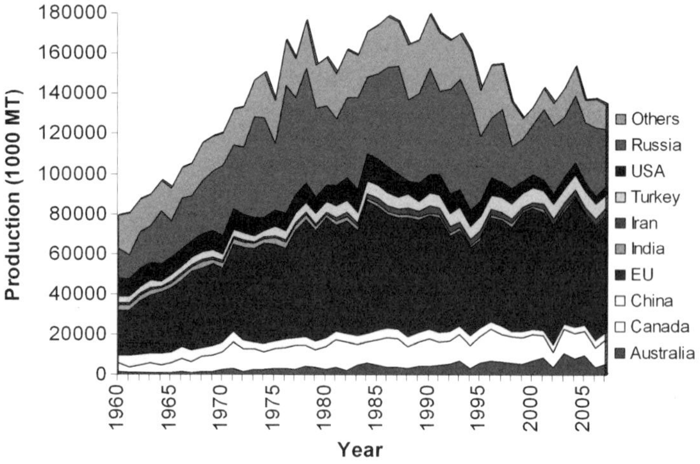

Fig. 1.6. World production of barley: total production, including major barley production countries

1.3 World Barley Trade

Barley trade accounted for about 7% of total production before the mid-1980s and increased to about 11% after the mid-1980s (Fig. 1.7). The total amount of barley traded had passed 20 million tonnes in 1990 and is now around 15 million to 18 million tonnes (Figs. 1.8, 1.9). The major exporters are EU countries, Australian, Canada, USA, and recently joined by Russia and former USSR countries which became the largest exporters in the last few years (Fig.

1.8). Saudi Arabia, China and Japan are currently the major barley importers. In the last two decades, the world barley trade market has been boosted by the huge increase in imports by Saudi Arabia, which comprises more than 40% of the total world barley imports. In 2005, Saudi Arabia imported nearly 7 million tonnes with 2 million tonnes from Ukraine, 1.4 million from Australia, 1.2 million from Germany and 1 million from Russia (USDA, GAIN report: SA6002). The EU countries used to be the major importers in the 1960s and 1970s but became the major exporters since the 1980s (Fig. 1.9). The significant decrease of world barley trade is mainly caused by the drought in the major exporting countries such as Australia. In 2006, its total exported barley falling from as much as 6 million tonnes in 2003 to 2 million tonnes (Fig. 1.8).

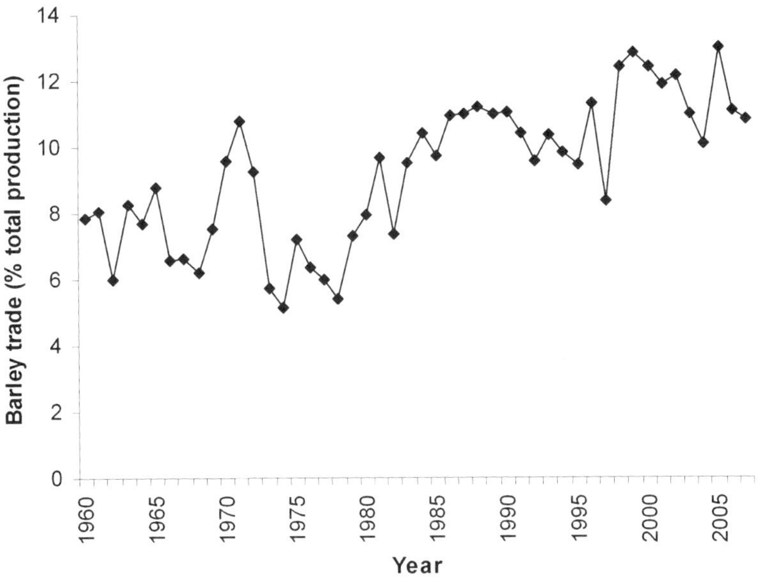

Fig. 1.7. World barley trade (average value of total imports and total exports)

1.4 World Barley Consumption

The total barley consumption in the world shows a similar trend to the total barley production (Figs. 1.6, 1.10). Barley is used commercially to manufacture malt, which is primarily used in beer production, for seed and human

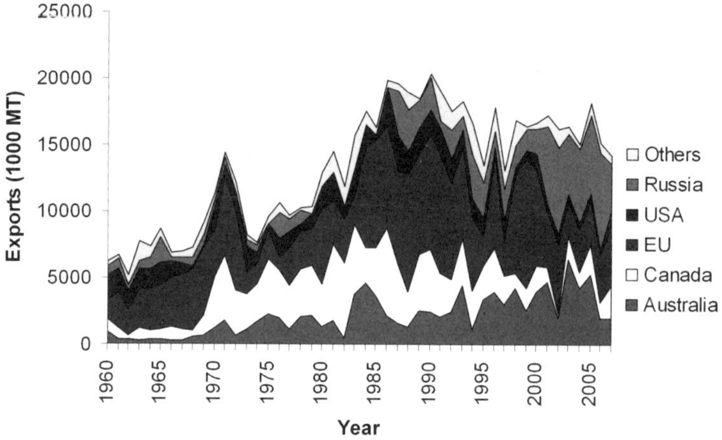

Fig. 1.8. World barley exports

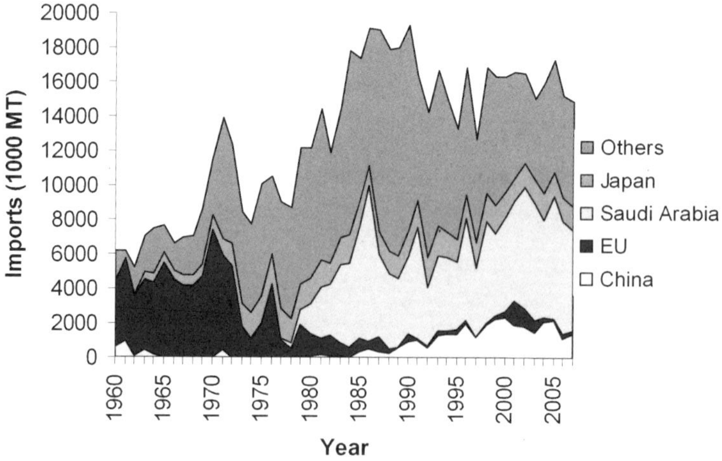

Fig. 1.9. World barley imports

food applications. Feed comprises about 70% of barley use. About 16% of barley is used for malting, seed or other industries, and 14% is used for food (Fig. 1.11). Most of the barley (more than 70%) consumed in China and the USA was used for malting production, while more than 85% of barley consumed in Canada is used as feed.

EU countries are the major barley consumers (Fig. 1.10). More than one third of world total barley was consumed in EU. Russia is another major barley consumer. It consumes nearly 20% of the barley. Other major consumers

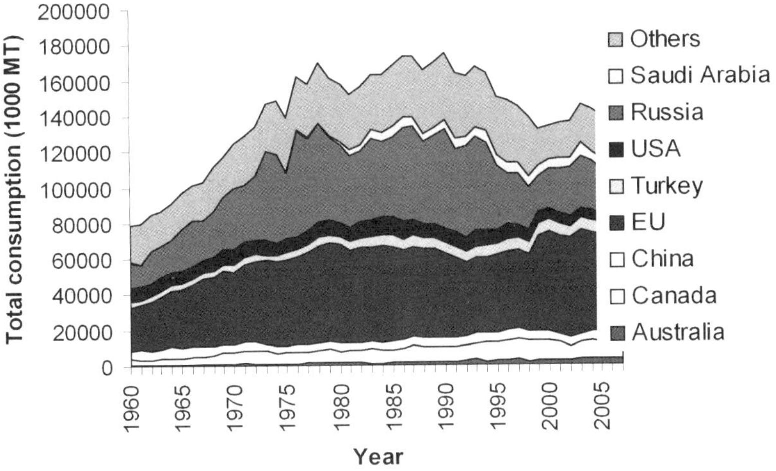

Fig. 1.10. World barley consumption

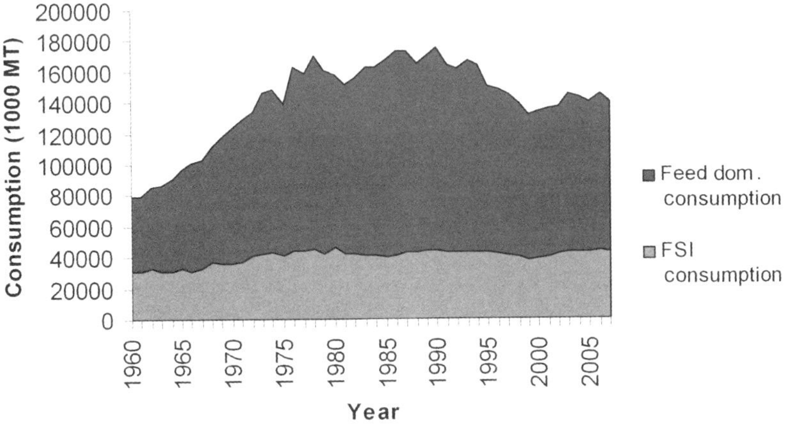

Fig. 1.11. World barley consumption

include Canada, Turkey, China, USA, Iran, Australia and India. In the last two decades Saudi Arabia showed a huge increase in the use of barley, from less than 0.1 million tonnes before the mid-1970s to more than 6 million tonnes since the mid-1980s. It becomes the fifth largest barley consumer after the EU, Russia, Canada and Turkey. Nearly all of the barley consumed in Saudi Arabia is for animal feed.

1.4.1 Feed Consumption

Barley is one of the four major feed grains (corn, barley, oats and wheat) and is widely used as a feed for livestock. The grain may be used as a major source of energy, protein and fibre for ruminants, and a major source of energy and protein for swine. In Australia, barley and wheat are the grains most common for Australian livestock industries which, when combined, represented around 60% of all cereal grains feed. Oats (20%), sorghum (10%) and triticale (10%) made up the other cereal grains used by the livestock industries. Approximately 40% of the barley was fed to feedlot cattle, 34% to dairy cows, 20% to pigs and 6% to grazing ruminants. Less than 1% was used for poultry. In the USA, corn is the most common feed grain, with the amount of tonnage being increased from 80 million in the 1960s to 140 million in the 2000s and comprising more than 90% of all cereal grain feed, followed by wheat, sorghum, and oats. The amount of barley used for feed has reduced from 5 million tonnes in the 1960s to 1 million tonnes in recent years.

The percentage of barley used for livestock feed in different countries varies from below 50% to as high as 90%. As shown in Fig. 1.12, EU countries and Russia are the major feed barley consumers, followed by Canada, Turkey and Saudi Arabia. Since the amount of barley used for food, seed and industrial consumption was relatively consistent, feed barley consumption was more closely related to the total barley production(Figs. 1.6, 1.12).

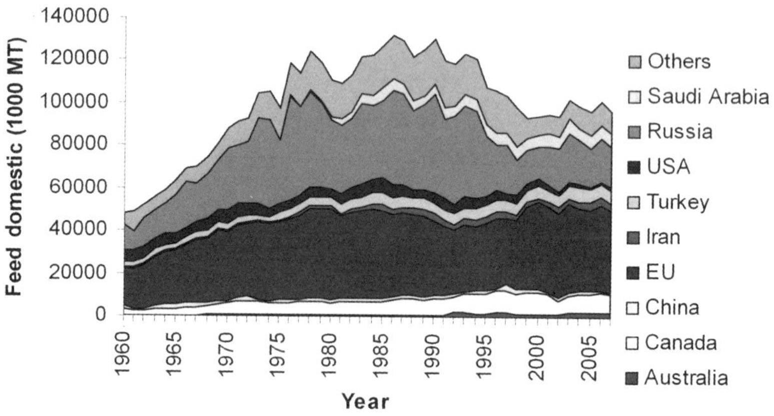

Fig. 1.12. World barley domestic consumption

1.4.2 Malting Barley Consumption

The second largest use of barley is for malt. Barley is the primary cereal used in production of malt. Historically this may have resulted from its availability compared to that of other cereals, but there are several sound reasons for its use. Although wheat and rye also produce α-amylase and β-amylase, which together are more efficient than either alone in hydrolysing starch, only barley has tightly cemented lemma and palea. These structures protect the embryo during grain handling as well as the coleoptile during malting, resulting in more uniform germination. In addition, the lemma and palea, or hulls, serves as an aid for filtering the brewing mash. Finally, the steeped barley kernel has a firmer consistency than wheat and rye, and it can be handled at high moisture levels with less risk of damage (Burger and LaBerge, 1985). The amount of barley for food, seed and industrial uses (FSI) is relatively consistent in the last thirty years compared to that for feed uses which largely depend on the total production (Fig. 1.13). More than half of the FSI consumption was for industrial uses. Barley ranked second in industrial use of grain, projected to a total of 27.4 million tonnes in 2008, against 26.5 million tonnes in the previous year (Fig. 1.14). Most industrially used barley is for malting. The total amount of barley for industrial uses is quite steady with only around 1%~3% annual increase, which is very small compared to the annual increase in the industrial uses of corn (more than 10% in recent years). However, it is expected that by 2010 global demand for malt is expected to increase by 27% to reach 19 million tonnes and world trade to expand by 8%. Most demand growth for malting barley and malt during the next five years will be in countries with rapidly expanding beer production, e.g. China (18% growth), Russia and Eastern Europe (17% growth), South America (17% growth) and Africa (21% growth) (http://www.pir.sa.gov.au/grains/grain_value_chains/malt_and_beverage).

The growth of malting barley demand has not been proportional to growth in beer production, especially in recent years. For example, in China the demand for barley malt is estimated at 2.62 million tonnes for the production of 291 Mhl of beer in 2004, while in 2000 2.64 MT of malt was needed for the production of 220 Mhl of beer. Two reasons are responsible for the lower usage of barley malt and malting barley. First, the substitution of an adjunct for barley malt has increased. Chinese breweries have the tradition of using rice or, to a less extent, corn as an adjunct in beer production. This creates a special taste favoured by local consumers. And at the same time, it reduces barley malt usage and input costs. In recent years when malting barley supplies were short, and malting barley prices were high relative to rice prices, breweries adjusted their production techniques to incorporate more rice in substitution for barley malt. In the last couple of years when rice prices increased more than malting barley prices, substantial amounts of corn and even grain syrups were used as a substitute for barley malt. Second, the original gravity of beer, defined as the amount of malt and adjunct as a percentage of water in wort, has decreased significantly from 11%~12% to 6%~7% in recent years. Thus more beer is produced from a given amount of malt and adjunct. Consequently, the ratio of barley malt to beer is estimated to have decreased from more than 13 kilograms of barley malt for one hectolitre of beer (kg/Hl) in the 1980s to 12 kg/Hl in the 1990s and 9 kg/Hl over the last four years (Agriculture and Agri-Food Canada, 2005).

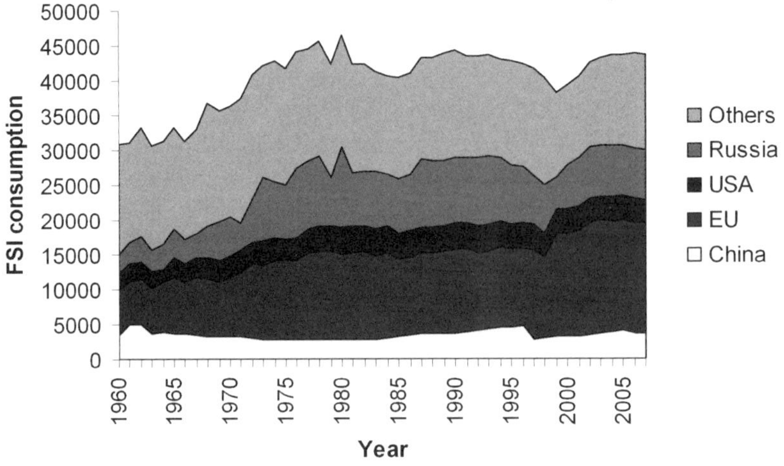

Fig. 1.13. World barley food, seed and industrial consumption

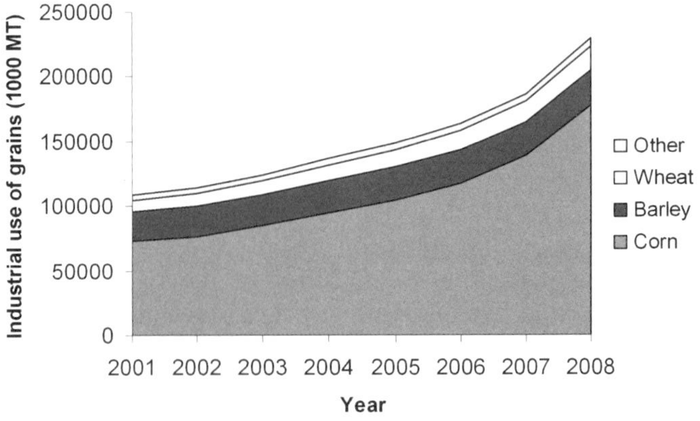

Fig. 1.14. World grain industrial consumption

1.4.3 Food Consumption

Barley is still a major staple food in several regions of the world: some areas of North Africa and the Near East, in highlands of Central Asia, the Horn of Africa, the Andean countries and the Baltic States. These regions are characterised by harsh living conditions and are home to some of the poorest farmers in the world who depend on low productive systems. At present very little barley is used as human food in developed countries. However, in the last two decades we have seen a rediscovery of food preparations with barley with a significant increase in the total amount of barley used for food, from 17.3 million tonnes in 1990 to 23.7 million tonnes in 2005 (Fig. 1.15). China has the biggest increase in barley used for food, from 2 million tonnes in 1990 to 4 million tonnes in 2005. Other countries saw a significant increase in the food use of barley including Russia, Ethiopia and Saudi Arabia, from nothing in 1990 to more than 1 million tonnes in 2005.

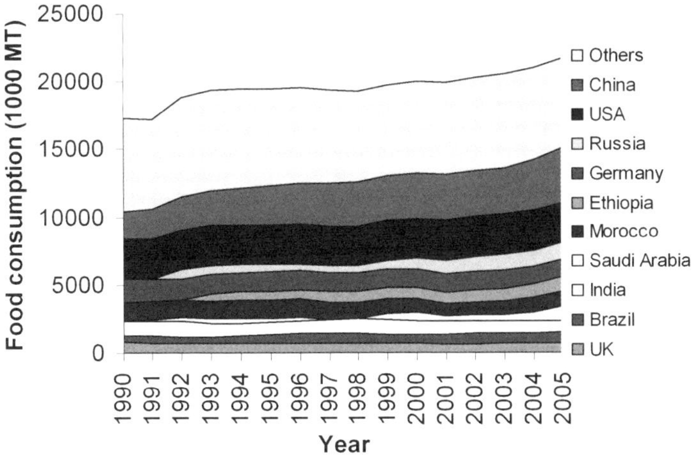

Fig. 1.15. World barley food consumption

From the FAO statistics in 2005 (Fig. 1.15), China is the largest consumer of barley as food (4.0 million tonnes), followed by USA (2.9 million tonnes), Russia, Germany, Morocco, Ethiopia and Saudi Arabia (1 million tonnes to 1.3 million tonnes). In China most of the barley was consumed in Tibet, where it accounts for 56% of the total food production, and about 2.1 million people consume barley. The main product of hull-less barley is a roasted barley flour known as *Tsangpa*;*Chang* brewed from hull-less barley is the major alcoholic beverage. Barley is also used in many other countries in traditional dishes such as *Kasha* in Russia and Poland, *miso* in Japan, and *sattu* or popped barley in India. In Korea barley is the second most important food crop after rice. In the West Asia-North Africa much of the barley is consumed as pearled grain in soups, flour in flat bread, and ground grain in cooked porridge. In Western countries small quantities of barley are used in breakfast cereals, soups, stews, porridge, bakery blends, and for baby foods.

Barley products utilized for traditional food preparations can be classified as: a) Whole grain; b) Cracked grain; c) Raw-grain flour (fine and coarse); d) Whole roasted grain; e) Roasted-grain flour (fine and coarse). Barley products are suitable in many food preparations, including different types of bread, pasta, rice extender, and baby foods, although most of the use has been largely confined to pot or pearled barley in soup and to flakes in breakfast cereals. Roasted barley can be used as coffee-substitute. 'Barley coffee' is very popular in Europe. In Italy, *Caffe d'orzo*, it is commonly used as a breakfast drink for children, often mixed with milk. Barley is eaten in soups or as porridge after it has been popped, blocked, or pearled and is sometimes used as a rice extender, which needs a longer cooking time than rice.

Among the 10 major consumers shown in Fig. 1.15, Saudi Arabia and Morocco have the highest average consumption (more than 35 kg per person), followed by Germany, Ethiopia (around 14 kg per person), UK, USA, Russia (about 10 kg per person), Brazil, China (3~4 kg per person). India has the lowest average annual consumption (0.7 kg per person).

There has been a renewed interest in food barley as healthy food in general and as a source of soluble fibre implicated in hypocholesterolemia and hypoglycaemia in non-insulin-dependent diabetes. The β-glucan content of whole-grain barley is equivalent to or greater than that found in whole-grain oats. In a weight-for-weight comparison, whole-grain barley is lower in fat, protein and calories, and higher in total dietary fibre than whole-grain oats. A comprehensive review of the scientific evidence suggests that increasing whole-grain consumption can reduce the risk of coronary heart disease and diabetes, and keep weight maintenance (http://www.health.gov/dietaryguidelines/dga2005/document/pdf).

The potential benefit of soluble dietary fibre such as the β-glucan in lowering the cholesterol level and postprandial blood glucose and insulin response has been reported by several authors (Brennan and Cleary, 2005). Tocols (tocopherols and tocotrienols) are also reported to lower the total cholesterol and the low density lipoprotein cholesterol (Wang *et al.*, 1993). Barley grain is a good source of both β-glucan and tocols.

1.4.4 Other Industrial Uses

Barley native starch and starch derivatives are extensively used in the food industry. The waxy starches from barley cultivars containing waxy or glutenous and high-amylose starches have special uses due to their high swelling power and colloidal stability (Greenwood, 1988). Barley has been used in some countries for the production of glucose, maltose syrups and beta-amylase. High-maltose syrup has special properties for use in food and pharmaceutical industries. β-glucans can be extracted and purified from barley or its β-glucan-enriched bran fraction (Bhatty, 1993). Isolated β-glucan has potential as a thickening agent in foods, as industrial hydrocolloid, and in pharmaceuticals. Hydrolyzed β-glucan has been suggested as a bulking agent for replacement of sucrose (Cowan and Mollgaard, 1988).

The total grain used for various industries has been increased significantly in the last several years, from 124.6 million tonnes in 2002 to 186 million tonnes in 2007 (Fig. 1.14, Table 1.1). The major increase was in the amount of grain used for biofuel production. The production of biodiesel has nearly doubled in the EU in the last 12 months, from 3.2 million tonnes in 2005 to 6.7 million tonnes in 2007 (Lyndon, 2007). In 2006, the US became the largest ethanol producer, turning out 20 billion litres, an increase of nearly 25% from the year before, and overtake Brazil, which produced an estimated 17 billion litres (Sosland, 2007). The ethanol production is expected to reach 6 billion gallons in the US in 2007 (Donley, 2007). Currently, China produces more than 1 million tonnes of ethanol and the new project would increase fuel ethanol production by more than 10 million tonnes (Biofeuls news digest, 2007). In contrast to the US, where most ethanol is produced from corn, EU countries use a wider range of materials, with half from cereal grains, mainly wheat, corn, barley and rye. Barley used for bioethanol production would represent 25% of the total grain production in Finland and 29% of barley production (www.biofeulreview.com). However, since regular hulled barley cannot be converted to fuel ethanol using a conventional corn-to-ethanol process without significant modifications, the use of this feedstock is problematic and current processes for conversion to fuel ethanol are not cost competitive today in the U.S. versus the use of corn. The abrasive nature of hulled barley, the high viscosity of barley fermentations, and the low starch and high fibre content lead to high production costs, low ethanol yields, and a distillers dried grains with soluble (DDGS) product that can't be fed to monogastric animals. A multidisciplinary research effort at the Eastern Regional Research Center, ARS, USDA in Wyndmoor PA, in cooperation with research partners, has been initiated to solve these technical problems. The research approaches include: 1) development of high-starch, hulless barley varieties specifically bred for ethanol production; 2) development of pre-fermentation barley fractionation processes to remove non-fermentables from hulled and hulless barley to produce a starch-rich feedstock for ethanol production; 3) use of new beta-glucanases to decrease viscosity of mashes, increase yields of ethanol, and decrease levels of residual β-glucans in DDGS.

There are lots of potential uses for the whole-crop of barley. Petersen and Munck (1993) illustrated the possible industrial uses of whole barley plant, which includes barley grain and straw. The industrial uses of barley grain have been discussed above. Barley straw can be used as ruminant feed. Straw can be chemically treated to improve the feeding value, in which treatments with ammonia or sodium hydroxide are commonly used. The treatment with ammonia increases the content of crude protein in addition to the digestibil-

Table 1.1. Global industrial use of grain (million tonnes) (Sosland, 2007)

	2002-2003	2003-2004	2004-2005	2005-2006	2006-2007
Ethanol	29.4	34.9	41.0	50.2	65.3
Starch	61.8	64.9	68.2	72.8	79.0
Brewing	27.6	29.2	29.7	30.7	31.3
Other	6.7	8.0	9.1	10.1	10.3
Total	124.6	137.0	147.9	163.6	186.0

ity (Petersen and Munck, 1993). Before the 19th century, most paper was produced from annual plants. A few straw-based cellulose factories still exist. Most of them are in the developing countries. According to Petersen and Munck (1993), 1 ha of barley crop producing 5 tonnes of grain and 5 tonnes of straw would potentially yield 3.1 tonnes of starch and 1.6 tonnes of cellulose. The yields of paper pulp would be about 2.4 tonnes. The internode chips of barley straw are ideal for the production of particle boards because of their high physical strength properties (Petersen and Munck, 1993).

References

Agriculture and Agri-Food Canada (2005) China: Beer and Malting Barley. Biweekly Bulletin, 18(10)

American Society of Agronomy, Crop Science Society of America, Soil Science Society of America, Publishers, Madison, Wisconsin D C, 367-401

Bhatty R S (1993) Extraction and Enrichment of (1→3), (1→4)-β-D-glucan from Barley and Oat Brans. Cereal Chem, 70: 397-402

Biofeuls News Digest (2007) World Grain, 25(1): 38-42

Brennan C S, Cleary L J (2005) The Potential Use of Cereal (1→3, 1→4)-β-D-glucans as Functional Food Ingredients. J Cereal Sci, 42: 1-13

Burger W C, LaBerge D E (1985) Malting and Brewing Quality. In: D C Rasmusson (eds.) Barley. Madison, Wisconsin: American Society of Agronomy, 367-401

Cowan W D, Mollgaard A (1988) Alternative uses of Barley Components in the Food and Feed Industries. In: Sparrow R C M, Lance, Henry R J (eds.) Alternative End Uses of Barley. DHB. Waite Agricultural Research Institute, Glen Osmond, Australia, 35-41

Donley A (2007) A New Way to Produce Ethanol. World Grain, 25(1): 33-35

Greenwood C T (1988) Barley Starches-basic Concepts. In: Sparrow R C M, Lance, Henry R J (eds.) Alternative End Uses of Barley. DHB. Waite Agricultural Research Institute, Glen Osmond, Australia, 67-72

Lyddon C (2007) Europe Races to Produce More "Green" Fuel. World Grain, 25(1): 28-32

Petersen P B, Munck L (1993) Whole-crop Utilization of Barley, Including Potential New Uses. In: MacGregor A W, Bhatty R S (eds.) Barley: Chemistry and Technology. American Association of Cereal Chemists Inc. St Paul, Minnesota, USA, 437-474

Sosland M I (2007a) Industrial Strength. World Grain, 24(8): 48-50

Sosland M I (2007b) Ethanol Powers Expansion in Industrial Use of Grains. World Grain, 25(8): 32-35

Wang L, Xue Q, Newman R K, Newman C W (1993) Enrichment of Tocopherol, Tocotrienol, and Oil in Barley by Milling and Pearling. Cereal Chem, 70(5): 499-501

2

Barley Germplasm and Utilization

D. F. Sun and X. Gong
Huazhong Agricultural University
Wuhan, China
sundongfa1@mail.hzau.edu.cn

Germplasm, genetic resources and gene resources are similar glossaries in crop breeding and genetics, which include cultivated plants, wild plants and any other life forms used in crop breeding. Crop breeders utilize selected gene resources for combination of new variations or defined agronomic traits like disease resistance commercial varieties. Novel germplasm are becoming increasingly important for developing crop varieities with high yield, quality and abiotic or biotic resistence. In this respect, original varieties, landraces and in particular, wild relatives of crops have been receiving increasing attention as gene donors. Earlier difficulties of sexual isolation in gene transfer using conventional methods can now be overcome by the development of new biological techniques which also make more distantly related species accessible for gene transfer. Cultivated barley, *Hordeum vulgare* L., one of the major cereals planted in the world, is a founder crop of the Old World Neolithic food production and one of the earliest crops domesticated. It is an important crop, ranking fifth in world crop production. In order of importance, barley is used for animal feed, brewing malts and human food. Barley is a short season, early maturing cereal with high yield potential. It may be found in widely varying environments, including extremes of latitude and altitude where other crops are not adapted. The number of barley germplasm recorded is up to 370,000, the second largest after wheat (von Hintum, 2000). This chapter aims to give a comprehensive overview of the genus *Hordeum*, relationships of the species, origin and evolution of cultivated barley, and utilization of barley germplasm in breeding programs, including some successful examples and potential for improvement of cultivated barley.

2.1 Origin and Evolution of Cultivated Barley

Cultivated barley, *H. vulgare*, a predominantly self-pollinated diploid crop ($2n=14$), including two-rowed and six-rowed barley, belongs to the tribe Triticeae in the grass family, Poaceae. Triticeae is a big plant group which contains economically important cereals, forages and about 350 wild species. Due to great possibilities for interspecific and intergeneric hybridization, all of these are of putative interest as gene donors in breeding.

2.1.1 The Taxonomy of Barley

The common temperate annual cereal crops, wheat (*Triticum aestivum*), rye (*Secale cereale*) and triticale, are also placed in this tribe together with several very important perennial forage grasses in many temperate areas of the world, such as Russian wild rye (*Psathyrostachys fragilis*), the crested wheatgrasses (*Agropyron cristatum* s.l.), Altai wild rye (*Leymus angustus*) and intermediate wheatgrass (*Thinopyrum intermedium*). Triticeae is found in most areas of the world, but the major center is in central and southwestern Asia. With the development of biological techniques and the increase basic research in breeding and genetic engineering on their taxonomy, distribution, variation patterns, species relationships, and cytogenetic behavior in Triticeae. Some genera in Triticeae are fairly well known, especially the species of *Aegilop, Triticum and Hordeum.* Most scientists involved in research of the Triticeae have noticed that conflicting opinions on generic circumscriptions have caused major disagreements among the taxonomists over the last couple of decades, and so far no universal acceptance of generic boundaries has been achieved. The genetic delimitation of *Hordeum* has been rather stable over the years and most scientists agree on the species content of the genus although there are some conflicting opinions. Some scientists suggest separating cultivated barley, *H. vulgare*, together with *H. bulbosum* into a genus of its own (*Hordeum* in a narrow sense) and all other species would then be referred to a separate genus, *Critesion* (Dewey, 1984). This view has not been widely accepted because it does not reflect the actual relationships of these species. Combining the research of *Hordeum* and the traditional view of the genus being widely accepted in this presentation, Bothmer(1992) considered that *Hordeum* can be divided into about 30 species or 45 taxa (Table 2.1).

Table 2.1 The species/taxa of Hordeum. The basic chromosome number is x=7. A, annual; P, Perennial* H. arizonicum is sometimes biennial or even with a shorter life cycle

Species/taxon	Ploidy level	Distribution	Life form
H. vulgare L. subsp. *vulgare*	2x(4x)	Cultivated	A
subsp. *spontaneum* (Koch) Thell	2x	SW Asia	A
H. bulbosurn L.	2x, 4x	Mediterranean	P
H. murinum L. subsp. *murinum*	4x	Mediterranean	A
subsp. *glaucum* (Steud.) Tzvelev	2x	Mediterranean	A
subsp. *leporinum* (Link) Arc	4x,6x	Mediterranean	A
H. marinum Huds. subsp. *marinum*	2x	Mediterranean	A
subsp. *gussoneanum* (Parl) Thell	2x,4x	Mediterranean	A
H. muticum Presl	2x	South America	P
H. cordobense Bothmer *et al.*,	2x	South America	P
H. chilense Roem Schult	2x	South America	P
H. stenostachys Godr	2x	South America	P
H. erectifolium Bothmer et al.,	2x	South America	P
H. flexuosum Steud	2x	South America	P
H. euclaston Steud	2x	South America	A
H. intercedens Nevski	2x	W USA	A
H. pusillum Nutt	2x	North America	A
H. jubatum L	4x	North America-East Asia	P
H. comosum Presl	2x	South America	P
H. pubiflorum Hook. f	2x	South America	P
H. lechler (Steud) Schenk	6x	South America	P
H. procerum Nevski	6x	South America	P
H. arizonicum Covas & Stebbins	6x	SW USA	A/P*
H. secalinum Schreb	4x	Europe-(North Africa)	P
H. bogdanii Wil	2x	Central Asia	P
H. roshevitzii Bowden	2x	Central Asia	P
H. brevisubulatum (Trin.) Link	2x,4x	E Asia	P
subsp. *brevisubulatum*	2x,4x	SW Asia	P
subsp. *violaceum* (Boiss & Hohen) Tzvelev	2x,4x	E Asia	P

(to be continued)

Species/taxon	Ploidy level	Distribution	Life form
subsp. *turkestanicum* (Nevski) Tzvelev	6x	Central Asia	P
subsp. *nevskianum* (Bowden) Tzvelev	4x	Central Asia	P
subsp. *iranicum* Bothmer	6x	W Asia	P
H. brachyantherum Nevski	4x,6x	W North America	P
H. californicum Covas & Stebbins	2x	W USA	P
H. depressum (Scribn & Sm) Rydb	4x	W USA	A
H. guatemalense Bothmer *et al.*,	4x	Central America	P
H. capense Thunb	4x	South Africa	P
H. parodii Covas	6x	South Africa	P
H. tetraploidum Covas	4x	South America	P
H. fuegianum Bothmer *et al.*,	4x	South America	P
H. pataonicum (Haum.) Covas subsp. *Patagonicum*	2x	South America	P
subsp. *mustersii* (Nicora) Bothmer *et al.*,	2x	South America	P
subsp. *santacrucense* (Parodi & Nicora) Bothmer *et al.*,	2x	South America	P
subsp. *setifolium* (Parodi & Nicora) Bothmer *et al.*,	2x	South America	P
subsp. *magellanicum* (Parodi & Nicora) Bothmer *et al.*,	2x	South America	P

*Cited from Bothmer, 1992

Wide crosses between *Hordeum* species showed that the intraspecific hybrids were in general fully fertile. For example between *H. marinum* and *H. brevisubulatum*, *H. patagonicum* (Landström *et al.*, 1984; Bothmer *et al.*, 1986b, 1989a), and all interspecific hybridizations were sterile for both seed and pollen except the tetraploid hybridization *H. jubatum* × *H. brachyantberum*, and the hexaploid hybridization *H. lechleri* × *H. procerum* which showed occasional seed set. Based on meiotic observations of Hordeum hybrids, *H. vulgare* and *H. bulbosum* are considered to share the common basic genome, called I (Kasha and Sadasivaiah, 1971), as the hybrid of the two species has high meiotic pairing. In general, this genome is considered to be not correlated with other genomes in *Hordeum*, because *H. bulbosum* occurs in two cytotypes , a diploid in the western part of the Mediterranean and a tetraploid in the eastern part (Jørgensen, 1982). The 4x cytotype is an autotetraploid shown by the high frequency of multivalents.

H. marinum (diploid and tetraploid) and *H. murinum* (diploid, tetraploid and hexaploid), the two annual Mediterranean species, are distantly related to each other and to the I genome species (*H. vulgare* and *H. bulbosum*), as well as to the other *Hordeum species* (Bothmer *et al.*, 1989a, b). Primarily, the genomes of the two annuals have been designated as "X" (*H. marinum*) and " Y" (*H. murinum*) by Bothmer *et al.* (1986c, 1987) because they do not to be closely related to the other genomes in Hordeum. Based on later research, the two species would be considered as the autoploid from the study of karyotypes and C-bands (Linde-Laursen and Bothmer, 1989), and isoenzyme variation (Jaaska and Jaaska, 1986), which showed differentiation within the genus (i.e. an alloploid origin), and from the cytogenetic studies, which showed a close relationship between the genomes of the polyploid cytotypes within *H. marinum* and *H. murinum*, respectively (Bothmer *et al.*, 1989a, b).

The general pattern of the relationships revealed by the genome studies is mainly confirmed by studies of the isoenzyme variation (Jørgensen, 1992), chloroplast DNA (Doebley *et al.*, 1992), C-banding patterns (Linde-Laursen *et al.*, 1980, 1989, 1990), and repeated sequences of nuclear DNA (Gupta *et al.*, 1989). The remaining taxa in Table 2.1 are all closely related, sharing the same basic genome (H), which is somewhat differentiated between American and Asiatic taxa (Bothmer *et al.*, 1986c). Nearly all the polyploids are segmental alloploids in which the genomes are closely related. The H genome and its differentiations constitute the basis (Bothmer *et al.*, 1989b, c). *H. brevisubulatum* is a polyploid complex, where the 4x and 6x cytotypes are of a true autoploid nature (Bothmer *et al.*, 1989b, c).

Based on the taxonomical research, Xu (1993) proposed a system for *H. vulgare* classification, in which *H. vulgare* can be divided into five sub-species, i.e subsp. *spontaneum* (the ear-uranium being brittle in mature and all side-spikelets being sterile), subsp. *agriocrithon* (the ear-uranium being brittle in mature and all side-spikelets being fertile), subsp. *distichon* (the ear-uranium being not brittle in mature and all side-spikelets being sterile), subsp. *vulgare* (the ear-uranium being not brittle in mature and all side-spikelets being fertile) and subsp. *intermedium* (the ear-uranium being brittle in mature and all side-spikelets being partly fertile).

2.1.2 Origin of the Cultivated Barley

Most scientists involved in studies of the origin and evolution of cultivated barley have noticed that conflicting opinions on the primary center of origin and evolution have caused major disagreements among scientists. Vavilov (1926, 1951) and Harlan (1968, 1971) discussed the problems involved in the centers of origin of cultivated barley in the Near East, China, Africa and America. Concerning ambiguities about the definition and boundaries of centers of origin, the primary center of origin and secondary center of origin are not clear up to now, and no universal verdict has been achieved. The following geographical areas are possible centers of origin of cultivated barley.

- The Near East Fertile Crescent Center:

This center comprises Asia-Minor, Caucasia, Iraq, Turkey, Jordan, Syria, Palestine, Lebanon and some Mideastan countries which are called "The Fertile Crescent". Vavilov (1926, 1951) proposed that "The Fertile Crescent" was the center of origin of *H. distichon* and *H. nutans*. The current view is that the earliest agriculture including barley started about 10,000 years ago in the area of the southwestern Asia called "The Fertile Crescent" (Harlan, 1968; Zohary, 1973). The main supporting evidence is as follows: (1) The Near East has centuries of cultural history in which settled agricultural society appeared about 10,000 years ago with the planting of cereal and crops and breeding of livestock; (2) The immediate progenitor of the crop, i.e. the two-rowed form *H. vulgare* subsp. *spontaneum*, is conspecific to cultivated barley (*H. vulgare* subsp. *vulgare*) and is still growing in this area in natural habitats as well as in ruderate ground (Nevo, 1992). This area is the primary center of origin of perennial wild barley species *H. sibericum*, *H. violaceum*, *H. turkestanieum* and *H. bogdani* (Rajhathy, 1964); (3) Archaeological studies support this, such as the carbide of grains and spikelets of barley before 7,000 B.C., found in Jarmo of Iraq, which seems like a type between annual wild barley subsp. *spontaneum* and cultivated barley subsp. *vulgare*, the carbide of barley before 8,000 B.C., found in Mureybat of Syria (Braidwood, 1958; Helbaek, 1959, cited by Xu, 1993), and so on.

- The Chinese Center:

This center comprises the mountainous area in the middle and west of China, and the area abutting China of India, Afghanistan, Pakistan. Nepal. Vavilov (1926, 1951) considered that China was the center of origin of the naked six-rowed cultivar of barley *H. hexastichon*. According to Xu (1987), China is the center of cultivated barley planted in China and Eastern nations. The supporting evidence is as follows: (1) China has a very long history of agricultural culture with settled agricultural society appearing over 10,000 years ago and the planting 5 cereal crops including barley; (2) The immediate progenitor of the six-rowed form *H. vulgare* subsp. *agriocrithon*, the two-rowed form *H. vulgare* subsp. *spontaneum*, and the intermedium *H. vulgare* subsp. *intermedium* are all found in Qing-Tibetan Plateau (Åberg, 1938; Friesleben, 1943; Schiemann, 1951; Shao, 1975, 1982; cited by Xu, 1993). The genetic diversity of annual wild barley in China is very rich. Currently 418 accessions of annual wild barley have been identified; (3) Archaeological studies also support this assumption, such as the carbide of grains and spikelets of barley before 5,000 B.C., found in Gansu, China, and they are similar to six-rowed naked cultivated barley; (4) A new species found in Inner-Mongolia, *H. innermongolicum*, which belongs to section biflora, as its middle spikelets often has a second flower, is being considered the oldest sect in *Hordeum* (Xu, 1993). The discovery of new species provided new evidence supporting the view that China is the original center of barley. Some scientists suggested that Inner-Mongolia is the original center of *Hordeum*, because the area is an ancient continent in which the species of *Hordeum* remained from antiquity (Xu, 1993). From this center, *Hordeum* was passed to America through Siberia and the Bering Strait, and to Europe and Africa through the Near East and the Middle East. After analyzing the differentiation and geographical distribution of barley germplasm based on RAPD markers, Strelchenkol *et al.* (1999) concluded that the westward distribution of barley is from west-Asia to Europe and the New World across Ethiopia and then the Mediterranean region.

- The African Center:

This center includes the northern and eastern nations of Africa, such as Ethiopia, Egypt, Tunis and Morocco. Ethiopia has been considered as the center of genetic diversity of barley (Xu, 1993), in which 25,000 barley accessions have been collected and many of them proved different from the barley collection from other areas in resistance, protein content and other agronomic traits. In addition, Egypt enjoys a long history in agricultural activities. Barley had been an important crop in the pyramid epoch in Egypt. The carbide of grains and spikelets of barley before 4,500 B.C., found in Foyum in Egypt, demonstrate that the barley in very early days in Egypt is similar to that in Ethiopia, and is significantly different from modern barley (Xu, 1993). In Tunis and Morocco, there exist some *H. vulgare* subsp. *spontaneum* plants, not found in Egypt. In Africa no perennial wild barley species have been found.

- The South American Center:

This center includes Argentina, Chile, Peru and Urugury. Zhukovsky suggested that South America was one of the original centers of *Hordeum*, in which some perennial wild barley species could be found, such as *H. comosum* found in Argentina and Chile, *H. chilense* found in Chile, *H. lechleri*, *H. procerum*, *H. stenostachys*, and *H. compressum* found in Argentina, *H. muticum* found in Peru (cited by Xu, 1993). Moreover, some annual wild barley species also exist in the area, such as *H. euclaston* found in Argentina. However, there are some conflicting opinions about whether South America is the original center of *Hordeum*. Bothmer (1985, cited by Xu, 1993) considered that South America and South-west Asia were the two original centers of Hordeum based on the fact that the continental perennial wild barley species *H. lechleri* and *H. pubiflorum*, containing many special ecotypes, could be found in these areas. The results from the research of cytogenetics and hybridization between species show that the three hexaploid perennial wild barley species on the continent, i.e. *H. arizanicum*, *H. lechleri* and *H. hexaploidum* , all have the same genome as that of *H. jubatum*.

2.1.3 Evolution and Domestication of Cultivated Barley

There is an inconsistent viewpoint on the evolution and domestication of cultivated barley. The controversy started in 1938 when Åberg found annual wild six-rowed barley, *Hordeum agriocrithon* in China. During the past 60 years a huge number of studies have been done in barley evolution and domistication, including ancestry and pre-domesticated genetic polymorphisms (Brown *et al.*, 1978a), distinct morphogenetic identity (Baum, 1983; Chakrabarti and Subrahmanyan, 1986; Jaradat, 1989a, b), and biological uniqueness. Several theories regarding the evolution and domestication of cultivated barley have been brought forward. The current view is that the earliest agriculture, including barley planting started about 10,000 years ago in the area of Southwestern Asia, called "The Fertile Crescent" (Harlan, 1968; Zohary, 1973). The immediate progenitor of cultivated barley, i.e. the two-rowed form *H. vulgare* subsp. *spontaneum*, is conspecific to cultivated barley (*H. vulgare* subsp.) and is still growing in this area in natural habitats as well as in ruderal ground (Nevo, 1992). It was suggested that among annual six-rowed wild barley, *H. agriocrithon* was the wild ancestor of cultivated barley, as annual six-rowed wild barley *H. agriocrithon* and intermedium wild barley, *H. paradoxon* have been found in Daofu county of Sichuan province and Linzi county of Tibet Autonomous Region (cited by Xu, 1993). In contrast, other investigators considered that six-rowed and two-rowed cultivated barleys were domesticated from six-rowed and two-rowed wild barley, respectively (cited by Xu, 1993).

2.2 Annual Wild Barley

H. spontaneum, *H. agriocrithon* and *H. paradoxon* are annual, diploid ($2n = 14$), brittle, two-rowed or six-rowed or inter-medium wild barley, respectively. They share the same genome as cultivated barley does, and are completely interfertile. Moreover, they are the only wild Hordeum species that are cross-compatible and fully interfertile with the cultivated barley. Hybrids of *H. vulgare*×*H. spontaneum* (or *H. agriocrithon* or *H. paradoxon*) show normal chromosome pairing and segregation in meiosis, and their progeny is fully fertile. However, the spikes of the three wild barleys are fragile at maturity and the grains shatter easily, while the spikes are tough and the grains persist on them for domesticated barleys. *H. spontaneum* and cultivated two-rowed distichum varieties also show many morphological similarities. The main differences between them are their modes of seed dispersal. The ears of *H. spontaneum* are brittle and at maturity disarticulate into individual arrow-like triplets. These are highly specialized devices which ensure the survival of the plant in nature. Under cultivation, this specialization was replaced by non-brittle mutants, which were immediately selected by man for reaping, threshing and sowing (Zohary and Hopf, 1988). Annual wild barley *H. spontaneum*, mainly exists in the Near East Fertile Crescent and *H. agriocrithon* or *H. paradoxon* is widely distributed over the Qing-Tibetan Plateau.

2.2.1 Annual Wild Barley in the Near East Fertile Crescent

In the Near East Fertile Crescent, *H. spontaneum* is the major annual wild barley and it is also a well studied wild taxon in *Hordeum* due to its large genetic variation and important agronomic characters, such as earliness, biomass and yield.

2.2.1.1 Distribution

H. spontaneum is widely distributed over the eastern Mediterranean basin and in western Asiatic countries, across a wide range of climates and soils. In the regions, central populations of wild *H. spontaneum* are often continuously and massively distributed and can be found in primary habitats, and secondary, segetal, man-made habitats, such as infesting cereal cultivation and fruit tree plantations. *H. spontaneum* constitutes an important annual component of open herbaceous formations, and it is particularly common in the summer-dry deciduous oak park-forest belt, east, north, and west of the Syrian Desert and the Euphrates basin, and on the slopes facing the Jordan Rift Valley (Zohary, 1973). From here, *H. spontaneum* spills over the drier steppes and semideserts. In the Near Eastern countries, wild barley also occupies a whole array of secondary habitats, such as opened-up Mediterranean maquis, abandoned fields and roadsides. Peripheral populations of *H. spontaneum* become increasingly sporadic, semi-isolated and isolated. This is true in the Aegean region, southern Israel, Mediterranean shore of Egypt, Cyrenaica and Morocco, and further east in northeast Iran, former Soviet central Asia, Afghanistan and India. In these peripheral regions *H. spontaneum* rarely builds large stands and seems to be largely restricted in most localities to segetal habitats, ruins, or to sites which have been drastically disturbed by human activity. In general, *H. spontaneum* in the Near East Fertile Crescent, is not tolerant of extreme cold temperature, and it is only occasionally found above 1,500 m in altitude, such as the elevated continental plateaus of Turkey and Iran.

H. spontaneum shows macrogeographic differentiation in population genetics. Nevo *et al.* (1979a, b; 1986a, b) analysed enzyme polymorpbisms encoded by 27 shared loci in 2125 individuals representing 52 populations of wild barley in Israel, Iran, Turkey and the Near East Fertile Crescent, where all isozyme loci are specified. *H. spontaneum* populations differ considerably in their allelic content (Nevo *et al.*, 1986c). Out of 127 alleles at 27 shared loci, 65 alleles (51%) were unique, that is to say they occurred only in one country. Out of these 65 unique alleles, the distribution in Israel, Turkey and Iran was 50, 7 and 8, and the percentage in each of the countries was 48.5%, 11.1% and 12.9%, respectively. Most of the unique alleles (51% out of 65%, or 78.5%) were not widespread but localized or sporadic. Remarkably, 40 out of the 65 unique alleles (61.5%) were common (frequency >10%). In the Fertile Crescent, *H. spontaneum* is very variable in genetic composition. Overall, 127 alleles at 27 loci were described: 103, 63 and 62 for Israel, Turkey and Iran, respectively. The number of alleles increases with the average range of annual rainfall in the three countries: 100, 19 and 21% of the total range for Israel, Turkey and Iran, respectively for climatic variation and allele frequencies at 27 loci in the three countries (Nevo *et al.*, 1986c). The high values of mean proportion of polymorphic loci per population and gene diversity show high levels and wide ranges of genetic diversity in comparison with mean genetic estimates in nature (Nevo *et al.*, 1984c; Nevo, 1988b).

2.2.1.2 Diversity of target characters in *H. spontaneum* and potential use in breeding

Barley breeders have explored germplasm collections and wild species as sources of favourable alleles for further yield improvement. It has been well demonstrated that the annual wild barley germplasm, *H. spontaneum* or *H. agriocrithonor*, *H. paradoxon*, is a valuable gene pool for barley improvement. It has been shown that the wild barley species *H. spontaneum* is rich in the alleles for resistance to biotic and abiotic stresses and thus has potential for use in improving barley yield. These genes are seldom found in the cultivated barley germplasm (Nevo, 1992; Ellis *et al.*, 2000; Fetch *et al.*, 2003; Williams, 2003; Vanhala, 2004). A large numbers of novel QTLs/genes were identified for abiotic and biotic stress tolerance, and for improvement of productivity and sustainability. A search for new alleles for malting quality has just started and *H. spontaneum* may provide novel genes for improving malting quality and other traits. AB-QTL analysis has the potential to improve yield- and quality-related characters. During the past two decades, many genes or QTLs have been detected from *H. spontaneum* (Table 2.2).

Table 2.2 A list of genes or QTLs detected from *H. spontaneum*

Traits	Chromosome location/genes	Donor Plant	References
Days to heading	1H,2H,4H,5H	Apex × ISR 101-23	Pillen *et al.*, 2003
Plant height	1H, 4H	Apex × ISR 101-23	Pillen *et al.*, 2003
Lodging at harvest	1H	Apex × ISR 101-23	Pillen *et al.*, 2003
Lodging at flowering	5H	Apex × ISR 101-23	Pillen *et al.*, 2003
Thousand-grain weight	2H,5H,7H	Apex × ISR 101-23	Pillen *et al.*, 2003
Water absorption	4H,5H,7H, 3HL	Apex × ISR 101-23	Pillen *et al.*, 2003 Suprunova *et al.*, 2007
Grain yield	4H	Apex × ISR 101-23	Pillen *et al.*, 2003

(to be continued)

Traits	Chromosome location/genes	Donor Plant	References
Powdery mildew resistance	1H	Scarlett × ISR42-8	Korff *et al.*, 2005
	Mla 29	110-4 × Sonja	Kinizios *et al.*, 1995
	Mla 32	142-29 × Dura	Kinizios *et al.*, 1995
	M1-i	RS42-8 × Oriol A	Jahoor & Fischbeck, 1987
	Mlf (7HL)	RS137-28*E	Schönfeld *et al.*, 1996
	Mlj (5HL)	HSY-78*A	Schönfeld *et al.*, 1996
	Mlt (7HS)	RS42-6*O	Schönfeld *et al.*, 1996
	Rbg-1H-1-3	OUH602 × Harrington	Yun *et al.*, 2005
	QPm.S42-1H	Scarlett × ISR42-8	Korff *et al.*, 2005
	Qpm.S42-2H	Scarlett × ISR42-8	Korff *et al.*, 2005
	QPm.S42-3H	Scarlett × ISR42-8	Korff *et al.*, 2005
	Lang1031Q Pm.S42-4H	Vada × 1B-87	Backes *et al.*, 2003
	QPm.S42-5H	Scarlett × ISR42-8	Korff *et al.*, 2005
	QPm.S42-7H	Scarlett × ISR42-8	Korff *et al.*, 2005
SSLB(Septoria speckled leaf blotch)	Rsp-2H-7-11	OUH602 × Harrington	Yun *et al.*, 2005
	Rsp-6H-10-14	OUH602 × Harrington	Yun *et al.*, 2005
Spot blotch	Rcs-7H-2-4	OUH602 × Harrington	Yun *et al.*, 2005
NTNB (net type net blotch)	Rpt-3H-4	OUH602 × Harrington	Yun *et al.*, 2005
	Rpt-4H-5-7	OUH602 × Harrington	Yun *et al.*, 2005
Leaf rust	Rph2 (5HS)	HJ198*/HS2310	Jin *et al.*, 1995
	Rph3 (7HL)	Bowman*4/PI466-324	Jin and Steffenson, 1994; Feuerstein *et al.*, 1990
	Rph10 (3H)	Clipper BC8 × *H. Spontaneum*	Feuerstein *et al.*, 1990
	Rph11 (6H)	Clipper BC67 × *H. Spontaneum*	Jin *et al.*, 1993

(to be continued)

Traits	Chromosome location/genes	Donor Plant	References
	Rph12 (5HL)	Triumph × I91-533-va	Feuerstein *et al.*, 1997
	Rph13 (5H)	Berac* 3 × HS2986	Feuerstein *et al.*, 1997
	Rph15 (2H)	Bowman*4/PI355-447	Ivandic *et al.*, 1998
	Rph16 (2HS)	L94 × HS084	Backes *et al.*, 2003
	QLr.S42-2H (2HL)	Vada × 1B-87, Scarlett × ISR42-8	Korff *et al.*, 2005
	QLr.S42-3H	Scarlett × ISR42-8	Korff *et al.*, 2005
	QLr.S42-4H	Scarlett × ISR42-8	Korff *et al.*, 2005
	QLr.S42-5H	Scarlett × ISR42-8	Korff *et al.*, 2005
	QLr.S42-7H	Scarlett × ISR42-8	Korff *et al.*, 2005
β-amylase	Bmy1-Sd3	*H. spontaneum*	Eglinton *et al.*, 1998
	Bmy1-Sd4	*H. spontaneum*	Eglinton *et al.*, 2001
	Bmy1-Sd5	*H. spontameum*	Eglinton *et al.*, 2001

(1) QTLs affecting yield in *H. spontaneum*:

Pillen *et al.* (2003) found some agronomically-related QTLs by using a BC2 and F2 population, which was derived from the initial cross Apex (*Hordeum vulgare* ssp. *vulgare*) × ISR101-23 (*H.* v. ssp. *spontaneum*). QTLs for days to heading were mainly on chromosomes 1H, 2H, 4H, 5H; QTLs for plant height were located on the chromosomes 1H and 4H; QTLs for lodging at harvest were located on chromosome 1H; QTLs for lodging at flowering were located on chromosome 5H; QTLs for grain weight were mapped to chromosomes 2H, 5H and 7H; six putative QTLs were located on chromosomes 4H, 5H and 7H for the quality-related trait, water absorption; QTLs for yield were located on all chromosomes except for 6H. The yield increase due to exotic *H. spontaneum* donor ranged from 2.7% to 10.9%, with an average of 7.7% under six tested environments.

(2) Genetic diversity of enzyme and cytoplasmic DNA:

The diversity of enzymes, such as esterase and α-amylases, are higher in wild barley than in cultivated barley. Nevo (1992) compared total allozymic diversity of 19 shared loci of wild barley in Iran, Turkey and Israel with that of barley landraces from Iran and a collection of European landraces, also including two accessions from Egypt and one from India, and two composite crosses. Allozymic diversity was lowest in the composite crosses, intermediate in the landraces, and highest in the wild germplasm. These results seem to hold even if the world collection of cultivated barley, which presumably represents the allelic diversity maintained in the domestic gene pool, is compared.

When this comparison is made for four esterase loci, the differences between the wild and cultivated gene pools in number of alleles are: Est1 (8,7); Est2 (17,12); Est4 (10,6); Est5 (6,6) (Nevo *et al.*, 1986c). These results emphatically demonstrate the potential role of the wild progenitor as well as that of the landraces (Brown and Munday, 1982) as valuable genetic resources for barley breeding. The α-amylases diversity of H. *spontaneum* in the Near East Fertile Crescent is higher than that of the cultivated barley varieties (Brown and Jacobsen, 1982). The cytoplasmic DNA diversities, such as chloroplast DNA (cpDNA) and mitochondrial DNA (mtDNA), are also higher in the wild barley than the cultivated barley. Clegg *et al.* (1984) found the cpDNA diversity in 11 lines of *H. spontaneum* was significantly higher than that in 9 lines of *H. vulgare*, suggesting that the level of cytoplasmic diversity was markedly restricted during barley domestication. Neale *et al.* (1988) also found the cpDNA diversity within and among populations of *H. spontaneum* from Israel and Iran, which displayed that ecogeographic correlation, and diversity was higher in *H. spontaneum* than in *H. vulgare.* By contrast, however, landraces of *H. vulgare* proved more polymorphic in cpDNA and in mitochondrial DNA (mtDNA) diversity than *H. spontaneum* (Holwerda *et al.*, 1986), which demonstrated that more genotypes and populations need to be studied for making the comparison meaningful.

(3) The variation of outcrossing rate:

The mating system of *H. spontaneum* is different from that of cultivated barley, and its variation of outcrossing rate in natural populations is much higher than the latter. Nevo (1992) used their allozyme data to estimate outcrossing rates across the range of 26 populations of wild barley (*H. spontaneum*) in Israel. The extent of cross-pollination was estimated quantitatively through using allozyme variation at 22 polymorphic loci. Outcrossing of individual *H. spontaneum* population varied from 0 to 9.6%. The overall average was a very low rate, about 1.6%. Outcrossing varied with ecological conditions, and was significantly higher in populations from mesic (2.1%) than from xeric regions (0.4%). The multilocus heterozygosity encountered at 22 loci displayed a very high degree of multiple-disequilibrium. In general, these results indicate that extensive genetic polymorphisms in the progenitor probably predate the domestication of the crop, and are maintained independently of introgression, presumably by natural selection in accordance with environmental heterogeneity.

(4) The variations of agronomic traits:

The wild gene pool of *H. spontaneum* in the Near East Fertile Crescent is generally considered to have more genetic diversity than cultivated barley and landraces (Nevo *et al.*, 1979a, b). However, Jana and Pietrzak (1988) reported different results showing that the landraces from the eastern Mediterranean countries Jordan, Syria, Turkey and Greece had wider genetic diversity than wild barley. Nevo (1992) suggested that this discrepancy might be derived from a lower number and a somewhat different set of tested gene loci. The reduction in genetic diversity of inbred lines has made cultivated barley increasingly vulnerable to new pathogens, pests and adverse climates.Therefore it is necessary to test the variation in quantitative traits of agronomic importance of *H. spontaneum* as potential genetic resources for breeding.

Nevo *et al.* (1992) tested the allozymes, disease resistance polymorphisms, quantitative phenotypic traits, germination, earliness, biomass and yield of the populations of wild barley *H. spontaneum.* The field experiments were conducted in two relatively standardized and contrasting environments: Mesic (Mount Carmel, Haifa) and Xeric (Avedat farm and Sede Boqer, both in the northern Negev desert). The results indicated that the characters studied are partly genetically determined. A dramatically genetic variation was found between and within populations in each site, whereas remarkable environmental variation, including genetic environmental interaction, was found between the Mesic and the Xeric sites, as well as between populations and years. They concluded that wild barley in Israel varies not only in allozyme polymorphisms and disease resistances, but also in quantitative traits of agronomic importance. Similar results were obtained by others (Giles and Lefkovitch, 1983). These characteristics are economically significant and can be utilized in barley breeding programs. It has been proved that *H. spontaneum* can provide a wealth of additional genetic traits of economic significance, such as earliness, high yield and high biomass. When parental genotypes of the progenitor are used as donors of alien genes for disease resistance to a cultivar, it is desirable that they also combine the genetic capacity for high yield and biomass to avoid yield depression in the resulting crosses.

(5) Tolerance to drought and salinity:

Stress tolerance can be described as the ability to maintain a stable yield under the stress environment. The wide ecological range of wild barley differs in water availability, temperature, soil type, altitude, and vegetation, generating a high potential for adaptive diversity to abiotic stresses. A significant proportion of diversity at the molecular level of stress responsive genes appears to be adaptive rather than neutral (Nevo *et al.*, 1998; Liviero *et al.*, 2002). Abiotic stress tolerance is a quantitative trait and controlled by genetic factors, and a lot of genes for abiotic tolerance in wild barley have been lost during domestication (Ivandic, 2000; Cattivelli *et al.*, 2002).

Drought is a major abiotic stress that limits plant growth and crop productivity. Drought tolerance in plants is a complex trait, which cannot be analysed genetically in the same way as monogenic resistance. When a crop is under stress, it can respond alternatively to the environmental changes. Wild barley *H. spontaneum* was demonstrated as a key genetic source for drought tolerance. Carbon isotope discrimination (CID), an indicator of stress tolerance, has been associated with drought tolerance in terms of water-use efficiency and yield stability in drought-prone environments (Teulat-Merah *et al.*, 2000). Ten QTLs for CID have been found on all the chromosomes except 4H (Teulat *et al.*, 2002).

Dehydrins (Dhns), peripheral membrane proteins, which functionally protect the cell from water deficit or temperature change, are among the most frequently observed proteins in plants under water stress (Suprunova, 2004). A total of 13 Dhn genes were found on four barley chromosomes (Choi *et al.*, 1999; Choi and Close, 2000; Rodrguez *et al.*, 2005). Dhn1, Dhn2 and Dhn9 (previously reported for Dhn4a) were mapped to the long arm of chromosome 5H; Dhn3, Dhn4, Dhn5, Dhn7, Dhn8 and Dhn12 were allocated to the long arm of chromosome 6H; Dhn6 and Dhn13 were allocated to 4HS; Dhn10 and Dhn11 were identified on 3HL. Dhn1 and Dhn2 are completely linked, and together with Dhn9 are located in the same QTL region of salt tolerance, freezing tolerance and ABA accumulation. In wild barley, Dhn genes (Dhn 1, 3, 5, 6, and 9) were also found and were not expressed in well-watered plants. The role of Dhn1 in drought tolerance is supported by several reports on co-localization of such QTLs with Dhn genes, e.g. QTLs for RWC (Relative Water Content) (Teulat *et al.*, 2003) and winter-hardiness (Pan *et al.*, 1994; Zee *et al.*, 1995) overlapping with a cluster of Dhn genes on chromosome 5H. High polymorphism with no geographic structure was found in Dhn5 in a collection of wild barley from the Mediterranean across the Zagros Mountains and into south west Asia, and moderate polymorphism associated with geographic structure was found in Dhn9 locus (Morrell *et al.*, 2003). Dhn1 showed earlier expression (after 3 h dehydration) and higher levels (after 12 h and 24 h) in resistant plants when compared with sensitive plants. Dhn3 gene was induced by drought stress, ABA, and salt tolerance (Choi and Close, 2000), and expression of Dhn3 and Dhn9 was detected in sensitive and resistant genotypes during 3–12 h. However, after 12 h and 24 h, they showed no clear difference. Dhn5 was expressed after 3 h of dehydration stress in all genotypes and increased after 12 h of dehydration, and a slightly higher level was found in the resistant genotypes than in the sensitive types. A low level of Dhn5 was detected in TR- and BA genotypes. It is interesting that Dhn6 had a higher expression in resistant genotypes after 3h of dehydration, and after 12 h and 24 h of dehydration the sensitive plants showed a higher expression level as compared to the resistant plants (Suprunova *et al.*, 2004). QTLs for agronomic traits related to drought resistance were detected in the cross "Arta" × *Hordeum spontaneum* 41-1. QTLs for the most important character "plant height" under drought stresses were localized on 2H, 3H and 7H. The "plant height" QTLs, specially the one on 3H, showed pleiotropic effects in traits such as days to heading, grain yield and biological yield (Baum *et al.*, 2003).

In addition to drought tolerance, the gene pool of wild relatives may represent a valuable source of new loci for salt tolerance. In wild barley *H. spontaneum*, salt tolerance is associated with drought environments (Pakniyat *et al.*, 1997). The natural populations of *H. spontaneum* showed rich genetic variation in resistance to physiological stresses, such as drought and salinity. This was highlighted by the tremendous variation both between and within populations observed in the extreme desert station at Sede Boqer (Nevo *et al.*, 1984a). Subsequently, they tested populations and genotypes of wild barley *H. spontaneum* for salt tolerance under controlled conditions using 150,

250 and 350 mM NaCl (equivalent to 25%, 40% and 60% sea water) with non-saline controls (Nevo, 1992). They found wide a difference in high salt tolerance within and between populations of *H. Spontaneum*, and identified wild barley genotypes with high tolerance from coastal Mediterranean populations (e.g. Caesarea), the Jordan Valley (Mehola) and the northern Negev desert (Makhtesh Gadol) and they could spike normally under 350 mM NaCl. Natural populations of wild barley extend southwards to the northern Negev, yet they only grow there in dry watercourses, thereby effectively utilizing the extra run-off water. However, under a water regime of 130 mm and highly saline soil conditions, some populations and superior genotypes have proved relatively outstanding in both growth and yield, despite these extremely harsh ecological conditions. Thus, the genetic resources of the progenitor display marked morphological and physiological performances, which are of potential economic importance for barley improvement.

Mano *et al.* (1998) investigated salt tolerance of wild barley species at germination and seedling stages. The results showed that seed germination tolerance in wild *Hordeum* species were generally lower than those found in cultivated barley (Mano *et al.*, 1996, 1998). Among *Hordeum* species, *H. agriocrithon*, *H. spontaneum*, other wild Hordeum species showed high, middle and low levels of NaCl tolerance, respectively. The higher tolerance level of *H. agriocrithon* probably originated from natural hybridization between *H. spontaneum* and cultivated barley (Briggs, 1978). At the seedling stage, the NaCl tolerant level in wild *Hordeum* species was generally higher than those found in cultivated barley (Mano and Takeda, 1995). Thus, wild Hordeum species as a donor will benefit from salt tolerant breeding at the seedling stage.

A number of QTLs affecting salt tolerance were detected on all chromosomes in a cross between *H. spontaneum* and *H. vulgare*, although several QTL clusters were present on chromosomes 1H, 4H, 6H, and 7H (Ellis *et al.*, 1997). In cultivated barley, QTLs controlling salt tolerance were mapped to chromosomes 1H, 4H, 5H, and 6H at germination and to chromosomes 1H, 2H, 5H and 6H at the seedling stage (Mano and Takeda, 1997). In a cross of Derkado × B83-12/21/5, the largest individual effects on salt tolerance were associated with the chromosomal regions around the two dwarfing genes sdw1 (3H) and ari-e.GP (5H). The sdw1 gene resulted in an overall yield increase, but was only detected as a secondary QTL (Ellis *et al.*, 2002).

Compared with cultivated barley, *H. spontaneum* has higher efficiency or ability in nitrogen utilization. Corke *et al.* (1988) measured the nitrogen contents and dry weights of leaf and "stem" (stem plus sheath) at anthesis, and the final grain size and protein content of 33 *H. spontaneum* and two *H. vulgare* genotypes. They found that H. spontaneum was generally higher in nitrogen content of leaves and stems, but lower in dry weight at anthesis. Consistent with previous reports, it showed that the *H. spontaneum* genotypes were considerably higher in grain protein content than the cultivated barleys. There was a wide variation between and within populations of *H.*

spontaneum, suggesting that accessions combining high vegetative nitrogen content, dry weight and grain protein content can be selected for breeding.

(6) Resistance to Disease:

Disease can affect the profitability of barley production. It either reduces final yield or lowers grain quality. Both of them result in a relatively lower financial return to growers. In the early 1980s, disease was considered as the key limitation on yield, and was the major focus of breeding programs. At least 30 diseases and pests are reported to affect barley. The major ones included leaf scald, leaf rust, yellow rust, spot form and net form of net blotches. Considerable effort has been made to identify and localize major disease resistance genes and QTLs from *H. spontaneum* chromosomes (Yun *et al.*, 2005). Genes from *H. Spontaneum* have proved to be crucial to the genetic improvement of disease resistance in cultivated barley. Currently, leaf scald and powdery mildew resistance QTLs were mapped on all chromosomes (Garvin *et al.*, 1997, 2000; Genger *et al.*, 2003a), however, only few of them were detected on certain wild barley chromosomes. Five QTLs for spot blotch resistance were localized on chromosomes 1H, 2H, 3H, 5H and 7H (Bilgic textit *et al.*, 2005). Four QTLs for SSLB resistance were mapped on chromosome 2H, 4H, 1H and 6H (Toubia-Rahme *et al.*, 2003; Yun *et al.*, 2005). Loci conferring resistance to NTNB were mapped to every barley chromosome (Richter *et al.*, 1998; Steffenson *et al.*, 1996; Yun *et al.*, 2005). The presence of the exotic QTL alleles reduced disease symptoms by a maximum of 51.5, 37.6 and 16.5% for powdery mildew, leaf rust and scald, respectively (Korff *et al.*, 2005).

The accessions of *H. spontaneum* from Israel have special potential in resistance to various pathogens, such as powdery mildew and rust. The reason is that Israel is one of the epicenters where active co-evolution of several crop ancestors and their pathogens occurs, including wild barley, wild wheat and their rust and powdery mildew pathogens. According to the theory and genetics of host-parasite co-evolution, the highest and broadest genetic immunity against devastation in the progenitors of some crops would be expected in the Fertile Crescent in general, and in Israel in particular. Israel is ideal for the exploitation of resistance genes, which could provide useful material for barley breeding, either for the short-term, or for the long-term solution to disease problems in agriculture (Segal *et al.*, 1980, 1982; Vanhala, 2004).

Powdery mildew (Esysiphe graminis hordei) is one of the most common diseases in barley. Yield reduction caused by powdery mildew can reach 30%, with an average of 5%~10% across all the regions. The disease can also deteriorate quality characteristics (Czembor, 2000). At least two genetically separable pathways control resistance to powdery mildew in barley (Jørgensen, 1994; Peterhänsel *et al.*, 1997). In the first pathway, resistance is mediated by recessive alleles at the *mlo* locus. The second resistance pathway can be induced by a number of race-specific resistance genes (R genes) like Mla, Mlg, Mlk (Jørgensen, 1994), and is almost invariably associated with the activation of rapid host cell death at attempted infection sites (Freialdenhoven *et al.*, 1994). Mla locus on barley chromosome 1HS is known to be complex. At present 31 alleles in the Mla locus were found by multi-copy RFLP, SCAR, gene-specific SNP and SSR markers (Kintzios *et al.*, 1995). The growing numbers of new alleles for the Mla locus are mainly originated from *H. spontaneum* lines from Israel (Kintzios *et al.*, 1995), and they were neither allelic nor linked to alleles of the Mla locus, and showed independent segregations. The resistant genes derived from wild barley also showed different powdery mildew resistant levels: mlt with 3 alleles showed a recessive mode of inheritance on chromosome 7HS. Mlf and Mlj localized on chromosomes 7HL and 5HL were semi-dominant resistance genes (Schönfeld, 1996). QTLs of powdery mildew resistance were mapped to all chromosomes. However, in wild barley they were mainly detected on 1H (Yun *et al.*, 2005).

In many countries a lot of resistant resources of barley to powdery mildew have been identified and used extensively in developing resistant cultivars (Moseman *et al.*, 1983). Many *H. spontaneum* accessions in Israel being identified are highly resistant to powdery mildew (Moseman *et al.*, 1980). Moseman and Craddock (1976) reported that 75% of 277 accessions of *H. spontaneum* collected in Israel were resistant to the two races of powdery mildew. Fischbeck *et al.* (1976a) reported that about 50% of 2700 unselected accessions of *H. spontaneum* were attacked only slightly or not at all by European field isolates of the pathogen.

Barley leaf scald was caused by the fungal pathogen *Rhynchosporium secalis*. The pathogen is highly variable among barley populations and can easily overcome newly developed resistance genes (Zhang *et al.*, 1987; McDonald *et al.*, 1999). Therefore it is very difficult to find a host that can permanently resist this pathogen. However, it is possible to find some resistant genotypes from its wild progeny. In cultivated barley, most genes for resistance to barley leaf scald were mapped either to the Rrs1 locus on the long arm of chromosome 3H, or the Rrs2 locus on the short arm of chromosome 7H (Schweizer *et al.*, 1995). However, resistant genes in wild barley populations were identified at 5 chromosome locations (1HS, 3H, 6HS, 7HL, and 7HS). Rrs12 was mapped to the short arm of chromosome 7H and may be allelic to Rh2 in cultivated barley (Abbott *et al.*, 1992); Rrs13 was mapped to the short arm of chromosome 6H (Abbott *et al.*, 1995); Rrs14 mapped to the short arm of chromosome 1H (Garvin *et al.*, 1997); Rrs15 was allocated on chromosome 7HL (Genger *et al.*, 2005), while Rrs16Hb was introgressed from *H. bulbosum* on 4HS (Pickering *et al.*, 2006). Up to date, four QTLs (Rrsq1, Rrsq2, Rrsq3 and Rrsq4) for resistance against scald were detected on chromosomes 2H, 3H, 4H and 6H, respectively. A resistant QTL QRh.S42-4H, two susceptible QTLs QRh.S42-2H and QRh.S42-3H were identified from a BC2DH S42 (a cross between spring cultivated barley "Scarlett" and a *spontaneum* accession "ISR42-8") (Korff *et al.*, 2005). Another QTL related to scald resistance, designated as Rrs-1H-1-4, was found from a cross between OUH602 and Harrington (Yun *et al.*, 2005).

Leaf rust caused by *Puccinia hordei Otth* is another important disease of barley in many regions of the world. It caused yield loss of up to 32% in susceptible cultivars (Griffey *et al.*, 1994). There are two pathways of resistance against *P. hordei* in barley: partial resistance and race-specific resistance. Controlled by several or many genes, partial resistance is often considered to be more durable than race-specific resistance (Kicherer *et al.*, 2000). To date, there are 17 identified leaf rust resistant genes on all chromosomes except 4H. Resistance in cultivated barley was shown to be very restricted and mainly limited to the genes Rph3 and Rph7 (Jin *et al.*, 1995). Four resistance genes (Rph10, Rph11, Rph15, Rph16) were found in *H. spontaneum*, and a large variability was found to exist in accessions from Israel. Moseman *et al.* (1990) identified the reactions to infection of 292 *H. spontaneum* accessions being collected at 16 sites encompassing the entire ecological range of *H. spontaneum* in Israel. Of the accessions tested, 52% were resistant to race Tel-Aviv, 67% were resistant to race 57. This confirmed the gene for gene hypothesis and furthermore demonstrated that the Near East represents a major centre for the evolution of resistance to *P. hordei Otth* (Moseman *et al.*, 1990). Rph1 and Rph16 were mapped to chromosome 2H (Ivandic *et al.*, 1998). Rph2, Rph9, Rph12 and Rph13 were located on chromosome 5H with a linkage of Rph9 and Rph13 (Borovkova *et al.*, 1997, 1998). Rph3 was detected on chromosome 7HL and found to have three alleles (Jin *et al.*, 1993), Rph19 was mapped to 3H as a major disease resistant gene (Park and Karakousis, 2002). Rph4 was identified on chromosome 1HS, and linked to the Mla mildew resistance locus (Tan, 1978); Chromosome 3H is the concentrate region of Rph genes including Rph5, Rph6, Rph7 and Rph10; Rph5 showed linkage to Rph7 on chromosome 3H (Tan, 1978; Brunner *et al.*, 2000; Manmmadov *et al.*, 2003). Rph11 was designated to 6H (Feuerstein *et al.*, 1990). Rph13, Rph14, and Rph15 were effective at the adult plant stage.

QTLs for leaf rust resistance were also detected on 2H, 3H, 4H, 5H and 7H. In cultivated barley, there were in total 6 QTLs related to leaf rust resistance. Rphq1 (1H), Rphq2 (2H) and Rphq3 (6H) were effective at seedling stage and contributed approximately 56% to the phenotypic variance. Rph2, Rphq3, Rphq4 (7H), and Rphq5 (4H) contributed approximately 63% of the phenotypic variance and were effective at adult plant stage. Rphq6 (2H), which is located in the same QTL region as Dh2, Ph2, was found to affect the latent period only at adult plant stage (Qi *et al.*, 1998). However, in wild barley the resistant QTLs were mapped on the same chromosome locations, especially at the QTL location QLr.S42-2H, exotic allele reduced disease symptoms by 25.9% (Korff *et al.*, 2005).

Resistance to other pathogens which have been studied so far include barley yellow dwarf virus (BYDV)yellow rust, stem rust and barley yellow mosaic virus (BYMV) in *H. vulgare* ssp. *spontaneum*. Schooler (1980, 1981) observed that there existed some resistant accessions in wild barleys to barley yellow dwarf virus (BYDV). Anikster *et al.* (1975) found that some of *H. spontaneum* germplasm had resistance to yellow rust. Steffenson *et al.* (2007)reported that wild barley exhibited a wide range of resistance to stem rust. The percentage of resistant accessions in wild barley was 5.7%~20.1% for stem rust, depending on the race used. Using RAPDs, Ordon *et al.* (1997)analyzed the genetic similarity to the barley yellow mosaic virus complex (BaMMV, BaYMV, BaYMV-2) between thirty-six *Hordeum vulgare* varieties and 12 *H. spontaneum* germplasm originating from different parts of the world. They found that the *H. spontaneum* germplasms and Hordeum vulgare varieties showed different reactions to the barley yellow mosaic virus complex (BaMMV, BaYMV, BaYMV-2). The results from their studies indicate that wild barleys are rich sources of these pathogens resistance genes for cultivated barley improvement.

(7) The variation in protein content:

According to Ladizinsky (1975) and Ahokas (1977, 1982), there were many significant differences in protein content between wild barley populations from different bio-geographical regions. Nove *et al.*, (1985) determined protein content of *H. spontaneum* in Israel. The study involved 195 genotypes of wild barley from 25 populations, 15 central and 10 marginal. The results indicated that protein content varies greatly within and between populations. Interestingly, the ten marginal populations exhibited higher protein contents (15.06%), but lower kernel weights (21.4 mg), as compared with the 15 central populations, which showed lower protein contents (13.92%), but higher kernel weights (36.0 mg). They concluded that natural populations of wild barley in Israel contain large amounts of potential genes for improvement of protein content.

(8) The variation in malting quality:

Fermentability is a crucial quality parameter for brewing, particularly in the systems using starch-based adjuncts. β-amylase (1,4-α-glucan maltohydrolase) catalyses the liberation of β-maltose from the non-reducing ends of 1,4-α-glucans and is a key enzyme in the degradation of starch during brewing. During the brewing process, the yield of degradation hydrolysates, i.e. fermentable sugars directly affects the level of alcohol produced. Temperatures in excess of 63°C are employed to gelatinize starch granules, which facilitate rapid and complete starch degradation. However, native barley β-amylase retains maximum activity up to 55°C, thus its stability decreases rapidly as the temperature increases above 55°C (Thacker *et al.*, 1992). Therefore improvement of β-amylase activity and its thermostability is quite important for increasing fermentability during malting. Bmy1 encodes β-amylase hydrolytic enzyme in malt barley (Zhang *et al.*, 2007). Currently, 23 Bmy1 alleles were identified, of which 16 were detected in cultivated barley, with 7 major types and 9 rare and unique types. The other 7 types were devoted to *H. spontaneum* accessions, which turned out to be unique. Noticeably, almost all the major allele types in cultivated barley were observed in wild barley except the B-I-8 allele, which was predominant in north European cultivated barley (Zhang *et al.*, 2007). Nearly the same result was found by Chalmers *et al.*(1992), that wild barley from Israel possesses seven isoforms of β-amylase enzyme, of which only two have been found in European cultivars.

Three β-amylase alleles, Bmy1-Sd3, -Sd4, and -Sd5 were identified in 154 accessions of *H. spontaneum*. The corresponding Sd4 and Sd5 enzymes showed a middle level of thermostability, similar to the Sd1 β-amylase in cultivated barley, while the Sd3 β-amylase exhibited greater thermostability than the other allelic forms of β-amylase and it can remain in a stable state in the temperatures of 60.8°C. Examination of the gene structure of the Bmy1-Sd3 allele revealed a 126bp deletion in intron III, consistent with elevated gene expression (Eglinton *et al.*, 2001). The Bmy1-Sd3 allele from *H. spontaneum* was used to develop commercial malting barley varieties through marker-assisted selection (Li *et al.*, 2004).

The mean activities of α-amylase, β-amylase, β-glucanase showed significantly lower values in the 257 *H. spontaneum* accessions when compared with those of 32 unimproved Finnish landraces and 5 cultivars (Ahokas *et al.*, 1996). Landraces have adapted in many ways to the local environment and are morphologically closer to the desired types than wild barley. Compared with the variation in wild barley (*H. spontaneum*) located in Finland (Ahokas and Naskali, 1990), the landraces have significantly higher mean activities for all the three enzymes. The variance of β-amylase is very significantly wider in the Finnish landraces than in the wild barleys (Ahokas and Poukkula, 1999).

β-amylase-less mutant lines had a naked seed character and various colors, and could give a new taste and flavour to beer and other liquors. Thus, use of the mutant is also expected to extend to enzyme technology. The β-amylase-less trait may be introduced into elite malting varieties by backcrossing. An enzyme assay showed that there were eight β-amylase-less mutant lines in Tibetan landrace barley. These mutants result from the same insert mutation in the β-amylase structure gene of Tibetan barley. Both RFLP and PCR analysis strongly suggest that the insertion occurs between 2071 bp and 3101 bp from the initiation codon and the structure gene causes the deficiency in enzyme activity (Kaneko *et al.*, 2000).

2.2.2 The Annual Wild Barley in Qing-Tibetan Plateau

Exploration, collection and evaluation of the annual wild barley in Qing-Tibetan Plateau was initiated in the 1980s, much later than the corresponding studies for the annual wild barleys in the Fertile Crescent. However, there are a lot of relevant researches in the recent two decades.

2.2.2.1 Genetic variation of some important characters

The annual wild barleys in Qing-Tibetan Plateau, including *H. spontaneum*, *Hagriocrithon* and *Hparadoxon*, are rich in genetic variation, notably in important agronomic characters, such as earliness, biomass, dwarf and yield (Xu, 1993), lysine and protein content, carbohydrate composition, hordein fractions and disease resistance. Furthermore, it was demonstrated that the wide biochemical, morphological and physiological diversity is not distributed randomly (Xu, 1993). In the past three decades, we have evaluated 27 accessions of annual wild barleys from Qing-Tibetan Plateau. These wild barleys exhibited distinct disease resistance, including yellow rust, powdery mildew, leaf rust and net type net blotch. We also found some accessions with very early heading, extremely short straw (30-40cm), tolerance to drought, salinity, at seedling stage or tolerance to waterlogging, and tolerance to low temperature (unpublished data). Recently we have tested the protein content, activities of alpha-amylase, beta-amylase, limit dextrinase and beta-glucanase in more than 200 accessions of annual wild barleys from Qing-Tibetan Plateau, and some genotypes showed great potential for improving malt, food and feeding quality (unpublished data). In earlier days we found some genotypes with potentially useful traits in the annual wild barleys, such as nuclear male sterility (Xu, 1993), nucleo-cytoplasmic male sterility (Sun, 1995) and high out-crossing ability.

A great genetic diversity exists in the annual wild species and cultivated barley in this region. We compared the variations in agronomic characters and DNA diversity using SSR molecular markers among 175 barley varieties or accessions, including 40 annual wild barley accessions from Qing-Tibetan Plateau, 46 European barley varieties, 81 American barley cultivars and 8 Australian barley varieties. The results indicated that there were dramatic differences among examined genotypes in agronomic characters. Clustering analysis with UPGMA illustrated that 175 accessions could be divided into five groups. Groups 1, 2 and 5 in total had 30 barley varieties, and were mainly composed of wild barley in the Qing-Tibetan Plateau. Group 3 consisted mainly of American barley cultivars. European and Australian barleys belong to the biggest group 4 (unpublished data).

2.2.2.2 Utilization in barley breeding

Annual wild barley germplasm in the Qing-Tibetan Plateau, including *H. spontaneum* and *Hagriocrithon*, have been successfully utilized in modern barley breeding programs. In 1984, Xu firstly used the wild relatives in barley breeding, and developed a new six-rowed naked variety, named Chuanluo No.1. After that, Xu and Sun have successively released 8 commercial barley varieties using the annual wild barley germplasm from the Qing-Tibetan Plateau. These varieties are Chuan Nong Da 1 (two-rowed malt barley, released in 1992), Chuan Nong Da 2 (two-rowed feed barley with high protein content, released in 1995), Huadamai 1 (two-rowed naked barley with high protein, released in 2001), Huadamai No.2 (two-rowed malt barley with low protein content, released in 2001), Huadamai 3 (six-rowed feed barley, released in 2002), Huadamai 4 (six-rowed feed barley with high yield potential, released in 2004), Huadamai 5 (two-rowed malt barley with low protein content, released in 2006) and Huadamai 6 (two-rowed malt barley with low protein content, released in 2008). It demonstrated that the annual wild barleys from Qing-Tibetan Plateau have great potential for genetic improvement of barley just like the annual wild barleys from the Near East Fertile Crescent.

2.2.3 Other Annual Wild Barley Species

Other annual wild barley species, such as *H. murinum*, *H. marinum* and *H. depressum*, also showed potential use for the genetic improvement of cultivated barley, although a huge difficulty exists in terms of interspecific gene transfer through sexual hybridization between cultivated barley and these wild species. It was reported that these annual wild barly species, including *H. murinum*, *H. marinum* and *H. depressum*, have a wide genetic diversity in their resistance to plant pathogens, such as powdery mildew, loose smut and greenbug. Giles & Barrett (1983) investigated 25 accessions of *H. murinum subsp* from 9 countries. In a greenhouse experiment, the accessions were exposed to 3 isolates of *Erysiphe graminis f.* sp. *hordei* from the UK. Most of accessions belonged to a resistant type and only two populations were susceptible. When exposed to powdery mildew of agricultural origin, *H. murinum* showed high levels of resistance, but variation existed between populations both in resistant type and in expressing time during plant development. In their second experiment they inoculated 20 individuals from each of four *H. murinum* populations with two different races of *Erysipbe graminis* originating from a bulk collection from Israel. The results showed that there was a great variation in resistance reactions, mainly between populations. When *H. murinum* was exposed to powdery mildew of wild origin, it exhibited higher levels of susceptibility than when exposed to isolates of cultivated origin. Isolates from cultivated barley have probably been exclusively selected for a crop performance whereas the Israeli isolates may have been exposed to a wider range of host species and hence may have a wider host-range adaptation (Wahl *et al.*, 1978). In an extensive study of resistance of wild species of *Hordeum* to *Erysiphe graminis f.* sp. *hordei*, Gustafsson & Claesson (1988) screened 155 populations representing 28 *Hordeum* species. They used four cultures of the powdery mildew and the isolates exhibited virulence to most known resistance genes. Apart from the two highly susceptible cultivars (cvs Tellus and Vada) and a susceptible accession of *H. marinum*, all populations of the wild species showed immune or resistant reactions. Loose smut is also frequently found attacking wild species of *Hordeum* under natural conditions in almost all parts of the world (Bothmer, 1992). Nielsen (1978, 1987) investigated resistance of 19 *Hordeum* species to the disease and found that *H. murinum*, *H. marinum*, and *H. depressum* had high resistance to smut fungi *Ustilago nuda* and *U.tritici*, and several other species showed resistance to one of the pathogens. During the evolution of *Hordeum*, many species retained susceptibility to both ancestral *U. tritici* and *U. nuda.* A current study at ICARDA in Syria showed that one accession of *H. murinum* subs.had moderate tolerance and *H. secalinum* had excellent tolerance to greenbug infections (Bothmer, 1992).

2.2.4 Prospective

The rapid increase in human population requires more and more food production. The requirement may be met through advanced crop improvement. The dramatic loss of plant genetic resources in the world due to rapid development of intensive agriculture, has brought limitaton and difficulties in crop genetic improvement. Narrowing genetic diversity has been accelerated during domestication, particularly over the last century, due to modern pure line breeding practices, which have increased genetic homogeneity in the cultivars. As a result, many important food crops have become increasingly susceptible to diseases, pests and environmental stresses. The present interdisciplinary review of the annual wild barley germplasms *H. spontaneum* in the Near East Fertile Crescent and the *H. spontaneum*, *Hagriocrithon* and *Hparadoxon* in the Qing-Tibetan Plateau, supports the idea that the greatest hope for future crop improvement lies in rationally and effectively exploring and exploiting the rich gene pool in wild relatives. In general, both annual wild barley germplasms *H. spontaneum* in the Near East Fertile Crescent and *H. spontaneum*, *Hagriocrithon* and *Hparadoxon* in the Qing-Tibetan Plateau, have great potential for the improvement of cultivated barley, although currently these genetic germplasm have been rarely used in most barley breeding programs. The idea may be at least supported by the following considerations. The first is that the annual wild barleys have wider genetic diversity than cultivars in many important agronomic characters, as described above, while genetic diversity of cultivated barley becomes narrow. Secondly, these wild barleys belong to the primary gene pool of the crop, have no biological barriers with the cultivated barley, and hybrids between them could occur even in natural conditions. Thirdly, many elite genes and relevant molecular markers have been detected in the annual wild barley germplasms, thus providing a good chance for transferring the alien genes into cultivated barleys by molecular marker assistant selection.

The development of an *in vitro* cultivation technique greatly increases the possibility to transfer the target genes of the annual wild barley species, which commonly show genetic barriers with cultivated barley caused by hybird sterility, such as *H. murinum*, *H. marinum* and *H. depressum*, into cultivated barley. *In vitro* cultivation technique can overcome genetic barriers, and may also induce various kinds of somaclonal variants through a cycle of *in vitro* cultivation. In the callus of primary sterile interspecific hybrids there may be possibilities for the induction of translocations or partial elimination of alien chromosomes with a subsequent establishment of substitution or addition lines. These variants should be useful in barley breeding.

2.3 Perennial Wild Barley Germplasm

During the past 30years, especially after the application of in vitro techniques to wild Hordeum species, there are a lot of studies on wild crosses between the perennial wild barley germplasm and cultivated barley. The interspecific aspect of wild crosses in barley has been quite extensive and the value of these perennial wild barley germplasms in barley breeding seems clear.

2.3.1 The Variation of Target Traits

As shown in Table 2.1, *H. vulgare* is one of the species in the genus *Hordeum* within the tribe *Triticeae*, which consists of 24 other species at three ploid levels. Among all perennial wild barley taxa, *H. bulbosum* is the most studied perennial wild species in *Hordeum*.

H. bulbosum, occurring as two cytotypes, diploid ($2n=2x=14$) and tetraploid ($2n=4x=28$), is a perennial outcrossing species found in the Mediterranean region, and is considered to be closely related to *H. vulgare*. The main reason for *H. bulbosum* becoming the most studied wild species in *Hordeum* is the utilization of its selective chromosome elimination system (so-called, bulbosum method), which has been widely used in breeding for the production of homozygous barley lines (doubled haploids). The other reason is that *H. bulbosum* has long been known to contain several interesting agronomic characters, especially resistant to powdery mildew, which is also the most studied pathogen in wild species of *Hordeum*, but no successful transfers to cultivated barley have been reported despite many attempts.

There is much research concerning the variation in resistance to powdery mildew and other pathogens in Hordeum, which show that *H. bulbosum* and other perennial wild barley taxa have genetic potential in improvement for resistance to the pathogens. Jones and Pickering (1978) studied the variation in resistance to powdery mildew in *H. bulbosum*. The isolates of mildew from two susceptible genotypes of *H. bulbosum* were found to be avirulent to more than 30 genotypes of cultivated barley. Some of the most susceptible barley lines (e.g. cvs Proctor and Golden Promise) were resistant to the *H. bulbosum* forms of the pathogens. One genotype of *H. bulbosum* was resistant to five isolates of powdery mildew of *H. vulgare*. Prasad *et al.* (1983) studied the resistance of six tetraploid strains of *H. bulbosum*, which also showed a large variation among individual plants in their resistance/susceptibility to three Japanese races of powdery mildew. Gustafasson and Claesson (1988) also made a study of the infection process in which they compared reactions in susceptible barley with those in more or less resistant or immune wild species. There was a significant difference in frequency of conidial germination among the species, from 35% in *H. brachyantherum* up to 79% in *H. jubtum*. A significant difference was also found between the susceptible cultivars of *H. vulgare*, in which up to 85% of the germinated spores formed well-developed primary haustoria. But the variation of primary haustoria in the wild species was only 2%~39%. Differences between compatible and incompatible reactions were most pronounced for penetration of secondary hyphae and formation of secondary haustoria. Very few abnormal haustoria were found in *H. bulbosum* and *H. procerum*. There were no secondary haustoria in all the other wild species.

Typhula blight, one of the major obstacles for enlarging cultivation of winter barley in more northern areas, is caused by two species of a parasitic fungus, namely *Typhula incarnata* and *T. ischicariensi*. It is thus a component of winter hardiness, which apart from cold tolerance also includes resistance to several different pathogens. Mielke (1980, unpublished, cited by Bothmer, 1992) studied the resistance against *T. incarnata* in eight *Hordeum* species, including *H. vulgare* cv. Dura of German origin as a standard, during two successive years. He found that *H. bogdanii*, *H. jubatum*, and *H. brevisubulatum subsp. violaceum* and *subsp, turkestanicum* were especially free of infection after 2 years and *H. vulgare subsp. spontaneum* had a similar degree of infection as the vulgare-line "Dura". Lundin (Hagberg, 1983, unpublished, cited by Bothmer, 1992) made a larger study in which 28 *Hordeum* species, with at least 2~3 populations of each, were screened for resistance to Typhula blight. He observed that there was particularly good resistance in some of the American species and the most prominent resistance reactions were found in some populations of *H. lechleri* and *H. pubiflorum* from Argentina.

There are some variations in resistance to loose smut, a world-wide disease in barley and wheat caused by the two embryo- infecting fungi (*Ustilago nuda* and *U. tritici*), in perennial wild barley germplasm. Nielsen (1978, 1987) investigated resistance to the disease in 19 *Hordeum* species, many of them with 2–3 accessions, and found good resistance to both species of *Ustilago* in *H. murinum*, *H. chilense*, *H. marinum*, and *H. depressum*, and several other species were resistant to one of the pathogens. Nielsen (1987) concluded that outside *Hordeum* only species of *Agropyron* (Elymus) are susceptible to both fungal pathogens. The perennial wild barley germplasms also have some potential for resistance to aphids and greenbug, which are regular and severe parasites of spring barley in many areas. Weibull (1987) studied resistance reactions against *Rhophalosiphum padii* in 27 accessions of 10 perennial Hordeum taxa and 6 inter-specific Fl hybrids between wild species and cultivated barley. Susceptibility was expressed as the total number of aphids per plant and per original female aphid after 16 days of exposure. The mean number of nymphs produced per female was only on average about 14% of that in the control barley cv. Tellus (206 nymphs per plant), with *H. bogdanii* being the most resistant species (4 nymphs per plant) and one accession of *H. lechleri* the most susceptible (58 nymphs per plant). The most resistant species were all diploids (*H. bogdanii*, *H. californicum* and *H. comosum*) with the inter-specific hybrids being more or less intermediate between the parents. A study at ICARDA in Syria showed that *H. secalinum* had excellent tolerance to greenbug infections (Bothmer, 1992).

2.3.2 Approaches for Utilization of Perennial Wild Barleys

A huge difficulty exists for inter-specific gene transfer through sexual hybridization between perennial wild barley and cultivated ones. The wide hybridization pulsing *in vitro* cultivation technique is the right way to overcome genetic barriers and transfer target genes from the wild relatives into cultivated barley. Although rather few investigations of *in vitro* cultivation techniques have been carried out on wild species of *Hordeum*, development of the technique in cultivated barley greatly increases the possibility of gene transfer from wild barley species into cultivated barley. In the process of distant hybridization and *in vitro* cultivation, some variants, having great potential for barley breeding, may be produced, such as haploids, somaclonal variations, spontaneous polyploidy and chromosomal variation (aneuploidy, chromosomes substitution translocations or addition lines, partial elimination, chromosomal rearrangements). *In vitro* cultures may be also a reasonable system for physiological study of stress tolerance or resistance. Haploids of barley were firstly obtained from *H. vulgare* × *H. bulbosum* due to the chromosome elimination of *H. bulbosum*. Utilization of the selective chromosome elimination system found in *H. bulbosum*, a method producing haploids, has been established, which has been widely used in breeding for the production of homozygous barley lines (doubled haploids). Some other haploids of barley were obtained from other perennial wild barley taxa. Orton (1980b) obtained some haploids of barley from a phase of callus culture of *H. vulgare* × *H. jubatum*. 5%∼10% of all regenerants were haploids resembling the *H. vulgare* parent. Some of the regenerants also exhibited activity of specific *H. jubatum* esterases, which indicated that chromosomal or sub-chromosomal substitutions or additions may occur *in vitro*. An interesting feature was that all *H. vulgare*-like regenerants (thus more or less barley haploids) could be maintained as perennial stocks, like the *H. jubatum* parent and in contrast to the annual *H. vulgare*. This showed that perennial characteristics of perennial wild barley taxa could be transferred into cultivated barley.

Orton (1980a) also observed a high frequency of chromosomal variation in the hybrid of *H. vulgare* × *H. jubatum*. There were a high frequency of spontaneous polyploidy, aneuploidy and chromosomal rearrangements in both callus and suspension cultures of the pure species and of the hybrid. The chromosomal aberrations were highly dependent on the differentiation stage, age and history of the culture. The genome interactions of the parental species were not responsible for the chromosomal variability. By applying colchicine in a low concentration, Orton and Steidl (1980) tried to induce the regeneration of chromosome doubled, fertile amphidiploids. Around 20% of the regenerated plants were aneuploid with a mean of $2n = 37$, and about 40% were stable and fertile amphidiploids of *H. vulgare* × *H. jubatum*. Jørgensen and Andersen (1989) studied the primary hybrids between cultivated barley and seven wild species (*H. brachyantherum*, *H. jubatum*, *H. roshevitzii*, *H. tetraploidum*, *H. brevisubulatum*, *H. lechleri* and *H. procerum*), one triple hybrid ((*H. lechleri*

× *H. procerum*) × *H. vulgare*) and one first backcross to cultivated barley of *H. lechleri* × *H. vulgare*. The phase of callus culture induced a wide variation of morphological and cytological variation as observed in the 1315 regenerated plants. Some regenerants looked more like *H. vulgare* than the primary hybrid, while others were more like the wild species parent. A significant increase in the proportion of variant plants (karyotypic and morphological variants) was observed after about 9 months in callus culture. The numerical chromosome changes comprised aneuploids with loss of individual chromosomes from either cultivated barley (chromosomes 5 or 7) or from the wild species parent. Seven amphiploids were regenerated from combinations with *H. jubatum* and *H. lechleri*. Structural rearrangements were identified including deletions, inversions (in *H. rosbevitzii* × *H. vulgare*), duplications and one possible case of translocation (*H. lechleri* × *H. vulgare*). The regenerated plants were pollen and seed sterile like the primary hybrids, except for a few plants, which were probably doubled haploids of *H. lechlen* arising from elimination of the barley chromosomes. Many regenerated plants were aneusomatic with normal cells together with aberrant ones. Of the cytologically studied materials, more than 50% of the plants had abnormal chromosome constitutions.

Picketing (1989) studied regeneration from callus and somaclonal variation in several inter-specific hybrids including *H. vulgare* × *H. bulbosum*. There was a large genotypic variation in regeneration capacity among the barley cultivars as well as among the *H. bulbosum* genotypes. He found 0%~33% of spontaneous doubling and all regenerated plants (except one amphidiploid plant) had chromosomes of $2n = 14$, thus being doubled haploids of cultivated barley. Eleven regenerants from vulgare-bulbosum (VB) callus had haploid and hybrid sectors. Fertility was restored after colchicine treatment. Apart from several cases of polyploidy and albinism, Picketing (1989) identified three variants, one of which showed maternal inheritance for yellow leaf striping, while the other two were controlled by single recessive genes.

The *in vitro* cultivation technique may be an ideal system for physiological studies of stress tolerance in crops. It may also be used as a rapid and relative cheap methods for screening and selection. Orton (1980c) used callus cultures for studies of salt tolerance in *H. vulgare* and *H. jubatum*. Growth and ion accumulation were measured under different salt concentrations. *H. jubatum* showed a considerably higher tolerance to high salt concentrations in comparison with cultivated barley. The tolerance is expressed at the tissue level and *H. jubatum* cultures were able to adjust to the salt stress up to 0.86 M, while the cultivated barley cultures did not recover after treatment with 0.17 M NaCl. *H. bulbosum* has been received with the greatest interest among wild species in *Hordeum* for utilization of the selective chromosome elimination system. Interest has also been focused on *H. bulbosum* for transfer of desirable genes. This species is the sole representative in the secondary gene pool of barley, and it is more compatible in crosses with cultivated barley than the remaining wild species. For this reason, *H. bulbosum* has been investigated in more details than the other *Hordeum* species. However, despite the high crossability and high meiotic pairing among the parental genomes, no fertile gamete has so far been reported in the primary hybrid between the two species at the diploid level (Picketing, 1988).

Martin (1988) reported that the South American diploid *H. chilense* was another considerably interesting wild species of *Hordeum* with a great agricultural potential. Both hexaploid (e.g. with *T. turgidum* and *T. dicoccoides*) and octaploid (with *T. aestivum* cv. Chinese Spring) fertile amphiploids have been produced, which all have higher protein content than the wheat parents and small but not shriveled kernels (Martin and Chapman, 1977; Martin and Cubero, 1981). Fernandez and Jouve (1988) reported that they got addition lines of *H. chilense* chromosomes to *T. turgidum* convardurum and found 50 monosomic or polysomic addition forms. Miller *et al.* (1981, 1985) reported that they got the addition and substitution lines in *T. aestivum* cv. Chinese Spring. Forster *et al.* (1990) reported that there existed a difference between *H. vulgare* and *H. chilense* in controlling tolerance to salinity.

2.4 Cultivated Barleys

Cultivated barleys, i.e. *H. vulgare*, mainly including primitive barleys or landraces, commercial varieties, mutants and some offspring derived from the process of genetic improvement, are the main genetic resource in barley breeding. Generally, cultivated barleys possess almost all the target genes we need in breeding. More exploration, collection and evaluation of cultivated barleys are needed for barley production and breeding.

2.4.1 Primitive Barleys or Landraces

Primitive barleys or landraces are the old cultivated barley varieties (*H. vulgare* subsp.) which were planted somewhere in the past. Now they can be only found in a barley breeder's experimental field or in the national and international gene banks. Generally, most of them are conserved in the gene banks. There are 9 countries in which the number of primitive barleys or landraces exceed 10,000, such as Canada, U.S.A, China and ICARDA in Syria. They have been screened for the sources of disease resistance, yield and quality improvement. Compared with the yield and quality characters, primitive barleys or landraces are considered to have a greater potential in abiotic and biotic resistance for barley breeding.

Many resistances or resistant genes to pathogens, such as barley leaf rust, yellow rust and powdery mildew have been identified in many primitive lines or populations of barley (*H. vulgare* L.). Some of them have been transferred by backcrossing into commercial varieties or advanced lines. In fact, all the powdery mildew resistances used commercially except one, the "spontaneum" resistance, are derived from primitive *H. vulgare*, mainly from West Asia,

Ethiopia and North Africa (Jørgensen, 1992). In many cases one or more additional resistant genes have been introduced, several of which are more or less tightly linked to the "main" gene. Eighteen different powdery mildew resistances and genes are recognized in this way (Jørgensen, 1992) in European spring and winter barley varieties. In addition, several more have been recognized very recently or are designated "unknown'. Another reason for the discrepancy between the about ten unique "sources" and the 18 resistances is that the "main" gene may have been introduced more than once, for example Mla 7 has been introduced from four different sources accompanied by four apparently different "additional" resistant genes.

The landraces of cultivated barley in Europe often have higher potential in malting quality and resistance due to their long malting history and the humid weather. The landraces of cultivated barley in China often have higher potential in early maturity, short straw, yield component, resistance to yellow rust and yellow mosaic virus, tolerance to waterlogging and acid soil due to the longer frost stage in the west area and humid weather in the east area. In China, three kinds of dwarf germplasms, i.e. semi-dwarf barley, such as Lixiahuangmimai (a six-rowed naked barley, from Xiaoshan city, Zhejiang province); dwarf barley, such as Qinghaizao 69, Ganzi813 (six-rowed naked barley, from Sichuan and Qinghai provinces, respectively); erectoide-dwarf barley, called ert-dwarf germplasm (found in Sichuan Province) have been found. Xu (1993) developed four naked barley varieties using ert-dwarf germplasm as breeding material. In the Middle East, barley landraces often have higher potential in tolerance to drought and pathogens, including powdery mildew and rust. In America, the landraces often have higher potential in early maturity, yield component and resistance to some pathogens, such as powdery mildew and rust.

2.4.2 Commercial Varieties

Commercial varieties are developed by breeders and planted now. Generally, they possess an appreciable level of resistance to several diseases, higher yield and quality potential than landraces.

There are quite a few reports about the resistance of commercial varieties in the world, especially in Europe. Jørgensen (1992) considered their resistance to pathogens mainly represented high or partial resistance accumulated during evolution and cultivation. In the case of powdery mildew, a considerable level of partial resistance is present in the northwest European barley germplasm, compared with the extreme susceptibility of barley lines/ varieties from northern Scandinavia or from more remote areas where the disease has not been prevalent for centuries. Knudsen *et al.* (1987) thought that the greater part of the resistance accumulated over centuries still appears to be present in many modern varieties. The use of race-specific resistance genes in barley breeding during the last few decades has relaxed the selection pressure necessary to maintain this level of partial resistance. Severe epidemics of net blotch, which was not prevalent on spring barley in northern areas in the early 1980s disclosed large varietal differences in resistance, from close to that known in recognized but primitive sources of resistance, to extreme susceptibility (Skou and Haahr, 1987a). Skou and Haahr (1987b) reported that the abolition of mercury compounds for seed-dressing encouraged extensive screening of old and new European barleys for resistance to the seed-borne barley leaf stripe [*Drechslera graminea* (Rabenh ex Schlecht) Shoe]. Subsequent studies of a cross of a highly resistant European variety and an extremely susceptible line revealed a single resistance gene (Haahr *et al.*, 1989), which turned out to be race- specific (Knudsen, 1986). In addition, the progeny that did not possess this gene exhibited a wide range in partial resistance originating from the European parent. The same range in partial resistance was seen in the screening of varieties. Arru *et al.* (2003) found some isolate-specific QTLs of resistance to leaf stripe in the "Steptoe" (partially resistant) "Morex" (susceptible) spring barley cross.

Barley breeders in the world, especially in Europe, benefit from a fortunate situation in which resistances or genes are "floating around" in commercial varieties, and thus probably also in their own breeding materials. Consequently, testing for susceptibility to pathogens such as net blotch and leaf stripe is advisable to eliminate breeding lines that are insufficiently resistant.

Commercial varieties have also constituted the main "source" of resistance in barley breeding for powdery mildew resistance in the world. Only about ten unique resistances have been introduced, mainly by individual scientists, but these resistances have been used extensively as resistance "sources" by practically all European barley breeders. In 1992, Jørgensen pyramidized these resistances to produce multigene-resistant varieties.

Malt quality of commercial varieties, from barley-exporting countries, such as Esterel and Alexi from France, Harrington and Stein from Canada, Schooner, Hamelin and Baudin from Australia, are better than those from barley-imported nations. Marquez-Cedillo *et al.* (2000) carried out QTL analysis of malt quality in barley based on the doubled-haploid progeny of two elite North American varieties representing different germplasm groups. They found that some QTLs were associated with seven grain and malt quality traits, such as percentage of plump kernels, test weight, grain protein content, soluble/total protein ratio, α-amylase activity, diastatic power and malt-extract percentage. The results showed that commercial barley varieties already possessed almost all good malting genes for barley breeding.

2.4.3 Mutants

Crop breeders expected that induced mutations would provide new genetic resources for improvement of disease resistance, quality and yield potential in the early phase of the "mutation era". They thought that mutants could be used to improve highly bred genotypes especially in resistance. However, the results of mutation in resistance have been disappointing. The crop breeders have got no new sources of resistance in cereals. Some scientists guess that this may be because the usually dominant gene-for-gene resistance genes all have a unique and highly specific function, whereas most mutations disrupt gene functions, leading to recessive genes with nonfunctional products. The production of a highly specific gene function by induced mutation is very unlikely. One outstanding exception is the mutagen-induced Mlo resistance to powdery mildew in barley, which is derived from a spontaneously occurring Mlo gene present in Ethiopian landrace barleys. Extensive studies on mutagen-induced Mlo resistance will increase the opportunity to use Mlo resistance in barley breeding and production in the world.

However, the results are exciting in inducing new sources of quality and some agronomical traits, such as high protein and lysine contents, dwarf and early maturity. Barley researchers have obtained many mutants, some of which have become commercial varieties or have been successfully used in barley breeding.

Hiproly (CI 3947), a mutant barley line with high crude protein and lysine contents, is a well-known mutant in the world. Munck *et al.* (1970, 1972, 1992) tested the environmental stability of high-lysine trait in relation to the crude protein content, which was established for Hiproly (CI 3947) grown at various localities and in the phytotrons, and was compared with its putatively isogenic sister line (CI 4362) as well as with commercial barley varieties. The lysine content of Hiproly in relation to its sister line was increased by 35%. In the pedigree material, the difference in lysine content based on protein between the high-lysine lines and the normal ones was reduced to 23% (Munck, 1972). The yield of Hiproly was lower than that of CI 4362 and only about 30% that of commercial barley varieties, due to smaller shriveled seeds. However, high-lysine plants from crosses between Hiproly and normal varieties were only 10% lower in yield compared with the normal lysine segregants from the same cross. The high-lysine seeds were plump although they are 15% smaller in size (Munck, 1972). The high lysine gene was preliminarily called *lys1* (Munck, 1972) and was found to be recessive and located near to the characters (short rachilla hairs) on the long arm of chromosome 7 (Karlsson, 1972). The *lys1* gene was associated with a hard endosperm which, however, could be modified into a soft high-lysine type found at a low frequency in crosses.

Mutant φ1508 is another very well known mutant, which was induced with ethylenimine in Bomi barley (Ingversen *et al.*, 1973). Mutant φ1508 displays a lysine increase in the barley protein of no less than 44%, approaching the nutritional value of milk protein, and associated increases in several other essential amino acids. The φ1508 mutant is highly variable in yield compared to Bomi, ranging in different years from a relative figure of 65% to 93%. It is about 10% smaller in seed size and the seed is shriveled. Jensen (1979) found that the gene *lys3a*, causing the high-lysine trait, was recessive and situated in the centromere region of chromosome 7. Two other alleles, mutant 18 (*lys3b*) and mutant 19 (*lys3c*), were found in Bomi material treated with sodium azide. A fourth allelic mutant, mutant 1460, was identified in a sodium azide-treated material of the high β-glucan variety "Minerva" by analyzing for low β-glucan content (namely, viscosity of acid barley extracts) by Aastrup from their research group in 1983 (Munck ,1992). Allison *et al.* (1979) observed that the φ1508 mutant was softer and gave less resistance in milling than the wild variety Bomi. The mutant has an opaque phenotype when compared with the wild variety in transmitted polarized light and, in contrast to *lys1*, it has a 50% increased level of free amino acids in the seed, as the opaque-2 mutant in maize. The prolamin fraction is reduced in mutant φ1508 by 66%, while the albumin and globulin fractions are increased correspondingly, and the glutelin fraction remains constant. Balasaraswathi *et al.* (1984) found that the *lys1* lines had 16% lower grain yield and 13% lower seed weight than the corresponding normal lines. Hordeins were reduced from 49.3% in the normal to 43.0% in the high-lysine lines. Using similar material, Doll and Koie (1978) observed that the *lys3a* gene caused yield reduction of 20% in the high-lysine condition. More than 90% of this reduction was attributed to lower carbohydrate production. The percentage of total DMSO-extractable carbohydrates including starch was on average 54.9% in grains from normal lines and 45.7% in high-lysine lines. Overall, the increase of lysine-rich endosperm proteins with a minimum interference with prolamin production will bring a negative effect on carbohydrate (starch) production and yield. Mutant φ1508 has been successfully used in barley breeding in China, with two commercial varieties being released (Chuan Nong Da 2, two-rowed feed barley with high protein content, released in 1995; Huadamai 1, two-rowed naked barley with high protein, released in 2001).

Mutant 7 is the only high-lysine barley mutant that does not have any reduction in seed size and starch content compared to the wild line (Koie and Doll, 1979). It has double recessive genes, i.e. *lys3a* and *lys1*. The mutant displays a low (15%) increase in lysine content. Compared with Bomi, the yield of mutant 7 only decreased by 10%, mainly due to a reduced seed number. The mutant also has a normal level of free sugars, differing from mutant φ1508. In contrast, it resembles mutant φ1508 in showing a decrease in prolamins, although only 20%, and a corresponding increase in non-prolamin proteins. Tallberg (1981) thought that both the Hiproly *lys1* gene and the mutant 7 gene were characterized by the fact that a significant part of the improvement in lysine content come from proteins in the glutelin fraction.

Tallberg and Eggum (1981) found that the double recessive between *lys1* and the high-lysine gene of mutant 7 was agriculturally much more promising with a lysine increase of about 35%. The mutant also displayed completely plump seeds, although with a 20% reduction in seed size. They observed that the *lys1*/mutant 7 double recessive had a reduction in available carbohydrates of only 5% and the hordein content was unchanged, compared with mutant 7. The increased glutelin fraction was the major contributor to the increased lysine content in the *lys1*/mutant 7 double recessive in spite of the fact that it had higher lysine content in the water-soluble protein fraction than even the Hiproly backcross line.

Doll (1976) created another theoretically interesting recessive mutant 56, which was induced in Carlsberg II barley by X-rays. Seed size is reduced by 10%, starch by 8% and prolamins by 23%, while lysine content is increased by 24%, compared with Carlsberg II (Koie and Doll, 1979). The changes in amino acid pattern of mutant 56 in comparison with normal barley show similarities to those of mutant 7 and *lys1*. The low content of the prolamin fraction in mutant 56 is due to a mutation at or near the Hor2 locus coding hordein-2 (B hordein) polypeptides in chromosome 5, making the mutant deficient in these polypeptides. Therefore, Doll (1980) called the mutation site '*hor2ca*', which was probably the first case to be identical to a structural locus for a protein (hordein).

Another outstanding example is the mutagen-induced barley variety Yanfuaizao No 3, an excellent commercial variety in China, which is derived from a commercial variety Zaoshu 3. Compared with the original variety, the mutant reduced the plant height by 20 cm, thus improving lodging resistance. In China this variety had been planted for more than 10 years and the total planting area was over 4 million ha. In addition, we have also found some mutants from the annual wild barleys of the Qing-Tibetan Plateau treated with sodium azide or X-rays, such as Huaai 11, characterized by short straw (35~45 cm) and early maturity, which are now being used in barley breeding programs.

References

Åberg (1938) Hordeum agriocrithon nova sp A Wild Six-rowed Barley. Ann Roy Agric Coll Sweden, 6: 159-216

Abbott D C, Brown A H D, Burdon J J (1992) Genes for Scald-resistance from Wild Barley (*Hordeum vulgare* ssp. *spontaneum*) and Their Linkage to Isozyme Markers. Euphytica, 61: 225-231

Abbott D C, Lagudah E S, Brown A H D (1995) Identification of RFLPs Flanking a Scald Resistance Gene on Barley Chromosome 6. J Heredity, 86: 152-154

Ahokas H (1977) Lysine and Tryptophan Content of Some High-protein Stocks and Species of Hordeum. Hereditas, 86: 143-145

Ahokas H (1982) Variation of Kernel Protein and Lysine in the Wild Progenitor of Barley. Hereditas, 96: 29-37

Ahokas H, Nashali L (1990) Variation of α-amylase, β-amylase, β-glucanase, Pullanase, Proteinase and Chitinase Activity in Germinated Samples of the Wild Progenitor of Barley. J Inst Brew, 96: 27-32

Ahokas H, Poukkula M (1999) Malting Enzyme Activities, Grain Protein Variation and Yield Potentials in the Displaced Genetic Resources of Barley Landraces of Finland. Genet Resour Crop Ev, 46: 251-260

Ahokas H, Uutela P, Erkkilä M J, Vähämiko S (1996) Another Source of Genes with High β-amylase Activity in Barley Grain: Finnish landraces. Barley Genet Newsl, 25: 36-40

Allison M J, Borzacki R, Cowe I A, McHale R (1979) Variation in Barley Collection for Endosperm Attributes that Relate to Malting Quality. J Inst Brew, 85: 86-89

Anikster Y, Moseman J G, Wahl I (1975) Parasite Specialization of *Puccinia hordei* Otth. and Sources of Resistance in *Hordeum sponluneum* C. Koch. Barley Genetics 3 (Proc. 3rd Int. Barley Genet. Syrup., Garching, 412-417

Arru L, Francia E, Pecchioni N (2003) Isolate-specific QTLs of Resistance to Leaf Stripe (Pyrenophora graminea) in the "Steptoe Morex" Spring Barley Cross. Theor Appl Genet, 106: 668-675

Backes G, Madsen L H, Jaiser H, Stougaard J, Herz M, Mohler V, Jahoor A (2003) Localization of Genes for Resistance against *Blumeria graminis* f. sp. *hordei* and *Puccinia graminis* in a Cross Between a Barley Cultivar and a Wild Barley (*Hordeum vulgare* ssp. *spontaneum*) line. Theor Appl Genet, 106: 353-363

Balasaraswathi R, Koie B, Doll H (1984) The Concentration and Yield of Hordein and Some Lysine-rich Proteins as Influenced by the lys Gene of Hiproly Barley. Hereditas, 100: 225-231

Baum B R (1983) Morphometric Relationships in *Hordeum-vulgare Triticae-Poaceae*. 2. *Hordeum agribcrithon*, *Hordeum-disticbum*, *Hordeum-lagunchliforme*, *Hordeum spontneum* and *Hordeum vulgare*. Can J Bot, 61: 2023-2031

Baum M, Grando S, Backes G, Jahoor A, Sabbagh A, Ceccarelli S (2003) QTLs for Agronomic Traits in the Mediterranean Environment Identified in Recombinant Inbred Lines of the Cross "Arta" x *H. spontaneum* 41-1. Theor Appl Genet, 107: 1215-1225

Bilgic H, Steffenson B J, Hayes P (2005) Comprehensive Genetic Analyses Reveal Differential Expression of Spot Blotch Resistance in Four Populations of Barley. Theor Appl Genet, 111: 1238-1250

Borovkova I G, Jin Y, Steffenson B J, Kilian A, Blake T K, Kleinhofs A (1997) Identification and Mapping of Leaf Rust Resistance Gene in Barley Line Q21861, Genome, 40: 236-241

Borovkova I G, Jin Y, Steffenson B J (1998) Chromosomal Location and Genetic Relationship of the Leaf Rust Resistance Genes Rph9 and Rph12 in Barley. Phytopathology, 88: 76-80

Bothmer R von, Flink J, Landström T (1986) Meiosis in Interspecific Hordeum Hybrids. I. Diploid Combinations. Can J Genet Cytol, 28: 525-535

Bothmer R von, Flink J, Landström T (1987) Meiosis in Interspecific Hordeum Hybrids. II. Triploid Hybrids. Evol Trends Plants, 1: 41-50

Bothmer R von, Flink J, Jacobsen N, Jørgensen, R B (1989a) Variation and Differentiation in Hordeum Marrinum. Nord J Bot, 9: 1-10

Bothmer R von, Flink J, Landström T, Thomas H M (1989b) Meiosis in Interspecific Hordeum Hybrids. V. Pentaploid Hybrids. Hereditas, 110: 217-226

Bothmer R von (1992) The Wild Species of Hordeum: Relationships and Potential Use for Improvement of Cultivated Barley. In: Peter R Shewry (ed.) Barley: Genetics, Biochemistry, Molecular Biology and Biotechnology. Wallingford, UK: CAB International, 3-18

Briggs D E (1978) The Origin and Classification of Barleys. In "Barley", London: Chapman and Hall, 76-88

Brown A H D, Nevo E, Zohary D (1978) Outcrossing Rates and Hetereozygosity in Natural Populations of *Hordeum spontaneum* in Israel. Heredity, 41: 49-62

Brown A H D, Jacobsen J (1982) Genetic Basis and Natural Variation of α-amlyase Isozymes in Barley. Genet Res (Cambridge), 40: 315-324

Brown A H D, Munday J (1982) Population Genetic Structure and Optimal Sampling of Land Races of Barley from Iran. Genetica, 58: 85-96

Brunner S, Keller B, Feuillet C (2000) Molecular Mapping of the Rph7 Leaf Rust Resistance Gene in Barley (*Hordeum vulgare* L.). Theor Appl Genet, 101: 783-788

Cattivelli L, Baldi P, Crosatti C, Fonzo N D, Faccioli P, Grossi M, Mastrangelo A M, Pecchioni N, Stanca A M (2002) Chromosome Regions and Stress-related Sequences Involved in Resistance to Abiotic Stress in Triticeae. Plant Mol Biol, 48: 649-665

Chakrabarti T, Subrahmanyan N C (1986) Analysis of DNA from Related and Diverse Species of Barley. Plant Sci, 42: 183-190

Chalmers K J, Waugh R, Watters J, Forster B P, Nevo E, Abbott R J, Powell W (1992) Grain Isozyme and Ribosomal DNA Variability in *Hordeum spontaneum* Populations from Israel. Theor Appl Genet, 84: 313-322

Clegg M T, Brown A H D, Whitefeld P R (1984) Chloroplast DNA Diversity in Wild and Cultivated Barley: Implications for Genetic Conservation. Genet Res (Cambridge), 43: 339-343

Choi D W, Zhu B, Close T J (1999) The Barley (*Hordeum vulgare* L.) Dehydrin Multigene Family: Sequences, Allelic Types, Chromosome Assignments, and Expression Characteristics of 11 Dhn Genes of Cv Dicktoo. Theor Appl Genet, 8: 1234-1247

Choi D W, Close T J (2000) A Newly Identified Barley Gene, Dhn12, Encoding YSK2 DHN, is Located on Chromosome 6H and has Embryo-specific Expression. Theor Appl Genet, 100: 1274-1278

Corke H, Atsmon D (1988) Amino Acid Resources of the Wild Relative of Barley *Hordeum spontaneum* in Israel. Plant Breeding, 101: 89-94

Corke H, Nevo E, Atsmon D (1988) Variation in Vegetative Parameters Related to the Nitrogen Economy of Wild Barley, *Hordeum spontaneum*, in Israel. Euphytica, 39: 227-232

Czembor J H (2000) Resistance to Powdery Mildew in Populations of Barley Landraces from Morocco. Aust Plant Pathol, 29: 137-148

Dewey D R (1984) The Genomic System of Classification as a Guide to Intergeneric Hybridization with the Perennial Triticeae. In: Gustafson J P (ed.) Gene Manipulation in Plant Improvement, (16th Stadler Genetics Symposium), 209-279

Doebley J, Bothmer R von, Larson S (1992) Chloroplast DNA Variation and Phylogeny in the Genus Hordeum (Poaceae). Am J Bot, 79: 576-584

Doll H (1976) Genetic Studies of High-lysine Mutants. In: Barley Genetics III. Verlag Karl Thiemig, München, Germany, 542-546

Doll H (1980) A Nearly Non-functional Mutant Allele of the Storage Protein Locus Hor2 in Barley. Hereditas, 93: 217-222

Doll H, Koie B (1978) Influence of the High Lysine Gene from Barley Mutant 1508 on Grain, Carbohydrate and Protein Yield. In: Seed Protein Improvement by Nuclear Techniques. IAEA, Vienna, Austria, 107-114

Eglinton J K, Evans D E, Brown A H D, Langridge P, McDonald G, Jefferies S P, Barr A R (2001) The Use of Wild Barley (*Hordeum vulgare* ssp. *spontaneum*) in Breeding for Quality and Adaptation. Proc 10th Aust Barley Tech Symp. Canberra ACT, Australian, 16-20, September

Eglinton J K, Langridge P, Evans D E (1998) Thermo-stability Variation in Alleles of Barley Beta-amylase. J Cereal Sci, 28: 301-309

Ellis R P, Forster B P, Waugh R, Bonar N, Handley L L, Robinson D, Gordon D C, Powell W (1997) Mapping Physiological Traits in Barley. New Phytol, 137: 149-157

Ellis R P, Foster B P, Robinson D, Handley L L, Gordon D C, Russell J R, Powell W (2000) Wild Barley: A Source of Genes for Crop Improvement in the 21st Century. J Exp Bot, 51: 9-17

Ellis R P, Forster B P, Gordon D C, Handley L L, Keith R, Lawrence P, Meyer R C, Powell W, Robinson D, Scrimgeour C M, Young G R, Thomas W T B (2002) Phenotype/genotype Associations of Yield and Salt Tolerance in a Barley Mapping Population Segregating for two *dwarfing* Genes. J Exp Bot, 53: 1163-1176

Evans D E, Wallace W, Lance R C M, MacLead L C (1997) Measurement of Beta-amylase in Malting Barley (*Hordeum vulgare* L.). II. The Effect of Germination and Kilning. J Cereal Sci, 26: 241-250

Fetch T G, Steffenson Jr B J, Nevo E (2003) Diversity and Sources of Multiple Disease Resistance in *Hordeum spontaneum*. Plant Dis, 87: 1439-1448

Fernandez J A, Jouve N (1988) The Addition of Hordeum Chilense Chromosomes to *Triticum turgidum* conv. durum. Biochemical, Karyological and Morphological Characterization. Euphytica, 37: 247-259

Feuerstein U, Brown A H D, Burdon J J (1990) Linkage of Rust Resistance Genes from Wild Barley (*Hordeum spontaneum*) with Isozyme Markers. Plant Breeding, 104: 318-324

Freialdenhoven A, Peterhänsel C, Kurth J, Kreuzaler F, Schulze-Lefert P (1996) Identification of Genes Required for the Function of Non-race-specific mlo Resistance to Powdery Mildew in Barley. Plant Cell, 8: 5-14

Fischbeck G, Schwarzbach E, Sobek F, Wahl I (1976) Mehltauresistenz aus Israelischen Populationen der Zwizeiligcn Wildgerste (*Hordeum spontaneum*). Ann Pflanzenzucht, 76: 163-166

Forster B P, Phillips M S, Miller T E, Baird E, Powell W (1990) Chromosome Location of Genes Controlling Tolerance to Salt (NaCl) and Vigor in *Hordeum vulgare* and *H. chilense*. Heredity, 65: 99-107

Freialdenhoven A, Scherag B, Hollricher K, Collinge D B (1994) Thordal-Christensen H, Schulze-Lefert P Nar-1 and Nar-2, two Loci Required for Mla12-specified Race-specific Resistance to Powdery Mildew in Barley. Plant Cell, 6: 983-994

Garvin D F, Brown A H D, Burdon J J (1997) Inheritance and Chromosome Locations of Scald-resistance Genes Derived from Iranian and Turkish Wild Barleys. Theor Appl Genet, 94: 1086-1091

Garvin D F, Brown A H D, Raman H, Read B J (2000) Genetic Mapping of the Barley Rrs14 Scald Resistance Gene with RFLP, Isozyme and Seed Storage Protein Markers. Plant Breeding, 119: 193-196

Genger R K, Brown A H D, Knogge W, Nesbitt K, Burdon J J (2003a) Development of SCAR Markers Linked to a Scald Resistance Gene Derived From Wild Barley. Euphytica, 134: 149-159

Genger R K, Williams K J, Raman H, Read B J, Wallwork H, Burdon J J, Brown A H D (2003b) Leaf Scald Resistance Genes in *Hordeum vulgare* and *Hordeum vulgare* ssp *spontaneum*: Parallels Between Cultivated and Wild Barley. Aust J Agric Res, 54: 1335-1342

Genger P K, Nesbitt K, Brown A D H, Abbott D C, Burdon J J (2005) A Novel Barley Scald Resistance Gene: Genetic Mapping of the Rrs15 Scald Resistance Gene Derived From Wild Barley, *Hordeum vulgare* ssp.*spontaneum*. Plant Breeding, 124: 137-141

Giles B E, Barrett J (1983) *Erysipbe graminis* Resistance in the *Hordeum murinum* Complex. Barley Genet Newsl, 13: 78-82

Griffey C A, Das M K, Baldwin R E, Waldenmaier C M (1994) Yield Losses in Winter Barley Resulting from a New Race of *Puccinia hordei* in North America. Plant Dis, 78: 256-260

Gupta P K, Fedak G, Molnar S J, Wheatcroft R (1989) Distribution of a *Secale cereale* DNA Repeat Sequence Among 25 Hordeum Species. Genome, 32: 383-388

Gustafsson M, Claesson L (1988) Resistance to Powdery Mildew in Wild Species of Barley. Hereditas, 108: 231-237

Haahr V, Skou J P, Jensen H P (1989) Inheritance of Resistance to Barley Leaf Stripe (*Drechslera graminea*) (Abstract) Vorträge für Pflanzenzüchtung, 15: 3-15

Harlan J R (1968) On the Origin of Barley. Barley: Origin, Botany, Culture, Winter Hardiness, Genetics, Utilization, Pests. USDA, Agricultural Handbook No.338, 9-31

Harlan J R (1971) On the Origin of Barley : A Second Look. In: Barley Genetics II, 45-49

Harlan J R (1976) Barley *Hordeum vulgare* (*Gramineae-Triticinae*). In: Simmonds N W (ed.) Evolution of Crop Plants. Longman, Essex, 93-98

Holwerda B C, Jana S, Crosby W L (1986) Chloroplast and Mitochondrial DNA Variation in *Hordeum vulgare* and *Hordeum spontaneum*. Genetics, 114: 1201-1292

Hintum T J L van, Menting F (2000) Barley Genetic Resources Conservation-now and Forever. In Barley Genetics VIII, 13-20

Ingversen J, Koie B, Doll H (1973) Induced Seed Protein Content of Barley Endosperm. Experientia, 29(1151): l-2

Ivandic V, Walther U, Graner A (1998) Molecular Mapping of a New Gene in Wild Barley Conferring Complete Resistance to Leaf Rust (*Puccinia hordei* Otth). Theor Appl Genet, 97: 1235-1239

Ivandic V, Hackett C A, Zhang Z J, Staub J E, Nevo E, Thomas W T B, Forster B P (2000) Phenotypic Responses of Wild Barley to Experimentally Imposed Water Stress. J Exp Bot, 51: 2021-2029

Jaaska V, Jaaska V (1986) Isoenzyme Variation in the Barley Genus *Hordeum*. I. Alcohol Dehydrogenase and Superoxide Dismutase. Biochemie und Physiologie der Pflanzen, 181: 301-320

Jaradat A A (1989a) Diversity Within and Between Populations of two Sympatrically Distributed *Hordeum-spp*. in Jordan. Theor Appl Gen, 78: 653-656

Jaradat A A (1989b) Ecotypes and Genetic Divergence Among Sympatrically Distributed Populations of *Hordeum-vulgare* and *Hordeum-spontaneum* from the Xeric Region of Jordan. Theor Appl Genet, 78: 857-862

Jana S, Pietrzak L N (1988) Comparative Assessment of Genetic Diversity in Wild and Primitive Cultivated Barley in a Center of Diversity. Genetics, 119: 981-990

Jensen J (1979) Location of a High-lysine Gene and the DDT-resistance Gene on Barley Chromosome 7. Euphytica, 24: 47-56

Jin Y, Statler G D, Franckowiak J D, Steffenson B J (1993) Linkage Between Leaf Rust Resistance Genes and Morphological Markers in Barley. Phytopathology, 83: 230-233

Jin Y, Steffenson B J, Bockelman H E (1995) Evaluation of Cultivated and Wild Barley For Resistance to Pathotypes of *Puccinia hordei* with Wide Virulence. Genet Resour Crop Ev, 42: 1-6

Jones I T, Pickering R A (1978) The Mildew Resistance of *Hordeum bulbosum* and its Transference into *H. vulgare* genotypes. Ann Appl Biol, 88: 295-298

Jørgensen R B (1982) Biosystematics of *Hordeum bulbosum* L. Nordic Journal of Botany, 2: 421-434

Jørgensen R B, Andersen B (1989) Karyotype Analysis of Regenerated Plants from Callus Cultures of Interspecific Hybrids of Cultivated Barley (*Hordeum vulgare* L.). Theor Appl Gen, 77: 343-351

Jørgensen J H (1992) Sources and Genetics of Resistance to Fungal Pathogens. In: Peter R Shewry (ed.) Barley: Genetics, Biochemistry, Molecular Biology and Biotechnology. Wallingford, UK: CAB International, 441-458

Jørgensen J H (1993) Durability of Resistance in the Pathosystems: Barley-powdery Mildew. In: Jacobs T H, Parlevliet J E (eds) Durability of Disease Resistance, Kluwer Academic Publishers, The Netherlands, 159-176

Jørgensen J H (1994) Genetic of Powdery Mildew Resistance in Barley. Crit rev plant sci, 13: 97-119

Kaneko T, Kihara M, Ito K, Takeda K (2000) Molecular and Chemical Analysis of β-amylase-less Mutant Barley in Tibet. Plant Breeding, 119: 383-387

Karlsson K E (1972) Linkage Studies on a Gene for High Lysine Content in Hiproly Barley. Barley Genet Newsl, 2: 34

Kasha K J, Sadasivaiah R S (1971) Genome Relationships Between *Hordeum vulgare* L. and *H. bulbosum* L. Chromosoma, 35: 264-287

Kicherer S, Backes G, Walther U, Jahoor A (2000) Localizing QTLs for Leaf Rust Resistance and Agronomic Traits in Barley (*Hordeum vulgare* L.). Theor Appl Genet, 100: 881-888

Kintzios S, Jahoor A, Fischbeck G (1995) Powdery Mildew Resistance Genes Mla29 and Mla32 in *H. spontaneum* Derived Winter Barley Lines. Plant Breeding, 114: 265-266

Knudsen J, Chr N (1986) Resistance to Barley Leaf Stripe. Zeitschrift für Pflanzenzüchtung, 96: 161-168

Knudsen J, Chr N, Dalsgaard H H, Jørgensen J H (1987) Partial Resistance to Barley Powdery Mildew. In: Yasuda S, Konishi T (eds) Barley Genetics V, Okayama, Japan, 645-650

Koie B, Doll H (1979) Protein and Carbohydrate Components in the Rise High Lysine Barley Mutants. In: Seed Protein Improvement Cereals and Gram Legumes. IAEA, Vienna, Austria, 205-215

Korff von M, Wang H, Leon J, Pillen K (2005) AB-QTL Analysis in Spring Barley. I Detection of Resistance Genes Against Powdery Mildew, Leaf Rust and Scald Introgressed from Wild Barley. Theory Appl Genet, 111: 583-590

Ladizinsky G (1975) Collection of Wild Cereals in the Upper Jordan Valley. Econ Bot, 29: 264-267

Landström T, Bothmer R von, Dewey D R (1984) Genomic Relationships in the *Hordeum brevisubulatum* Complex. Can J Genet Cytol, 26: 569-577

Li C D, Lance R, Tarr A, Broughton S, Harasymow S, Appels R, Jones M (2004) Improvement of Barley Malting Quality using a Gene from *Hordeum spontaneum*. VI. Intl Barley Genet Symp, Czech Republic

Linde-Laursen I, Bothmer R von, Jacobsen N (1980) Giemsa C-banding in Asiatic Taxa of Hordeum Section Stenostachys with Notes on Chromosome Morphology. Hereditas, 93: 235-254

Linde-Laursen I, Bothmer R von (1989) Giemsa C-banded karyotypes of *Hordeum marinum* and *H. murinum*. Genome, 32: 629-639

Linde-Laursen I, Bothmer R von, Jacobsen N (1990) Giemsa C-banded Karyotypes of South and Central American (Poaceae). II. 6 Polyploid Taxa, 112: 93-107

Liviero L, Maestri E, Gulli M, Nevo E, Marmiroli N (2002) Ecogeographic Adaptation and Genetic Variation in Wild Barley:Application of Molecular Markers Targeted to Environmentally Regulated Genes. Genet Resour Crop Ev, 49: 133-144

Mammadov J A , Zwonitzer J C, Biyashev R M, Griffey C A, Jin Y, Steffenson B J, Saghai M A (2003) Molecular Mapping of Leaf Rust Resistance Gene Rph5 in Barley. Crop Sci, 43: 388-393

Mano Y, Nakazumi H, Takeda K (1996) Varietal Variation in and Effects of some Major Genes on Salt Tolerance at the Germination Stage in Barley. Breeding Sci, 46: 227-233

Mano Y, Takeda K (1995) Varietal Variation and Effects of some Major Genes on Salt Tolerance in Barley Seedlings. Bull Res Inst Bioresour Okayama Univ.3: 71-81

Mano Y, Takeda K (1997) Mapping Quantitative Trait Loci for Salt Yolerance at Germination and the Seedling Stage in Barley (*Hordeum vulgare* L.). Euphytica, 94: 263-272

Mano Y, Takeda K (1998) Genetic Resources of Salt Tolerance in Wild *Hordeum* Species. Euphytica, 103: 137-141

Marquez-Cedillo L A, Hayes P M, Jones B L, Kleinhofs A, Legge W G, Rossnagel B G, Sato K (2000) QTL Analysis of Malting Quality in Barley Based on the Doubled-haploid Progeny of two Elite North American Varieties Representing Different Germplasm Groups. Theor Appl Genet 101: 173-184

Martin A (1988) Tritordeum: The First Ten Years. Rachis, 7: 12-15

Martin A, Chapman V (1977) A Hybrid between *Hordeum Chilense* and *Triticum Aestivum*. Cereal Res Commu 5: 365-368

Martin A, Cubero J I (1981) The *Hordeum Chilense* in Cereal Breeding, Cereal Res Commu, 9: 317-323

McDonald B A, Zhan J, Burdon J J (1999) Genetic Structure of *Rhynchosporium Secalis* in Australia. Phytopathology, 89: 639-645

Mielke H (1980) Untersuchungen Fiber Die Anfiilligkeit Verschiedener Wildgerstenarten Gcgenfiber Typhu/a Incarnata. Nacbricbten Blatt Deutches PJranzenschutzdienst (Brawnschwetg), 32-39

Miller T E, Reader S M, Chapman V (1981) The Addition of *Hordeum Chilense* Chromosome to Wheat. in: Induced Variabilitym Plant Breeding, (Int. Symp. Eucarpia, Pudoc, Wageningen), 79-81

Miller T E, Reader S M, Ainsworth C C (1985) A Chromosome of *Hordeum Chilense* Homoeologous to Group 7 of Wheat. Can J Genet Cytol, 27: 101-104

Moseman J G, Craddock J C (1976) Genetic Basis for Barley Germ-plasm Conservation. in: Gaul, H. (ed.), Barley Genetics III. Verlag Karl Thieming, Munich, 51-57

Moseman J G, Baenziger P F, Kilpatrick R A (1980) *Hordeum spontaneum*, An Overlooked Source of Disease Resistance. Europe and Mediterranean Cereal Rust Foundation, 91-93

Moseman J G, Nevo E, Zohary D (1983) Resistance of *Hordeum Spontaneum* Collected in Israel to Infection with *Erysiphe Graminis Hordei.* Crop Sci, 23: 1115-1119

Moseman J G, Nevo E, EI-Morshidy M A (1990) Reactions of *Hordeum Spontaneum* to Infection with two Cultures of Puccinia Hordei from Israel and United States. Euphytica, 49: 169-175

Munck L, Karlsson K E, Hagberg A, Eggum B O (1970) Gene for Improved Nutritional Value in Barley Seed Protein. Science, 168: 985-987

Munck L (1972) Improvement of Nutritional Value in Cereal, Hereditas, 72: 1-128

Munck L (1992) The Case of High-lysine Barley Breeding. in: Peter R Shewry (ed.) Barley: Genetics, Biochemistry, Molecular Biology and Biotechnology. CAB International, Wallingford, UK. 441-458

Neale D B, Saghai-Maroof M A, Allard R W, Zhang Q, Jorgensen R A (1988) Chloroplast DNA Diversity in Populations of Wild and Cultivated Barley. Genetics, 120: 1105-1110

Nielsen J (1978) Host Range of the Smut Species Ustilago Nuda and Ustilago Trztici in the Tribe Triticeae. Can J Bot, 56: 901-915

Nielsen J (1987) Reaction of *Hordeum* Species to the Smut Fungi Ustilago Nuda and U.tritici. Can J Bot, 65: 2024-2027

Nevo E, Brown A H D, Zohary D (1979a) Genetic Diversity in the Wild Progenitor of Barley in Israel. Experientia, 35: 1027-1029

Nevo E, Brown A H D, Zohary D (1979b) Genetic Diversity and Environmental Associations of Wild Barley *Hordeum Spontaneum* in Israel. Evolution, 33: 815-833

Nero E, Beiles A, Gutterman Y, Storch N, Kaplan D (1984a) Genetic Resources of Wild Cereals in Israel and Vicinity: II Phenotypic Variation within and between Populations of Wild Wheat, *Hordeum Spontaneum.* Euphytica, 33: 737-756

Nevo E, Belles A, Ben-Shlomo R (1984b) The Evolutionary Significance of Genetic Diversity: Ecological, Demographic Anti Life History Correlates. Lecture Notes Biomathematics, 53: 213-213

Nevo E, Atsmon D, Beiles A (1985) Protein Resources in Wild Barley *Hordeum Spontaneum* in Israel: Predictive Method by Ecology and Allozyme Markers. Plant Syst Ev, 150: 205-222

Nevo E, Beiles A, Kaplan D, Storch N, Zohary D (1986a) Genetic Diversity and Environmental Associations of Wild Barley, *Hordeum Spontaneum* (Poaceae), in Iran. Plant Syst Ev, 153: 141-164

Nevo E, Zohary D, Beiles A, Kaplan D, Storch N (1986b) Genetic Diversity and Environmental Associations of Wild Barley, *Hordeum spontaneum* in Turkey. Genetica, 68: 201-213

Nevo E, Beiles A, Zohary D (1986c) Genetic Resources of Wild Barley in the Near East: Structure, Evolution and Application in Breeding. Biol J Linn Soc, 27: 355-380

Nevo E, Beiles A, Krugman T (1988b) Natural Selection of Allozyme Polymorphisms: A Microgeographic Climatic Differentiation in Wild Emmer Wheat (*Triticum Dicoccoides*). Theor Appl Genet, 76: 737-752

Nevo E (1992) Origin, Evolution, Population Genetics and Resources for Breeding of Wild Barley, Hordeum Spontaneum, Cultivated Barley: In the Fertile Crescent. in: Peter R. Shewry (ed.) Barley: Genetics, Biochemistry, Molecular Biology and Biotechnology. CAB International, Wallingford, UK, 19-43

Nevo E, Baum B, Beiles A, Johnson D A (1998) Ecological Correlates of RAPD DNA Diversity of Wild Barley, *Hordeum Spontaneum* in the Fertile Crescent. Genet Resour Crop Ev, 45: 151-159

Orton T J (1980a) Chromosomal Variability in Tissue Cultures and Regenerated Plants of Hordeum. Theor Appl Genet, 56: 101-112

Orton T J (1980b) Haploid Barley Regenerated from Callus Cultures of *Hordeum Vulgare* × *H. Jubatum*. J Hered, 71: 280-282

Orton T J (1980c) Comparison of Salt Tolerance Between *Hordeum Vulgare* and *H. Jubatum* in Whole Plants and Callus Cultures. Zeit Schrifi Fiir Pfianzenphysiology, 98: 105-118

Orton T J, Steidl R P (1980) Cytogenetic Analysis of Plants Regenerated from Colchicine-treated Callus Cultures of an Interspecific Hordeum Hybrid. Theor Appl Genet, 57: 89-95

Ordon F, Schiemann A, Friedt W (1997) Assessment of the Genetic Relatedness of Barley Accessions (*Hordeum Vulgare* S.l.) Resistant to Soil-borne Mosaic-inducing Viruse (BaMMV, BaYMV, BaYMV-2) Using RAPDs. Theor Appl Genet, 94: 325-330

Pakniyat H, Powell W, Baird E, Handly L L, Robinson D, Scrimgeour C M, Nevo E, Hackett C A, Caligari P D S, Forster B P (1997) AFLP Variation in Wild Barley (Hordeum Spontaneum C. Koch) with Reference to Salt Tolerance and Associated Ecogeography. Genome, 40: 332-341

Pan A, Hayes P M, Chen F, Chen T H H, Blake T, Wright S, Karsai I, Bedo Z (1994) Genetic Analysis of the Components of Winter Hardiness in Barley (*Hordeum Vulgare* L.). Theor Appl Genet, 89: 900-910

Pakniyat H, Powell W, Baird E, Handly L L, Robinson D, Scrimgeour C M, Nevo E, Hackett C A, Caligari P D S, Forster B P (1997) AFLP Variation in Wild Barley (*Hordeum Spontaneum* C. Koch) with Reference to Salt Tolerance and Associated Ecogeography. Genome, 40: 332-341

Park R F, Karakousis A (2002)Characterization and Mapping of Gene Rph19 Conferring Resistance to *Puccinia Hordei* in the Cultivar “Reka 1” and Several Australian Barleys. Plant Breeding, 121: 232-236

Peterhänsel C, Freialdenhoven A, Kurth J, Kolsch R, Schulze-Lefert P (1997) Interaction Analyses of Genes Required for Resistance Responses to Powdery Mildew in Barley Reveal Distinct Pathways Leading to Leaf Cell Death. Plant Cell, 9: 1397-1409

Pickering R, Ruge-Wehling B, Johnston P A, Schweizer G, Ackermann P, Wehling P (2006) The Transfer of a Gene Conferring Resistance to Scald (*Rhynchosporium Secalis*) from *Hordeum Bulbosum* into *H. Vulgare* Chromosome 4HS. Plant Breeding, 125: 576-579

Picketing R A (1988) The Attempted Transfer of Disease Resistance from *Hordeum Bulbosum* L. to *H. Vulgare*. Barley Genet Newsl, 18: 5-8

Picketing R A (1989) Plant Regeneration and Variants from Calli Derived from Immature Embryos of Diploid Barley (*Hordeum vulgare* L.) and *H. Vulgare* L. × *H. Bulbosum.* L. Crosses. Theor Appl Genet, 78: 105-112

Pillen K, Zacharias A, Leon J (2003) Advanced Backcross QTL Analysis in Barley (*Hordeum vulgare* L). Theor Appl Genet. 107: 340-352

Prasad G, Yasuda S, Konishi T (1983) Reaction of *Hordeum Bulbosum* L. to Japanese Races of Powdery Mildew (Erysiphe Graminis F. sp. Hordez). Barley Genet Newsl, 13: 60-62

Qi X, Niks R E, Stam P, Lindhout P (1998) Identification of QTLs for Partial Resistance to Leaf Rust (*Puccinia Hordei*) in Barley. Theor Appl Genet, 96: 1205-1215

Rajhathy T, Morrison J W, Symko S (1964) Interspecific and Intergeneric Hybrids in Hordeum. Barley Genet, 1: 195-212

Richter K, Schondelmaier J, Jung C (1998) Mapping of Quantitative Traits Loci Affecting *Drechslera Teres* Resistance in Barley with Molecular Markers. Theor Appl Genet, 97: 1225-1234

Rodríguez E M, Svensson J T, Malatrasi M, Choi D W, Close T J (2005) Barley Dhn13 Encodes a KS-type Dehydrin with Constitutive and Stress Responsive Expression. Theor Appl Genet, 110: 852-858

Schooler A B (1967) A form of Male Sterility in Barley Hybrids. J Hered, 58: 207-211

Schooler A B (1980) Intetgeneric and Interspecific Barley Hybrids Show Tolerance to Barley Yellow Dwarf Virus. NC Dakota Farm Res, 38: 19-21

Schooler A B, Franckowiak J D (1981) Registration of North Dakota 497 and 586 Barley Germplasm. Crop Sci, 21: 352

Schönfeld M, Ragni A, Fischbeck G, Jahoor A (1996) RFLP Mapping of three New Loci for Resistance Genes to Powdery Mildew (*Erysiphe Graminis F.* sp. Hordei) in Barley. Theor Appl Genet, 93: 48-56

Schweizer G F, Baumer M, Daniel G, Rugel H, Rder M S (1995) RFLP Markers Linked to Scald (*Rhynchosporium secalis*) Resistance Gene Rh2 in Barley. Theor Appll Genet, 90: 920-924

Segal A, Manisterki J, Fischbeck J, Wahl I (1980) How Plant Populations Defend Themselves in Natural Ecosystems. in: Horsfall J G, Cowling E B (eds) Plant Disease, Vol V. Academic, New York, 75-102

Segal A, Manisterski J, Browning J A, Fischbeck J, Wahl I (1982) Balance in Indigenous Plant Populations. in: Hcybrock H M, Stephan B R, Van Weissenberg K (eds) Resistance to Disease and Pests in Forestnes. Center for Agriculture Publishing and Documentation (Pudoc), Wageningen, 361-370

Skou J P, Haahr V (1987a) Screening for and Inheritance of Resistance to Barley Leaf Stripe (*Drechslera Graminea*). Rise Report No.554, Rise National Laboratory, Denmark, 96

Skou J P, Haahr V (1987b) Field Screening for Resistance to Barley Net Blotch. Annu Appl Biol, 111: 617-627

Skou J P, Jörgensen J H, Lilholt U (1984) Comparative Studies on Callose Formation in Powdery Mildew Compatible and Incompatible Barley. Phytopathologische Zettschrift, 109: 147-168

Strelchenkol P, Kovalyoval O, Okuno K (1999) Genetic Differentiation and Geographical Distribution of Barley Germplasm Based on RAPD Markers. Genet Resour Crop Ev, 46: 193-205

Steffenson B J, Hayes P M, Kleinhofs A (1996) Genetics of Seedling and Adult Plant Resistance to Net Blotch (*Pyrenophora Teres* F. Teres) and Spot Blotch (*Cochliobolus Sativus*) in Barley. Theor Appl Genet, 92: 552-558

Steffenson B J, Jin Y, Griffey C A (1993) Pathotypes of *Puccinia Hordei* with Virulence for the Barley Leaf Rust Resistance Gene Rph7 in the United States. Plant Dis, 77: 867-869

Steffenson B J, Olivera P, Roy J K, Jin Y, Smith K P, Muehlbauer G J (2007) A Walk on the Wild Side: Mining Wild Wheat and Barley Collections for Rust Resistance Genes. Aust J Agric Res, 58: 532-544

Sun D F, Xu T W, Jiang H R, Zhao L (1995) Selection and Genetic Analysis Of Nucleo-cytoplasmic Male Sterile Line 88BCms in Barley. Scientia Agricultra Sinica, 28: 37-42

Suprunova T, Krugman T, Distelfeld A, Fahima T, Nevo E, Korol A (2007) Identification of a Novel Gene (Hsdr4) Involved in Water-stress Tolerance in Wild Barley. Plant Mol Biol, 64: 17-34

Tallberg A, Eggum B O (1981) The Nutritional Value of High Lysine Barley Genotypes. Plant Foods Hum Nutr, 31: 151-161

Tan B H (1978) Verifying the Genetic Relationships between Three Leaf Rust Resistance Genes in Barley. Euphytica, 27: 317-323

Teulat-Merah B, Rotter B, Francois S , Borries C, Souyris I, This D (2000) Stable QTL for Plant Water Status, osmotic Adjustment and Co-location with QTLs for Yield Components in a Mediterranean Barley Progeny. Barley Genet, 8: 246-248

Teulat B, Merah O, Sirault X, Borries C, Waugh R, This D (2002) QTLs for Grain Carbon-isotope Discrimination in Field-grown Barley. Theor Appl Genet, 106: 118-126

Teulat B, Zoumarou-Wallis N, Rotter B, Ben Salem M, Bahri H, This D (2003) QTL for Relative Water Content in Field-grown Barley and Their Stability Across Mediterranean Environments. Theor Appl Genet, 108: 181-188

Thacker S P, Ramamurthy V, Kothari R M (1992) Characterization of Barley β-amylase for Application in Maltose Production. Starch, 44: 339-341

Toubia-Rahme H, Johnston P A, Pickering R A, Steffenson B J (2003) Inheritance and Chromosomal Location of *Septoria Passerinii* Resistance Introgressed from *Hordeum Bulbosum* into *Hordeum Vulgare*. Plant Breeding, 122: 405-409

Vanhala T K, Rijn C P E, Van Buntjer J, Stam P, Nevo E, Poorter H, Van Eeuwijk F A (2004) Environmental, Phenotypic and Genetic Variation of Wild Barley (*Hordeum Spontaneum*) from Israel. Euphytica, 37: 297-309

Vavilov N I (1926) Studies on the Origin of Cultivated Plants. Bulletin of Applied Botany and Plant Breeding, 26: 1-248

Vavilov N I (1951) The Origin, Variation, Immunity and Breeding of Cultivated Plants (Translated from the Russian by K. Starr Chester). Chronica Botanica, 13: 1-364

Wahl I, Eshed N, Scgal A, Sobel Z (1978) Significance of Wild Relatives of Small Grains and Other Wild Grasses in Cereal Powdery Mildews. In: Spencer, D M (ed.) The Powdery Mildews. Academic Press, London, 83-100

Weibull J (1987) The Resistance of *Hordeum Bulbosum* Padi (L.) 2. *Hordeum* Species and Interspecific Hybrids, Euphytica, 36: 571-576

Williams K J (2003) The Molecular Genetics of Disease Resistance in Barley. Aust J Agric Res, 54: 1065-1079

Xu T W (1987) Origin and Phylogeny of Cultivated Barley in China. In: Barley Genetics V, Bejing, 91-99

Xu T W (1993) The Headway of Research in the Origin and Taxonomy of Barley. In: Liu H L (ed.) The Headway of Research in Crop Breeding. Chinese Agricultural Publishing Press, 17-35

Yang J L, Yen C (1985) Distribution of Weedrace Barley in China and the Center of Diversity. Sveriges Utsadesforenings Tidskrift, 95: 71-79

Yun S J, Gyenis L, Hayes P M, Matus I, Smith K P, Steffenson B J, Muehlbauer G J (2005) Quantitative Trait Loci for Multiple Disease Resistance in Wild Barley. Crop Sci, 45: 2563-2572

Zee K, Chen F Q, Hayes P M, Close T J, Chen T H H (1995) Cold-specific Induction of a Dehydrin Gene Family Member in Barley. Plant Physiol, 108: 1233-1239

Zhang Q, Webster R K, Allard R W (1987) Geographical Distribution and Associations between Resistance to Four Races of *Rhynchosporium secalis*. Phytopathology, 77: 352-357

Zhang W S, Li X, Liu J B (2007) Genetic Variation of Bmy1 Alleles in Barley (*Hordeum vulgare* L.) Investigated by CAPS Analysis. Theor Appl Genet, 114: 1039-1050

Zohary D (1973) The Origin of Cultivated Cereals and Pulses in the Near East. Chromosomes Today, 4: 307-320

Zohary D, Hopf M (1988) Domestication of Plants in the Old World. Clarendon, Oxford

3

Chemical Composition in Barley Grains and Malt Quality

G. P. Fox
Department of Primary Industries & Fisheries
Queensland Grains Research Laboratory
PO Box 2282
Toowoomba Qld 4350 Australia
glen.fox@dpi.qld.gov.au

3.1 Introduction

Barley is used for a wide range of traditional and novel end-uses (Edney 1996; Sparrow *et al.*, 1988). In most countries, the major portion of barley is fed to animals, particularly cattle and pigs. Human food uses of barley are more limited, although recent trends in the use of barley varieties, high in dietary fibre, have been identified. A significant high-value use is to produce malt as a raw material for the brewing industries, including beer and whiskey.

Improved barley production requires barley breeding programs to provide varieties with the combination of reliable and efficient production characteristics and grain quality attributes suited to these uses. Selection of varieties, with the correct range of traits, is a difficult process. It is necessary to produce high quality malt for beer and other fermented beverages. However, a number of factors influence the final quality of the barley grain as well as the malt. Testing of end product quality for each breeding line is not only expensive but requires the availability of larger quantities of barley than is available from single plants or lines at an early stage in barley breeding. Biochemical or molecular tests that predict likely feed or malting and brewing quality are therefore needed to allow rapid development of barley cultivars.

Research continues to unravel the biochemistry behind areas of grain development (gene to protein to product) which relates to end product quality. Research into well-established components such as starch and protein, improves our knowledge of the relation between barley, malt and beer. However, the genetic, chemistry and synergistic relationship between some components that are important in malt and beer quality are still unknown.

Many of the traditional testing and evaluation methods aim to ensure a consistency of quality without a link between the attribute measured and the end-use quality being known. This results in attributes of uncertain value being assessed to reduce the risk of adopting or using barley that causes difficulties in processing or end product quality. This process may discriminate against barley with superior processing traits and will only be overcome by improved understanding of the basis of barley quality. This is highlighted when comparing our understanding of barley carbohydrates or proteins and the enzymes that degrade them being tested in conditions that do not duplicate the conditions of the in-situ reaction. A number of these key enzymes have inhibitors and some aspects of the role and function of these inhibitors also are still unclear.

A number of quality attributes for malt barley are critical for the identification and release of malt varieties. These are hot water extract, viscosity, Kolbach index, wort β-glucan, fermentability, and diastatic power. Additional parameters include α-amylase, β-amylase, free amino nitrogen, friability, and β-glucanase. However, the value of parameters currently used to measure malt quality has been questioned (MacGregor, 1996; Palmer, 1983). The limitations of some of these traits in predicting brewery performance have led to suggestions that specific brewery tests were also needed to adequately describe malt quality (Axcel, 1998). More recently, Evans *et al.* (2007) demonstrated that the sum of individual starch degrading enzymes (as well as the thermostability of these enzymes) was a better indicator of diastatic power and that these enzymes also correlated better to fermentability.

The barley grain is comprised of three main components, namely the germ or embryo, the outer layers (husk, pericarp, testa and aleurone layers), and the endosperm. Fig. 3.1 shows the overall structure and individual components of the grain, with the endosperm being the dominant component. However, the main living organ in the grain is the embryo, with the aleurone layers (also a living tissue) surrounding the endosperm.

This review covers recent research in barley structure and chemistry as well as highlighting new research from a number of excellent reviews published over the last two decades (Duffus and Cochrane, 1992; Duffus and Cochrane, 1993; Swanston and Ellis, 2002). Also, for a number of barley components such as starch or protein, more detailed reviews are provided in later chapters of this book.

3.2 Physical Structure

Some physical structures affecting germination are associated with malt quality. Here grain size, dormancy and hardness will be in detail discussed.

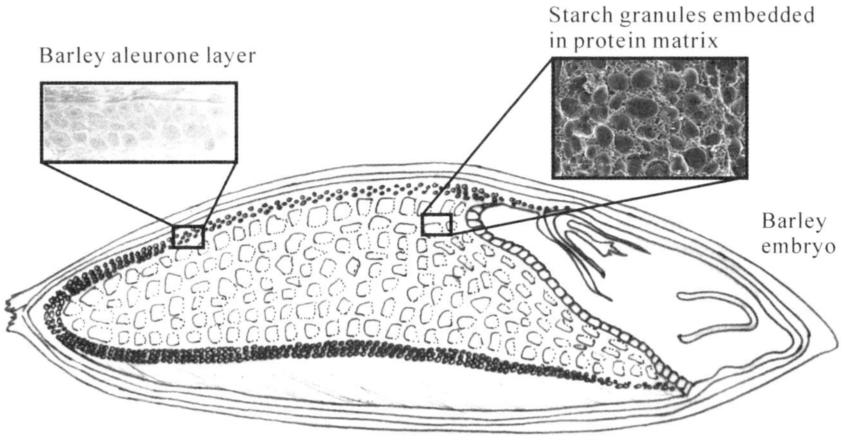

Fig. 3.1. Structure of barley

3.2.1 Grain Size

Grain size is an important descriptive trait based on the physiology of the grain. The final grain size is determined by environmental effects, which affect the biochemical components within the grain itself (Coventry *et al.*, 2003b). For thousands of years, when grain was used specifically for human consumption, the grain was selected based on size. Improvement in grain size and weight has been shown to have occurred through human selection rather than through evolutionary events (Ferrio *et al.*, 2006). Within the last century, barley breeders have continued to target large grain cultivars in association with improved yield and grain quality attributes. In Australia, the measurement of grain size is generally based on 4 fractions: <2.2 mm (screenings), >2.2 mm, >2.5 mm, and >2.8 mm. Fox *et al.* (2006a) demonstrated the genetic and environmental effects for improving grain size. Selection for increased plump grain (>2.8mm) provided a useful strategy in increasing overall grain size. Increased grain size also provided an increase in the important malt quality trait of diastatic power (Agu *et al.*, 2007). However, there were no corresponding protein determinations to ascertain if the increase in DP was related to protein content. Emebiri *et al.* (2007) reported a strong correlation between increased grain size and yield when using genotypes with low protein content alleles. Passarella *et al.* (2005) demonstrated the impacts of temperature during grain fill on the level of screenings with high temperature reducing the overall size. Fox *et al.* (2006a) also showed that from breeding trials, where sites suffered from terminal moisture and/or heat stress, grain size was significantly reduced. The negative impacts of heat and drought stress on grain size and weight have also been shown in Australian barleys under controlled experiments (MacNicol *et al.*, 1993; Savin and Nichols, 1996; Savin *et al.*, 1999; Savin and Nichols, 1999). The negative effects on heat during grain filling on grain quality and starch synthesis was reported by Wallwork *et al.* (1998a 1998b).

Industry standards on large grain are based on the amount of grain > 2.5mm. Smaller grain generally has lower starch and higher protein levels, thus reducing the extract/feed potential. Large grains conversely have increased levels of starch and therefore more potential extract. However, excessively large grain could impact on malt quality particularly on the rate of water hydration and modification during malting.

3.2.2 Dormancy

Dormancy is not a biochemical component such as starch or protein that can be isolated and measured from grain. However, an understanding of the level of dormancy in barley is critical in terms of malt quality. Malting barley is one of the few grains where the seed is required to germinate for product development, ie. production of malt. The failure of barley to germinate at an acceptable level, i.e. <95%, has the potential to cause problems during the malting process.

The physiological and biochemical components of dormancy are very complex. Han *et al.* (1999) reported on four possible gene regions related to dormancy, with Seed Dormancy (SD) 1 and 2 on chromosome 5H, SD 3 on 7H and SD4 on 4H. Li *et al.* (2003) confirmed genetic associations of dormancy on chromosome 5H. The role of hormones and enzymes in dormancy has been documented in recent reviews by Briggs and Woods (1993) and Benech-Arnold (2002), with the latter author describing the most significant and relevant aspects of dormancy and resistance to pre-harvest sprouting. Several biochemical mechanisms have been associated with dormancy, including an antagonistic effect between abscisic acid (ABA) and gibberellic acid (GA). For example, Weidner *et al.* (1993) demonstrated a relationship between phenolic acids and barley dormancy with an increase in total and free phenolic acid content coinciding with the pattern of dormancy after flowering and post ripening. In addition, it was demonstrated that ferulic and sinapic acid retarded germination of ripening barley embryos, and Wang *et al.* (1995) showed that an increased level of hydrogen peroxide reduced the inhibitory effect of ABA. The biochemistry of dormancy is complex and covered in a wide range of comprehensive literature.

Dormancy refers to a live grain not germinating, even under ideal conditions. However, for malting barley it is preferable for the grain to go into its dormant state prior to harvest as this dormant state can help in preventing the grain from sprouting in the head prior to harvest. Rain at harvest can cause the grain to sprout, resulting in a condition known as pre-harvest sprouting (PHS). This condition results in malt barley being downgraded to feed. Two methods currently can be used to assess the level of any pre-harvest sprouting, namely the Falling Number test and the Rapid Visco Analyzer (RVA) test. Both of these tests relate to the starch in a ground barley sample being gelatinized and the level of resistance from the gelatinized starch being an indication of how "sound" the grain is.

Recent studies have linked genetic regions for resistance to sprouting and dormancy (Ullrich *et al.*, 2005; Hori *et al.*, 2007) with a high level of correlation between tests that indicate pre-harvest sprouting resistance and grain dormancy (measured by germination tests). Controlling dormancy would contribute to the control of pre-harvest sprouting. Lin *et al.* (2008) also demonstrated varietal differences in dormancy which could be correlated to differences in PHS resistance. The differences were also linked to the level of α-amylase as well as measured by the Stirring Number method (similar method to RVA).

3.2.3 Grain Hardness

Grain hardness is a trait normally considered when looking into the structure and composition of barley. It is not routinely measured when evaluating barley grain quality. However, hardness in wheat is related to the texture of the grain endosperm where starch granules are either readily separated from the protein matrix (soft) or the granules resist separation (hard). In general, malting barley cultivars can be classified as soft, whereas non-malting or feed cultivars are classified as hard (Alison *et al.*, 1976). Hardness has also been associated with the level of modification of malt, which would imply that grain components within the endosperm directly affect modification. Specific proteins that interact with starch granules such as hordoindolines, including friabilin, have also been implicated in grain hardness (Darlington *et al.*, 2000).

Milling energy (ME) has been used as a measure of barley grain hardness. Finished malt may be analysed either for ME or for "friability", thereby providing an indication of malt endosperm modification. Recent studies have demonstrated the relationships between barley hardness (milling energy) and grain and malt quality parameters (Swanston *et al.*, 1995). Grain protein and β-glucan have been positively correlated with hardness (Henry and Cowe, 1990), with malt extract and endosperm modification correlated negatively to hardness (Swanston and Taylor, 1988; Swanston *et al.*, 1995). B and C hordein have been shown to be associated with milling energy, where an increase in C hordein along with a decrease in β-glucan corresponded to a decrease in ME. The reverse was reported for B hordein. However, that study was conducted with a limited dataset (Molina-Cano *et al.*, 1995). The effects of environmental impacts on grain components such as glucan and protein have been associated with changes in ME (Henry, 1985; Swanston and Cowe, 1989; Henry and Cowe, 1990).

Another aspect of grain hardness associated with endosperm structure is the steely or mealy appearance (glassy or chalky). A number of studies have highlighted this effect and how it related to protein content, composition and other parameters and its impact on malting quality. The effect is mostly by influenced growing environment; however, some varieties have been shown to be more prone to expressing this effect (Chandra *et al.*, 1999; Broadbent and Palmer, 2001; Koliasou and Palmer, 2003; Halopainen *et al.*, 2005).

Recent studies have investigated the role of grain hardness for malt and/or feed quality. These studies have used a different technology from the ME one. The measurement of particle size after milling, as well as the measurement of the resistance to crushing of single kernels (the Single Kernel Characterisation System (SKCS)), offered alternative hardness determinants (Osborne *et al.*, 2003; Fox *et al.*, 2007b; Psota *et al.*, 2007; Osborne *et al.*, 2007). While there were genetic and environmental effects on hardness, there were also correlations to malt and feed quality traits. Results from a recent study using SKCS also showed a strong relationship between endosperm cell wall composition, namely β-glucan, and barley hardness.

In wheat, there is a strong relationship between hardness and a single gene locus (two genes). However, in barley at the gene level, three genes encode the hardness proteins (hordoindolines). These three genes are *hina*, *hinb1* and *hinb2*. A number of researchers have explored the allelic variation between genotypes with reports that, unlike wheat, polymorphisms in the gene sequence did not relate to functional changes in malt or feed quality (Beecher *et al.*, 2000; Darlington *et al.*, 2001; Beecher *et al.*, 2002; Caldwell *et al.*, 2004; Fox *et al.* 2007a). Fox *et al.* (2007a) detailed the homology between a number of domesticated genotypes and wild accessions and land races, based on literature and database (Genbank) entries. There was a considerable diversity in the wild accessions and land races, which offers opportunities to exploit possible hordoindoline variation and functionality.

The indoline proteins have a tryptophan rich motif, which has been shown to bind lipids (see review Douliez *et al.*, 2000). While at this stage there is no comprehensive data in barley, it is thought that these proteins are involved in binding the endosperm storage proteins, lipid and starch granules. Further evidence of this lipid binding capacity has been shown through the wheat indolines binding to yeast membrane lipids (Evrard *et al.*, 2007). Douliez *et al.* (2000) also highlighted the similarity of indolines with Lipid Transfer Proteins.

3.3 The Internal Structure

The internal structure of barley grains associated with malt quality is involved in embryo, aleurone and husk.

3.3.1 The Embryo

The barley embryo is the most important living tissue within a barley grain. A comprehensive review of the formation and structure of the embryo was provided by Duffus and Cochrane (1992) and Duffus and Cochrane (1993). The embryo is the fertile part of the barley seed, which also contains starch, protein and lipids. These are used during the development of the embryo after fertilisation and as an initial food source when the harvested seeds commence germination. The dormancy level and viability of the embryo is critical for the next generation of seeds. A number of factors can have a negative impact on dormancy and viability including:

- environmental conditions after anthesis and during grain fill, i.e. heat (heat wave) or cold (frost) stress;
- plant moisture (avoiding drought conditions);
- harvest conditions (i.e, excessive abrasions during threshing in the harvester or pre-harvest rain);
- storage conditions (excessive heat in storage or high grain moisture can reduce germination potential).

Further, the development of hulless barley provides a number of positive characteristics for malting, food and feed end-uses. However, in hulless barleys, as the husk is removed during the harvesting process, physical damage can occur to the germ.

The germination process is initiated by the embryo imbibing water, from which a cascade of biological systems is initiated. The sole purpose of the embryo is to generate a plant which will produce new seeds. The biological systems initiate the breakdown of the endosperm reserves to produce sugars, amino acids and fats for the growing embryo to consume. These systems trigger the development of complex enzyme pathways (including the development of hormones) that will produce proteins (enzymes) for the breakdown of the endosperm cell walls, storage proteins and starch granules and subsequent transport of those products to the embryo.

Most of these systems are controlled by growth hormones including gibberellic acid, abscisic acid and salicylic acid. Variation in the levels of these hormones will impact on the expression of the enzymes. This variation can be influenced by the development of genetic variants that alter the level of the hormones produced. For example, α-amylase is one of the most important enzymes in the breakdown of starch during germination. Studies have shown that through a change in the gibberellic and/or abscisic acids' pathways the α-amylase can be up or down regulated. In terms of malt quality, most of the important components produced by the embryo are related to the breakdown of endosperm reserves. However, two enzymes that can have a negative impact on beer quality are lipoxygenase 1 and 2 (LOX 1 and LOX2). LOX has been related to beer staling with the production of trans-2-nonenal (cardboard flavour). A number of reviews have been published on LOX with Morrison (1993) providing a comprehensive summary. Both LOX isoforms have been identified during the grain fill period (Schmitt and van Mechelen 1997). While early work showed that LOX1 was present in resting barley embryos, both LOX1 and LOX2 were present in germinating embryos and the specificities of these enzymes on lipid substrates were different (Holtman *et al.*, 1997). Wu *et al.* (1997) showed there was genetic variability for LOX1 and

LOX2 although the ratios were similar. While Hirota *et al.* (2006) reported it was possible to produce germplasm free of LOX (Hirota *et al.*, 2006).

Yan *et al.* (1999) clearly showed the level of effect that the embryo, endosperm and the interaction between the two tissues had on malt quality. As a consequence they noted it was important to have an understanding of the impact of the embryo and endosperm for future developments in barley quality improvement.

3.3.2 The Aleurone

The thickness of the barley aleurone layer varies around the kernel depending upon the tissue its covers. The aleurone layer is two to three cells thick and encases the endosperm. The thickness of the aleurone is more variable in the crease and is approximately one cell thick at the germ. The aleurone cells are living tissue and contain protein, starch and lipids. The cell walls of the aleurone contain β-glucan, arabinoxylan and phenolic acids. However, the ratio of β-glucan to arabinoxylan is 25:75, which is the reverse for the endosperm cell walls. The sub-aleurone which varies in the number of cells is located between the aleurone and endosperm. The sub-aleurone also contains protein embedded in starch granules (Macewicz *et al.*, 2006). Macewicz *et al.* (2006) proposed that some of these proteins were isoforms of a storage protein (hordein B3) (see section on storage proteins below) as well as a serine proteinase inhibitor (serpin) protein.

A comprehensive description of the aleurone is presented in Duffus and Cochrane (1992, 1993). The aleurone plays a critical role in the expression of endosperm degrading enzymes during germination, including:

- β-glucanases (no inhibitors reported at this stage);
- Proteinases (Jones 2005a), and peptidases;
- α-amylase;
- Limit Dextrinase (Sissions *et al.*, 1992);
- α-glucosidase;
- Inhibitors (proteinases (Jones 2005b), trypsin/amylase, limit dextrinase (Macgregor *et al.*, 2002b; Stahl *et al.*, 2007).

A number of important transcript factors that regulate the synthesis of α-amylase and other enzymes include gibberellin oxidase (GAox) and GAMYB which are expressed in aleurone cells. Up-regulation of GAMYB results in increased expression of α-amylase and $(1\rightarrow3,1\rightarrow4)$-$\beta$-Glucan endohydrolase (Murray *et al.*, 2006), while inhibition of GA-oxidase proteins can stop the expression of amylase and other enzymes. Similar responses can also be observed when transcription factors in the abscisic acid pathway are affected.

The growing environment affects aleurone cell integrity. Excessive heat stress can induce the development of some proteins while having little effect on others. Heat Shock Proteins (HSP) are expressed in high temperatures.

These proteins can in turn impact on the expression of other proteins (Harju *et al.*, 2003).

In addition, important trace elements including zinc and iron, are stored in aleurone cells (Brinch-Pedersen *et al.*, 2007).

3.3.3 The Barley Husk

Barley is one of only four commercial species of grass that retains a husk after harvest. Rice, oats and millet are the other "covered" grains. The hull amounts to approximately 13% of grain weight but can range between 7%~25% depending upon cultivar, growing environment and grain size (Evers *et al.*, 1999). Variation occurs between row types, with six-row having a larger proportion than two-row barley (Evers *et al.*, 1999). Additionally, winter barleys have higher husk content than spring barleys and the husk content increases in regions closer to the equator (Evers *et al.*, 1999). Thin hulled barley cultivars exist but these can suffer from "skinning" during harvest. Hulless barley cultivars are also commercially available. For these barleys, the hull is removed during the harvesting process.

The hull plays an important role before and after harvest. During the later stages of grain ripening, the hull has been considered to have a role in grain dormancy and therefore pre-harvest sprouting resistance (Benech-Arnold *et al.*, 1999). During harvest, the husk acts to protect the germ during the abrasive threshing process in the harvester (Olkku *et al.*, 2005). Post harvest, the husk plays a role in processing for the malting, brewing and feed industries. In terms of the malting industry, the husk aids during the malting process by protecting the germ from physical abrasion during handling and preventing the growing acrospire from being damaged during germination and kilning (Meredith *et al.*, 1962). In the brewing process, the husk aids during filtration of the brewers extract from the lautering process. Huskless barley can "gum" up filters and reduce filtration rate, thereby adding to the cost of production and potentially impacting on the quality of beer (Edney *et al.*, 1998; Evans *et al.*, 1999b). However, Edney and Langrell (2004) recently demonstrated that good quality malt could be produced from hulless barley if the "appropriate malting conditions were used".

The intensive livestock industry also benefits from using a hulled barley grain. The hull aids in holding crushed or pressed grain together. A thinner hull is desirable as thicker hulls will result in higher levels of deleterious compounds, in particular lignin, which have been shown to have a negative impact on feed performance in ruminants (Kaiser, 1999).

3.4 The Barley Endosperm

Endosperm is a major part of the barley grain, and its chemical composition is directly related to malt quality.

3.4.1 Barley Carbohydrates

Barley carbohydrate composition has been one of the most studied aspects in terms of barley quality and its relation to feed, malt, and beer quality. Henry (1988) reviewed the current knowledge of barley carbohydrate composition in terms of malt quality, concluding that despite many years of research, our knowledge of barley carbohydrate chemistry was incomplete. Further work was required to fully understand barley carbohydrates before significant genetic gain could be reached. More recent reviews have covered the topic in similar detail (Duffus and Cochrane, 1992; Duffus and Cochrane, 1993). In this section we will cover the basics of barley carbohydrates as well as review the most recent research.

3.4.1.1 Starch

Starch is the most abundant component of the endosperm , comprising around 60% of total grain weight. Starch consists of two polymers, amylose and amylopectin. Amylose is a linear polymer made up of glucose molecules linked via α-(1-4) glucosidic bonds. AmylopectinindexAmylopectin. is the larger polymer with α-(1-4) glucosidic and α-(1-6) glucosidic linkages, which form a branched structure (Hough, 1985). The ratio of amylopectin to amylose is around 3:1 (Palmer, 1983). Both amylose and amylopectin polymers are present in the barley endosperm starch granules. The large granules, designated A type, are round in shape and contain 70%~80% amylopectin. The small, spherical, B type granules, contain 40%~80% amylose (May and Buttrose, 1959; Evers *et al.*, 1999).

The formation of starch is a complex pathway with amylose and amylopectin being formed simultaneously (Fig. 3.2 Rahman *et al.*, 2000). The genes responsible for particular proteins are expressed shortly after anthesis. For example, starch branching enzymes I and II (SBEI and II) have been shown to express 12 days post anthesis (Mutisya *et al.*, 2003).

Dosage of starch synthesis (SS) IIb has been shown to affect amylose synthesis in wheat (Konik-Rose *et al.*, 2007), while mutation of the SSIIa gene in barley resulted in a novel high amylose phenotype. The SSIIa gene is located near the barley sex6 mutation (Morrell *et al.*, 2003).

Recently, the gelatinisation properties of starch were reviewed by Evers *et al.* (1999). For barley, gelatinisation temperature plays an important role in the quality of malt and hot water extract. The temperature at which gelatinisation occurs varies between 55°C and 65°C. MacGregor *et al.* (2002) presented data on the effects of gelatinisation temperature on normal, high, low, and zero amylose starches. Each starch type was found to vary, with the high amylose and waxy starches having gelatinisation temperatures higher than that for the normal starch, whereas high amylose starch showed a high level of resistance to enzymic attack in a slow ramp mash. Ellis *et al.* (1979) had previously demonstrated that high amylose barley had an increased gelatinisation temperature. In "waxy" barley, the amount of amylopectin increases to >90% (Evers *et al.*, 1999). However, waxy barleys generally have lower hot water extract values, and higher β-glucan content and slower modification rates (Ullrich *et al.*, 1986; Swanston, 1996). These factors complicate any interpretation of the impact of changes on starch properties. The main requirement seems to be for a high level of starch, which is not lost during malting and was readily convertible to fermentable sugars in the brewery (low gelatinisation temperature).

Limited starch breakdown occurs during malting, although Allosio-Ouarnier *et al.* (2000) reported increased levels of maltose, maltotriose, and maltotetraose during germination. Starch was degraded more in the mashing process by the hydrolytic enzymes α-amylase, β-amylase, β-glucosidase, and limit dextrinase. High temperature infusion mashes readily solubilise the starch but limit the activity of thermolabile enzymes, in particular α-glucosidase and β-amylase (Osman *et al.*, 1996a). Oliverira *et al.* (1994) related the effect of starch properties, namely granule volume and size, to hot water extract.

3.4.1.2 Non-starch polysaccharides

The major constituent of barley endosperm cell walls is β-D-(1→3), (1→4)-glucan (75%), with a minor component identified as arabinoxylans (20%) (Fincher, 1975; Fincher and Stone, 1986; Henry, 1987). The arabinoxylan fraction is usually referred to as pentosan. The solubility of β-glucan in beer varies according to the number and arrangement of (1→3) and (1→4) linkages (Izawa *et al.*, 1993) as well as the size of the molecules. The range in barley for β-glucan is 2%~10% of total grain weight (Henry, 1987). Cultivar and environmental influences on the content of β-glucan have been reported (Henry, 1987; Stuart *et al.*, 1988; Zhang *et al.*, 2001; Li *et al.*, 2008).

Recent studies have presented models of the structure of endosperm cell walls. Autio *et al.* (1996) suggested that the cell walls of the endosperm have a layered structure. Bamforth and Kanauchi (2001) presented a model whereby the outer cell wall was made up of xylan, arabinose, and ferulic and acetic acids, with the inner layer composed of β-glucan. These results indicate that enzymic hydrolysis of the arabinoxylan layer would be critical to ensure that β-glucanase could hydrolyse the β-glucan layer. Vietor *et al.* (1991) and Han and Schwartz (1996) reported that arabinoxylan survives into malt and beer, albeit in smaller oligosaccharide forms. Similarly, it was shown that xylanases could release all pentosan as well as a limited amount of β-glucan from cell walls (Kanauchi and Bamforth, 2002). During germination or malting, the "modification" of the endosperm relates to the breakdown of the cell wall and endosperm components. Although endosperm modification is measured through the solubilisation of endosperm protein reserves, access to those reserves is only possible after cell wall breakdown. Henry (1987) measured the breakdown of β-glucan during malting. The results indicated that by day 4 of germination, barley β-glucan had decreased from 5% to around 1%. Similarly, Allosio-Ouarnier *et al.* (2000) reported an increased level of sugars derived from β-glucan and arabinoxylan components during malting. Although these studies analysed the complete malting process, Walker *et al.* (2001) suggested that by day 2 of germination, the β-glucan level could be used to indicate if a cultivar had desirable extract potential. Stewart *et al.* (1998, 2000) demonstrated that both β-glucan and pentosan impact on wort viscosity and beer filtration rates. Thus, the enzymic breakdown of β-glucan and pentosan during malting is critical for efficient brewing. Mashing temperature has an influence on the solubility of β-glucan. Palmer and Agu (1999) demonstrated the difference in solubility of β-glucan when malt was mashed at 45°C or 65°C, with an increased level of solubilisation at the latter temperature.

The level of β-glucan has been shown to have a relationship with other malt quality traits. Importantly, high β-glucan levels may not result in higher or lower extract but relate to other malt quality traits such as Kolbach Index (ratio of soluble to total protein), viscosity or the speed of filtration (Evans *et al.*, 1999). Views differ on the relationship between β-glucan and foam stability (Lusk *et al.*, 2001).

3.4.2 Grain Protein

Barley protein accounts for 8%~13% (dry basis) of the total composition of malt barley, with a more desirable range of 10%~11% for brewers using a liquid adjunct or for those using a low protein solid adjunct such as rice in which the protein content could be 12%~13%. Of total protein content, around 50% is the prolamin (hordein) protein, with the other three protein fractions, albumins, globulins and glutelins making up the remaining protein. In this review, the hordein protein component will be the main focus. However, structural proteins will be mentioned in the context of their interactions with specific components of barley.

The synthesis of protein commences quite early after anthesis, although there are different reports on the actual timing of these protein components. The hordein component and the individual fractions that make up this component have received the most attention (Shewry, 1992). In addition, a number of the enzymes responsible for the synthesis of the barley proteins have been identified and, most recently, the individual genes and allelic variation have been identified.

Protein accumulates during grain fill at the same time as other grain components. However, there are a number of factors that influence the final protein content and composition, including genotype, growing environment and fertilizer. Nitrogen fertilizer at particular growth stages has been shown to affect final protein content and composition (Sardana and Zhang, 2005; Zhao *et al.*, 2007). In addition, sulphur fertilizer also affects the total protein content and composition (Zhao *et al.*, 2006). Shewry *et al.* (2001) noted that there was little data describing the impact of sulphur fertilizer on malt or beer quality.

However, Zhao *et al.* (2006) reported effects on a range of malt parameters including germination, modification and beta-glucan levels, on malt produced from sites where there was yield response to sulphur fertilizer.

There are reported genetic variations in the starch protein relationship with some genotypes maintaining high protein content regardless of the final starch content whereas others can express low protein levels, which also seem independent of the starch content. This is potentially an important factor for barley breeders to consider. Barley protein has a complex interaction with quality. High protein is undesirable because of the strong correlation with low carbohydrate (starch) levels and thus low extract values (Bishop, 1930). However, if the protein content of malt is too low, brewing performance may be impaired through poor yeast amino acid nutrition. Protein, peptides and amino acid levels in packaged beer are also considered important, positively enhancing foam stability but negatively influencing shelf life by contributing to chill hazes. Many proteins have been identified with specific functions in terms of grain and malt quality, whereas a number have yet to have their function clearly defined.

3.4.2.1 Low protein barleys

Protein content has been shown to be controlled by environment as well as genotype. In addition, low protein content has been shown to be an inherited trait. Karl, a six-rowed variety was released in 1979, and was clearly shown to have low protein content across a range of environments. See *et al.* (2002) demonstrated a negative correlation between low protein and yield, as well as a number of genome regions for yield, maturity and protein. These relationships were also reported by Emebiri *et al.* (2003) using two-rowed breeding lines, one with Karl as a parental relative. Emebiri and Moody (2004) demonstrated the low protein content effect when eight varieties were grown on dryland and irrigated sites, using a range of N fertilizer applications. Again there was a strong relationship between yield and low protein content. Emebiri *et al.* (2004) further confirmed the close linkage between low protein content and agronomic traits, yield, maturity and plant height. Further, Moralejo *et al.* (2004) also identified genomic regions between low protein content, high yield and maturity (earliness). The authors concluded that the low protein content could be related to nitrate reductase genes and not hordein genes. Emebiri *et al.* (2007) reported a strong low protein Genotype x Environment x Management for improved yield, and concluded that through selection of low protein genotypes and higher nitrogen fertilizer application, significant yield gains could be achieved.

Low protein will have a strong effect on malt quality (see below). However, based on the relationship between yield and low protein content, it would be desirable to select for low protein content, while increasing grain yield (and grain size). Selection for low protein content would be a consideration when selecting for malt quality in that low protein content relates to a high extract.

A number of studies have explored the relationship between low protein content breeding and malting quality. Goblirsch *et al.* (1996) identified it was possible to select for low protein and kernel discolouration but the cost was a decrease in diastatic power (DP) using Karl as a parent. Similarly, Canci *et al.* (2003) also reported a close linkage between low protein content and kernel discolouration resistance. Emebiri *et al.* (2004) reported a negative correlation between protein and extract and a positive correlation between protein and diastatic power, using a low protein breeding population. Further, a number of genomic regions common for these traits were identified.

3.4.2.2 Storage proteins

Storage proteins exist in all cereals. This protein component forms a matrix around starch granules in the endosperm and provides a source of nitrogen for the growing embryo if germination occurs. These proteins are generally rich in the amino acids proline and glutamine (hence the term prolamine). In barley, the major storage protein is called hordein, and this comprises 40%~50% of total grain protein. This component is soluble in aqueous alcohol and comprises 4 fractions designated D, C, B, and A. The diversity in the hordein family has made the analysis of these fractions very useful in varietal identification. This diversity can be explained through differences within the B and C hordeins that occur between cultivars, as well as grain protein levels and environments (Benetrix *et al.*, 1994; Molina-Cano *et al.*, 2000b). Several hordein proteins have been purified and sequenced. However, the roles of the specific subunits remain undefined for B-, C- and D-hordeins.

Of these hordeins sub-groups, the A-hordein has been investigated for its possible role as a trypsin/amylase inhibitor and a possible protein involved in beer haze formation. First reports on the existence of five components of chloroform/methanol (CM) extracted proteins was in the early 1980s (Saledo *et al.*, 1980; 1982; Aragoncillo *et al.*, 1981; Paz-Ares *et al.*, 1983; Lazaro *et al.*, 1985). These five proteins have been identified as CM (a~e). Saledo *et al.* (1984) first reported on the genetics of these CM proteins and its close homology to A-hordein. Subsequent studies have confirmed the role of some of these individual proteins and their inhibitory role on trypsin as well as insect α-amylase. Moralejo *et al.* (1993) reported genetic variability in at least one specific protein, namely CMe. Additional roles have been identified for these proteins, including inhibiting serine proteinases in malting (see review by Jones 2005b).

3.4.2.3 Hordeins, malting and brewing quality

For over 100 years researchers have attempted to improve our understanding of the relationship between protein and hot water extract. Smith (1990), Shewry (1993) and Tatham and Shewry (1995) provided excellent reviews of previously published results on this subject, although initially Bishop (1930) concluded that there was a negative relationship between protein and hot water extract. Details from these reviews showed that each hordein group had some relationship to extract or final beer quality. In particular, aggregation of the sulfur-rich B and D groups forms gels which cause filtration problems in brewing. Individual B and C (sulfur-poor) groups have variable effects on extract (Skerritt and Janes, 1992; Janes and Skerritt, 1993). Initial studies by Marchylo *et al.* (1986) demonstrated the difference in hordein breakdown during the malting of two cultivars of differing quality. The malting cultivar was reported to give a higher level of modification than the non-malting. This was reflected in the hordein analysis where the malting cultivar exhibited a higher level of hordein breakdown. Recently, several studies have described a negative correlation between D hordein and hot water extract (Howard *et al.*, 1996; Molina-Cano *et al.*, 2000b). In contrast, Brennan *et al.* (1998) found that with D hordein isogenic lines there was no effect on extract, although the presence or absence of a D hordein allele had an impact on gel protein formation. Further, Brennan *et al.* (1998) related grain hardness, in the form of milling energy, to specific B and C hordein alleles. However, two cases noted a positive relationship between individual hordein fractions and malt quality. Janes and Skerritt (1993) reported a B hordein fraction with a positive effect on hot water extract, whereas Molina-Cano *et al.* (1995) suggested that a C hordein had a positive effect on water uptake during malting. Many of these observations may be explained by indirect effects of the levels of one group of proteins on total protein or the levels of other specific protein components.

Several studies have examined the effect of hordein breakdown products on beer quality. Using monoclonal antibodies, Kauffman *et al.* (1994) identified the presence of D- and B-hordein components in lager foam, whereas C-hordein components could not be detected. ELISAs (The Enzyme-Linked ImmunoSorbent Assay) have also been used to identify polypeptides derived from B, C, and D hordein in beer and beer foam (Sheehan and Skerritt, 1997). Further, a member of the A-hordein sub-group, a CMe protein has been identified as being involved in beer haze formation (Robinson *et al.*, 2005), where malts produced from varieties without the CMe protein failed to produce a haze under forced aging.

Although large proline-rich polypeptides influence chill haze formation (Asano *et al.*, 1982), smaller hydrophobic polypeptides have positive effects on foam formation. Hence, several important factors should be considered when investigating beer and beer foam quality, including the initial barley hordein profile, malt hordein profile, level of modification, mashing conditions, and the possible interaction of polyphenols and additional proteins such as protein Z and lipid transfer proteins (LTPs). Silva *et al.* (2007) also monitored hordeins (and other proteins) during malting, wort and beer production. There was a significant reduction in hordeins but other proteins (LTP and Protein Z) survived the malting process.

3.5 Malt Quality

Malt quality is derived from the impact of the malting process on components and enzymes within the barley grain. However, a number of these cannot be predicted from the analysis of the barley grain.

3.5.1 Diastatic Power

Diastatic Power (DP) is the term used to describe the collective activity of starch degrading enzymes in malt. Four enzymes, α-amylase, β-amylase, limit dextrinase and α-glucosidase, have been identified during malting and mashing (Osman *et al.*, 1996). β-amylase is the only enzyme in resting barley. The remaining enzymes are synthesised during germination. There are genetic and environmental effects on α-amylase, β-amylase and limit dextrinase (Arends *et al.*, 1995). Industry methods used to measure diastatic power vary considerably in a number of aspects including substrate, pH and assay temperature. These variations may impact on one or more enzymes. Most methods only provide data solely on the enzyme potential under those conditions which are far removed from industrial mashing conditions (Henry, 1984).

3.5.2 α-amylase

α-amylase is an endohydrolase that randomly cleaves α-(1-4) glucosidic bonds in starch. The level of this enzyme is not detectable in barley grain, however it increases once germination commences (Bathgate and Palmer, 1973). In most cereals, the α-amylase I form appears shortly after anthesis. The level declines during grain maturation. When germination commences, a second form, α-amylase II, appears. In the first hours of germination, α-amylase II is released from the scutellum. After the first day of germination the aleurone becomes the main source of α-amylase (Munck *et al.*, 1981). The level of α-amylase II formation is highly dependant upon GA. Studies have shown that this form was synthesized in an embryo-less kernel with GA as a stimulant (Freeman, 1984). In the presence of GA, both enzymes continue to be secreted by the aleurone (MacGregor, 1987). Multiple forms of amylase I and II exist and these forms can vary between varieties (MacGregor, 1987).

A third form of α-amylase (Ⅲ) has been reported. It is a complex between amylase II and an amylase inhibitor protein (MacGregor 1987). However, recent evidence on the location of expression of the inhibitor gene suggests that the binding of this inhibitor to α-amylase is probably not significant *in vivo* (Furtado *et al.*, 2002).

The pH optima for α-amylase is also below mashing pH. α-amylase II is highly dependant upon the level of calcium ions present. During mashing the enzyme was shown to be highly active, albeit a number of variables influenced the total activity. The optimal temperature for α-amylase II is around

65°C(Briggs *et al.*, 1981; Hoseney, 1986) which would allow the enzyme to perform efficiently under most mashing conditions. However, consideration has been given to improving its thermostability in barley by introduction of an alien genetic form of the enzyme from bacteria (Vickers *et al.*, 1996). The preliminary results suggested that mashing could be carried out at temperatures as high as 75°C. In a 65°C infusion mash, the activity of the bacterial enzyme was found to be 1.5 times that of the barley enzyme (Vickers *et al.*, 1996).

3.5.3 β-amylase

β-amylase is an exoenzyme that cleaves the disaccharide, maltose, from the non-reducing end of amylose and amylopectin. The enzyme activity alone catalyses the hydrolysis of approximately 70% of the amylose and 50% of the amylopectin fractions of barley starch (Hoseney, 1986). Two forms (free and bound) are present in barley. The bound form is linked to the hordein matrix and results suggest that the bound form was released by proteinase activity during germination (Sopanen and Lauriere, 1989; Grimes and Briggs, 1996; Buttimer and Briggs, 2000a; 2000b). A cellular pH of 5.3 is optimal for the enzyme. However, conditions during mashing of > 6.0 are not optimal and impact significantly on its action in commercial mashes. The thermostability of β-amylase decreases rapidly at temperatures above 55°C (Hoseney, 1986). For decoction style mashes (a low initial mash-in temperature followed by rapid heating) β-amylase remains active until the mash temperature exceeds 55°C. In comparison, for a mashing style where the infusion temperature is greater than 65°C, the activity of β-amylase is reduced to a few minutes. Fix (1989) has shown that in high temperature mashes, with a low grist liquor ratio, the maltose level in the final wort is higher. This suggests substrate protection of the enzyme within a thick mash.

An increase in the thermostability of β-amylase would provide brewers with opportunities to increase efficiencies in wort production. Two forms (SD1 and SD2) have been observed in commercial barley varieties. A third form (SD3) with increased thermostability has recently been identified in wild barley types (Eglington *et al.*, 1998). This enzyme has been shown to remain active at temperatures greater than 60°C under simulated commercial mashing conditions.

3.5.4 Limit Dextrinase

Limit dextrinase catalyses the hydrolysis of the α-(1-6) glucosidic linkages of amylopectin. The enzyme produces an increased number of smaller linear oligosaccharide chains that subsequently are rapidly hydrolysed by α-amylase and β-amylase. Limit dextrinase has been extracted from ungerminated barley (Manners and Yellowlees, 1973; Yamada, 1981; Lenior *et al.*, 1984; MacLeary, 1992). However, its level increases during germination, with maximum activity being obtained after eight days. The observed increase in total limit dextrinase activity is due to a bound form being released by the action of proteinase (Longstaff and Bryce, 1993). Purified limit dextrinase has been found to have an optimal pH of 5.5 and temperature at 50°C (Sissions *et al.*, 1992). Stenholm and Home (1999) have shown that malt limit dextrinase, under programmed mashing conditions, has a similar pH optimum but higher temperature optimum of between 60°C to 63°C. These results have been confirmed by Osman *et al.* (1996). Malt limit dextrinase has been observed to have a positive influence on wort fermentability. This is presumably due to an increase in the production of linear oligosaccharides, allowing α-amylase and β-amylase to carry out their roles (Stenholm and Home, 1999). It is also proposed that limit dextrinase inhibitors play a role in malt quality (MacGregor *et al.*, 2002) as well as starch synthesis (Stahl *et al.*, 2004).

3.5.5 α-glucosidase

α-glucosidase (EC 3.2.1.20) is the fourth enzymic activity involved in hydrolysis of starch during mashing. The enzyme catalyses the release of single glucose from maltose and higher sugars. Like α-amylase, α-glucosidase is synthesised during germination and dependant upon GA. The role of α-glucosidase in malting and mashing has not been clearly defined. Agu and Palmer (1997) reported an increase in the level of α-glucosidase and also in glucose levels during malting. Osman *et al.* (1996) demonstrated that α-glucosidase hydrolyses oligosaccharides preferentially over starch polymers. This suggests that α-glucosidase would rely on the availability of oligosaccharide substrates released by the prior action of the other three starch degrading enzymes. The pH optima for α-glucosidase appears to be dependant upon substrate, namely 4.5–4.6 for maltose substrate but 5.0 for starch as a substrate (Osman *et al.*, 1996a; Agu and Palmer, 1997). Under mashing pH conditions, α-glucosidase activity is reduced, suggesting that activity would be limited during mashing and dependant upon the activity of α- and β-amylases (Osman *et al.*, 1996a). The amount of α-glucosidase activity present in a hot water extract was found to be considerably lower than the activity of other starch degrading enzymes (Osman *et al.*, 1996b). Overall, the efficiency of α-glucosidase during mashing is dependant upon temperature, pH and the availability and form of substrate (Osman *et al.*, 1996a; 1996b; Agu and Palmer, 1997).

3.5.6 β-glucanase

β-(1→3), (1→4)-glucan-4-glucanhydrolases (EC 3.2.1.73) or β-glucanase has received considerable attention by researchers over the past 50 years. The function of β-glucanase is to hydrolyse β-glucan during germination. Two isoenzymes, EI and EII, have been identified and their functional properties reported (Woodward and Fincher 1982). β-Glucanase is produced during grain germination in the aleurone and scutella, with EII and EI produced in the aleurone and EI produce from the scutella (Stuart *et al.*, 1986). The level of this enzyme increases upon the addition of GA.

This first stage of endosperm modification is the breakdown of cell walls. This is one of the critical steps in producing good quality malt. The hydrolysis of β-Glucan during germination has a significant impact on the final malt quality. The mechanism and enzymes involved in cell wall modification remain controversial (Bamforth and Barclay, 1993). However, it is clear that an increase in malt β-glucanase levels results in reduced levels of β-glucan in wort. Slow and/or incomplete breakdown of barley β-glucan has been shown to have a negative impact on hot water extract (Henry, 1986; Stuart *et al.*, 1988) as well as causing viscosity and filtration problems in the brewhouse (Stewart *et al.*, 1998, Stewart *et al.*, 2000).

β-glucanase, as with most enzymes in cereals, was not designed to be active at high temperatures. Woodward and Fincher (1982b) reported the optimal temperatures for EI and EII to be approximately 37°C and 45°C, respectively. β-glucanase activity is considerably reduced during the kilning process, and in the initial stages of high temperature mashing (Woodward and Fincher, 1982b; Loi *et al.*, 1988). All activity is lost after approximately 15 minutes under these conditions. Hence, under industry commercial brewing conditions either high levels of β-glucanase are required through addition of enzymes or an increase in thermostability of the native enzymes.

β-glucan hydrolysis still remains an important target in barley breeding programs. High levels of β-glucanase would be required for high temperature mash breweries as well as breweries where unmalted barley is used as an adjunct. The development of varieties processing high levels of β-glucanase or increased thermostabilities will remain an important breeding target.

3.5.7 Proteinase

The influence of storage protein within the grain and malt on beer quality has been well documented (Jones and co workers), with Jones providing a comprehensive review recently (Jones, 2005a). The mechanisms and control of the breakdown of storage proteins has only been partially unravelled. Early studies were limited to general proteinase activity. Recently, assays have been developed that can more precisely describe the specificity of individual proteinases under germination or mashing pH and temperature optima (Schmitt and Budde, 2007) which could finally help breeders in developing varieties with optimal storage protein degradation.

Four classes of proteinases namely, cysteine, serine, aspartic and metallo, have been identified in barley. The most abundant are the cysteine enzymes (Zhang and Jones, 1995a, 1995b). Studies on individual proteinase classes have been limited and the roles of the various classes have not yet been clearly defined. Jones and co-workers have carried out a number of studies attempting to characterise the role of these enzymes during malting and mashing (Poulle and Jones, 1988; Worbel and Jones, 1992; Worbel and Jones, 1993; Zhang and

Jones, 1995a; 1995b; Fontanini and Jones, 2001; Jones and Marinac, 2002). Most of the work previously reported used non barley protein substrates. In contrast, Osman *et al.* (2002) characterised the major endoproteinase activities in malted barley using isolated native substrates, specifically hordein and glutelin. These activities were all found to be of the cysteine class. The temperature optimum for endoproteinases with glutelin and hordein substrates was 50°C and 40°C respectively. Osman *et al.* (2002) further investigated the proteinase activity in a range of genotypes. Obvious differences were observed between genotypes with the activity again found to be dependant upon the type of substrate. However, no statistical data was provided to indicate any significance between genotypes.

In the most recent review, Jones (2005a) showed that of the four proteinase classes, cysteine- and metallo-proteinases were the only two involved in storage protein breakdown during germination of mashing. However, Schmitt and Marinac (2007) reported that serine proteinase may be involved in the degradation of the hydrolytic enzyme β-amylase using gel studies on green malt.

3.5.8 Hot Water Extract

The hot water extract (HWE) of wort, commonly called malt extract (ME), is the most important trait whether selecting potential new malting varieties or trading malt. The quality of the extract is influenced by four factors. The first is the quality of barley which is affected by environmental factors such as growing conditions, temperature, fertiliser, available nitrogen or moisture. These factors do not impact on extract directly. Their effect is on content and compositions of traits that influence extract, particularly protein and starch.

The second is a number of physical and biochemical components that influence the final level of extract such as two or six rowed type, husk thickness, grain size, protein, starch, non-starch polysaccharides and enzyme production. Varieties with the optimum combination of these traits will consistently produce higher extract even with minor negative effects from either environmental or physiological factors.

The third factor is the malting process. Nischwitz *et al.* (1999) outlined the effects of malting on final beer quality. Most aspects of grain modification affect final beer quality, including important aspects such as clarity and foam stability. During malting, enzymes which have an impact on the degradation of substrates, are either synthesised or cleaved from their bound forms. The range of enzymes produced includes those that degrade cell wall components, proteins and starch. The process of malt production varies between countries with four day germination schedules in Australia and five to six day germination schedules in most overseas countries. The objective for most maltsters is to maintain high extract levels and yet somehow achieve relatively low protein modification levels ($< 50\%$).

Mashing is the fourth factor that influences extract. Within the mashing process, there are a number of physical factors that affect the resultant extract.

These are pH, mash time, mash temperature, grist/particle size and grist to liquor ratio. Cook (1964) and Briggs *et al.* (1981) presented reviews detailing previous and current knowledge and technology available on mashing. While all of the above aspects of mashing would influence the quality of the final extract, most result from the genetic attributes of the starting barley. For example, high DP varieties produce high levels of malt DP under optimal conditions. This in turn impacts on starch hydrolysis and the final fermentable sugar profile. Low DP barley varieties only produce low to moderate levels of starch degrading enzymes which affect the fermentable sugars profile.

A number of studies have presented results detailing the relationship between grain and malt physical attributes and extract. Some of these attributes include protein levels and the type of protein fractions (Smith, 1990; Howard *et al.*, 1996; Brennan *et al.*, 2000; Molina Cano *et al.*, 2001), hardness and milling energy (Ellis *et al.*, 1979; Alison, 1986; Swanston and Taylor, 1988), starch properties (Glennie-Holmes, 1995a,b,c,d), non-starch polysaccharides (Henry, 1985; Molina Cano *et al.*, 1995) and husk thickness (Roumeliotis *et al.*, 2000). The specifics of the relationships between extract and these attributes have been discussed in the sections above.

3.6 Feed Barley Quality Traits

The major portion of the Australian barley grain crop (around 70%) is used as feed grain. This is made up of a combination of poor quality malting cultivars and specific feed cultivars. A high portion of barley is used as the feed by the animal industries in developed countries. Overall, feed cultivars are those that either do not the pass industry malt quality standards or were bred for specific improved agronomic performance. These are simply designated as feed cultivars, not necessarily possessing desirable feed quality attributes.

Current understanding of true feed quality is somewhat limited. However, recent studies indicate that a number of grain traits in barley are relevant to feed quality. A number of reviews have listed specific traits that could be related to feed quality, including a high starch level, low protein, acid detergent fibre (ADF) and neutral detergent fibre (NDF) (Overnell-Roy *et al.*, 1998a,b,c;). A special edition of the Australia Journal of Agricultural Research (Volume 50, 1999) published a comprehensive set of reviews on the current knowledge of feed quality. Areas covered were grain composition and attributes that contribute to feed quality. These included both processing (Rowe *et al.*, 1999; Kaiser, 1999) and methodology (Petterson *et al.*, 1999; Wrigley, 1999). The general consensus was that there were critical differences between grain types and feed quality which have an impact on animal performance (both ruminants and monogastrics). Hussien *et al.* (2004) identified a number of genetic regions associated with hardness, fibre and an *in-sacco* dry matter digestibility method.

Starch, protein and fibre (ADF, acid detergent fibre; NDF, neutral detergent fibre) are the main grain attributes in barley that have a direct impact on feed potential. All of these traits are influenced by both cultivar and environment. Thus it can be assumed that it would be possible to genetically select for feed quality. However, from a breeding perspective it is still unclear as to the definitive range of values for grain traits that would have a direct and positive influence on animal performance. To date, only one "fee" quality cultivar (Valier) has been released with claims of a direct improvement in animal performance (Bowman *et al.*, 1999). Particle size, ADF and starch as well as rate and extent of fermentation were considered to be the four most significant factors influencing performance (Bowman *et al.*, 2000).

Within the malting industry, processing of the grain has a direct impact on malt and beer quality. Similarly, processing of grain in the feed industry would also appear to impact on feed quality. Barley hardness is proposed associated with improved feed grain quality. Bowman *et al.* (1996) have reported a negative correlation between particle size index (hardness), acid detergent fibre and dry matter digestibility (DMD), but a positive correlation to daily live weight gain. Rowe *et al.* (1999) has also found that particle size has a direct impact on animal performance. Dehghan-Banadaky *et al.* (2007) detailed the effects of processing on particle size and cattle feed quality. Further, treatment of the grain via heating or cooking, as well as the use of supplements, has been found to have a positive effect on animal performance.

Limited data has been published on the relationship between feed and malt quality. Crosbie and Portman (1977) have reported on the relationships between EBC style malt extract with fibre. Significant ($p < 0.05$) negative correlations were obtained between extract and both ADF and NDF. Further analysis is required on well designed sample sets to define the relationships between barley traits for malt and feed quality. These, in turn, would be useful for breeding programs to select for feed quality cultivars. It has been proposed to use the EBC extract as an indicator of feed quality (Molino-Cano *et al.*, 1997) in a combined feed/malt quality testing program within a barley breeding program (Fox *et al.*, 2008).

Overall, the feed industry has yet to define quality in any terms that can be used practically for improved breeding selection. Until then, feed cultivars will remain second class quality grain.

3.7 Conclusion

Barley remains one of the world's economically important crops for its uses in food, feed and beverage production. While the main components are quite well understood, the minor components such as enzymes and inhibitors, their functions and roles and the synergistic interactions within the grain and when used in making value-added products remain to be completely unravelled.

The opportunity to exploit wild and exotic barley germplasm, as well as possibly genetically enhanced barleys, will have two effects. This will allow researchers to continue to improve barley cultivars and gain future knowledge of the chemistry and biochemical processes of barley.

Another aspect in gaining further understanding of barley will be the use of the most modern technologies, such as Nuclear Magnetic Resonance, Infrared Technologies, improved gene studies, i.e. Single Nucleotide Polymorphsims and Diversity Array Technologies.

References

Agu R C, Palmer G H (1997) α-Glucosidase Activity of Sorghum and Barley Malts. J Inst Brew, 103: 25-29

Agu R C (2003) Some Relationships between Malted Barleys of Different Nitrogen Levels and the Wort Properties. J Inst Brew, 109: 106-109

Agu R C, Brosnan J M, Bringhurst T A, Palmer G H, Jack F R (2007) Influence of Corn Size Distribution on Diastatic Power of Malted Barley and its Impact on Other Malt Quality Parameters. J Agr Food Chem, 55: 3702-3707

Alison M J, Cowe I, McHale R (1976) A Rapid Test for the Prediction of Malting Quality of Barley. J Inst Brew, 82: 166-167

Alison M (1986) Relationships between Milling Energy and Hot Water Extract Values of Malts from Some Modern Barleys and Their Parental Cultivars. J Inst Brew, 92: 604-607

Allosio-Ouarnier N, Quemener B, Bertrand D, Boivin P (2000) Application of High Performance Anion Exchange Chromatography to the Study of Carbohydrate Changes in Barely during Malting. J Inst Brew, 106: 45-52

Arends A M, Fox G P, Henry R J, Marschke R J, Symons M H (1995) Genetic and Environmental Variation in the Diastatic Power of Australian Barley. J Cereal Sci, 21: 63-70

Asano K, Shinagawa K, Hashimoto N (1982) Characterization of Haze-formation Proteins of Beer and Their Roles in Chill Haze Formation. J Am Soc Brew Chem, 40: 147-154

Australian Lot Feeders Association (2001) Lotfeeding, Annual Report, September 14-17

Autio K, Mannonen L, Pietila K, Koskinen M, Siika-Aho M, Linko M, Morgan A (1996) Incubation of Barely Kernel Sections with Purified Cell Wall Degrading Enzymes. J Inst Brew, 102: 427-432

Axcell B (1998) Malt Analysis: Prediction or Predicament, Tech Q, 35: 28-30

Bamforth C W, Kanauchi M (2001) A Simple Model for the Cell Wall of the Starchy Endosperm in Barley. J Inst Brew, 107: 235-240

Bason M L, Wrigley C W, McLeary B V (1993) Modelling the Effects of Sprouting Damage on Malting Barley, Based on Viscometric or Amylase Analysis. In: Walker-Simmons M K, Reid J L (eds.) Preharvest Sprouting in Cereals. American Society of Cereal Chemists, St Paul, 417-425

Bathgate G N, Palmer G H (1973) the *in vivo* and *in vitro* Degradation of Barley and Malt Starch Granules. J Inst Brew, 79: 402-406

Beecher B, Smidansky E D, See D, Blake T K, Giroux M J (2001) Mapping and Sequence Analysis of Barley Hordoindolines. Theor Appl Genet, 102: 833-840

Beecher B, Bowman J, Martin J M, Bettge A D, Morris C F, Blake T K, Giroux M J (2002) Hordoindolines Are Associated With a Major Endosperm-texture QTL in Barley (*Hordeum Vulgare*). Genome, 45: 584-591

Benech-Arnold R L, Giallorenzi M C, Frank J, Rodriguez V (1999) Termination of Hull-imposed Dormancy in Developing Barley Grain Is Correlated in Change With Embryonic ABA Levels and Changes. Seed Sci Res, 9: 39-47

Benech-Arnold R (2002) Bases of Pre-harvest Sprouting Resistance in Barley: Physiology, Molecular Biology and Environmental Control of Dormancy in the Barley Grain. In: Slafer G A, Molina-Cano J L, Savin R, Araus J L, Romagos I (eds.) Barley Science: Recent Advances from Molecular Biology to Agronomy of Yield and Quality. Food Products Press, New York, 481-502

Benetrix F, Sarrafi A, Autran J C (1994) Effects of Cultivar and Nitrogen Nutrition on Protein Aggregates in Barley. Cereal Chem, 71: 75-82

Bhullar S S, Jenner C F (1996) Effects of Temperature on the Conversion of Sucrose to Starch in the Developing Wheat Endosperm. Aust J Plant Physiol, 13: 605-615

Bowman J G P, Blake T K, Surber L M M, Habernicht T K, Daniels T K, Daniels J T (1996) Genetic Factors Controlling Digestibility of Barley for Ruminants. Proc West Sec Am Soc Anim, 47: 257-260

Bowman J G P, Blake T K, Surber L M M, Habernicht D K, Bockelman H (2001) Feed Quality Variation in the Barley Core Collection of the USDA National Small Grains Collection. Crop Sci, 41: 863-870

Brennan C S, Harris N, Smith D, Shewry P R (1996) Structural Differences in the Mature Endosperms of Good and Poor Malting Barley Cultivars. J Cereal Sci, 24: 171-177

Brennan C S, Smith D B, Harris N, Shewery P R (1998) the Production and Characterisation of Hor3 Null Lines of Barley Provides New Information on the Relationship of D Hordein to Malting Performance. J Cereal Sci, 28: 291-299

Briggs D E, Hough J S, Stevens R, Young T W (1981) Malting and Barley Brewing Science, Volume 1, 2nd Edition, Chapman and Hall, London

Briggs D E, Woods J L (1993) Dormancy in Malting Barley: Studies on Drying, Storage, Biochemistry and Physiology. HGCA Project Report 84, 146

Briggs D E, Woods J L, Favier J F (1994) Drying and Storage Treatments for Overcoming Dormancy in Malting Barley. J Inst Brew, 100: 271-278

Briggs K G (1974) Combining Ability for Kernel Plumpness in a Diallel Cross of Five Canadian Barley Cultivars. Can J Plant Sci, 54: 29-34

Brinch-Pedersen H, Borg S, Tauris B, Holm P B (2007) Molecular Genetic Approaches to Increasing Mineral Availability and Vitamin Contents of Cereals. J Cereal Sci, 46: 308-326

Broadbent R E, Palmer G H (2001) Relationship between α-amylase Activity, Steeliness, Mealiness, Nitrogen Fractions on the Barley Grain. J Inst Brew, 107: 349-354

Buttimer E T, Briggs D E (2000a) Characterisation of Solubilized Forms of Bound Amylase Released By Various Agents. J Inst Brew, 106: 71-82

Buttimer E T, Briggs D E (2000b) Mechanisms of the Release of Bound Amylase. J Inst Brew, 106: 83-94

Caldwell K S, Langridge P, Powell W (2004) Comparative Sequence Analysis of the Region Harbouring the Hardness Locus in Barley and its Collinear Region in Rice1. Plant Physiol, 136: 3177-3190

Cakir M, Poulsen D, Galwey N, Ablett G, Chalmers K, Platz G, Park R, Lance R, Panozzo J, Read B J, Moody D, Barr A, Johnston P, Boyd W J R, Grime C R, Appels R, Li C, Jones M G K, Langridge P (2003) Mapping and QTL Analysis in the Barley Population Tallon X Kaputar. Aust J Agric Res, 54: 1155-1162

Canadian Wheat Board (2004) Barley - Malting Barley. Http://www.cwb.ca/en/growing/barley/12-malting_barley.jsp

Chandra G S, Proudlove M O, Baxter E D (1999) the Structure of Barley Endosperm - An Important Determinant of Malt Modification. J Sci Food Agric, 79: 37-46

Choct M, Selby E A D, Cadogan D J, Campbell R G (2004) Effects of Particle Size, Processing and Dry or Liquid Feeding on Performance of Piglets. Aust J Agric Res, 55: 237-245

Chowdhury S L, Wadlaw I F (1978) the Effect of Temperature on Kernel Development in Cereals. Aust J Agric Res, 29: 205-23

Cowe I A, Cuthbertson D C, Swanston J (1989) the Effect of Moisture and Nitrogen Levels on Milling Energy of Barley. J Inst Brew, 95: 423-425

Crosbie G B, Portman P A (1977) A Comparison of Screening Tests for Feed and Malting Quality for Barley. J Aust Inst Agri Sci, 43: 160-161

Dehghan-Banadaky M, Corbett R, Oba M (2007) Effects of Barley Grain Processing on Productivity of Cattle. Anim Feed Sci Tech, 137: 1-24

Dahl S W, Rasmussen S K, Hejgaard J (1996) Heterlogous Expression of Three Plant Serpins With Distinct Inhibitory Specificities. J Biol Chem, 271: 25083-25088

Darlington H F, Tesci L, Harris N, Griggs D L, Cantell I C, Shewery P R (2000) Starch Granule Associated Proteins in Barley and Wheat. J Cereal Sci, 32: 20-21

Darlington H, Rouster J, Hoffman L, Halford N G, Shewry P R, Simpson D J (2001) Identification and Molecular Characterisations of Hordoindolines from Barley Grain. Plant Mol Biol, 47: 785-794

Doan D N P, Fincher G B (1992) Differences in the thermostabilities of Barley (cv Clipper)(1$\rightarrow$3, 1$\rightarrow$4)β-glucanases Are only Partly Determined By N-glycosylation. FEBS Lett, 309: 265-271

Douliez J P, Michon T, Elmorjani K, Marion D (2000) Structure, Biological and Technological Functions of Lipid Transfer Proteins and Indolines, the Major Lipid Binding Proteins from Cereal Kernels. J Cereal Sci, 32: 1-20

Duffus C M, Cochrane M P (1992) Grain Structure and Composition. In: Shewry P R (eds.) Barley: Genetics, Biochemistry, Molecular Biology and Biotechnolog. CAB International Oxon, 291-318

Duffus C M, Cochrane M P (1993) Formation of the Barley Grain - Morphology, Physiology and Biochemistry. In: MacGregor A W, Bhatty R S (eds.) Barley-Chemistry and Technology. AACC, Minnesota, 31-72

Edney M (1996) Barley. In: Henry R J, Kettlewell P S (eds.) Cereal Grain Quality. Chapman & Hall, the University Press, Cambridge, 113-146

Edney M J, Mather D E (2002) Quantitative Trait Loci Affecting Germination Traits and Malt Friability in a Two-rowed By Six-rowed Barley Cross. J Cereal Sci, 39: 283-290

Edney M J, Langrell D E (2004) Evaluating the Malting Quality of Hulless CDC Dawn, Acid-dehusked Harrington and Harrington Barley. J Am Soc Brew Chem, 62: 18-22

Eglinton J K, Langridge P, Evans D E (1998) Thermostability Variation in Alleles of Barley Beta-amylase. J Cereal Sci, 28: 301-309

Ellis R P, Swanston J S, Bruce F M (1979) A Comparison of Some Rapid Screening Tests for Malting Quality. J Inst Brew, 85: 282-285

Emebiri L C, Moody D B, Panozzo J F, Chalmers K J, Kretschmer J M, Ablett G A (2003) Identification of QTLs Associated With Variations in Grain Protein Concentration in Two-row Barley. Aust J Agric Res, 54: 1211-1221

Emebiri L C, Moody D B (2004a) Potential of Low-protein Genotypes for Nitrogen Management in Malting Barley Production. J Agric Sci, 142: 319-325

Emebiri L C, Moody D B, Horsely R, Panozzo J, Read B J (2004b) the Genetic Control of Grain Protein Content Variation in a Double Haploid Population Derived from a Cross between Australian and North American Two-rowed Barley Lines. J Cereal Sci, 41: 107-114

Emebiri L C, Moody D B, Panozzo J F, Read B J (2004c) Mapping of QTL for Malting Quality Attributes in Barley Based on a Cross of Parents With Low Grain Protein Concentration. Field Crop Res, 87: 195-205

Emebiri L C, Moody D B, Black C, Van Ginkel M, Hernandez E (2007) Improvement in Malting Quality Barley Grain Yield By Manipulation of Genes Influencing Grain Protein Content. Euphytica, 156: 185-194

Evans E, Stenholm K, Vilpola A, Home S, Hughes G (1998) Pilot Scale Evaluation of Mash Filter Performance of Malts With Variable Quality. Tech Q, 35: 189-195

Evans D E, Hejgaard J (1999) the Impact of Malt Derived Proteins on Beer Foam Quality Part I: the Effect of Germination and Kilning on the Level of Protein Z4, Protein Z7 and LTP1. J Inst Brew, 105: 159-169

Evans D E, Sheehan M C, Stewart D C (1999a) the Impact of Malt Derived Proteins on Beer Foam Quality Part I: the Influence of Malt foam-positive Proteins and Non-starch Polysaccharides on Beer Foam Quality. J Inst Brew, 105: 171-177

Evans D E, Vilpola A, Stewart D C, Stenholm K, Poyri S, Washington J M, Barr A R (1999b) Pilot Scale Investigation of the Importance of the Barley Husk for Mash Filtration. Tech Q, 36: 375-382

Evers A D, Blakeney A B, O'Brien L (1999) Cereal Structure and Composition. Aust J Agric Res, 50: 629-650

Evrard A, Lagarde V, Joudrier P, Gautier M F (2007) Puroindoline-a and Puroindoline-b Interact With the Saccharomyces Cerevisiae Plasma Membrane Through Different Amino Acids Present in Their Tryptophan-rich Domain. J Cereal Sci, 48: 379-386

Ferrio J P, Alonso N, Voltas J, Araus J (2006) Grain Weight and Changes Over Time in Ancient Cereal Crops: Potential Roles of Climate and Genetic Improvement. J Cereal Sci, 44: 323-332

Fincher G B, Stone B A (1986) Cell Walls and Their Components in Cereal Grain Technology. In: Pommaranz Y (ed.) Advances in Cereal Science and Technology. American Society of Cereal Chemists, St Paul, 8: 207-295

Fincher G B (1975) Morphology and Chemical Composition of Barley Endosperm Cell Walls. J Inst Brew, 81: 116-122

Fontanini D, Jones B L (2001) Study of Metallopeptidase Isoenzymes from Malted Barley (*Hordeum Vulgare* Cv Morex). J Agric Food Chem, 49: 4903-4911

Fox G P, Kelly A M, Poulsen D M E, Inkerman P A, Henry R J (2006a) Genetic and Environmental Effects on Selecting Improved Barley Grain Size in Dry Environments. J Cereal Sci, 43: 198-208

Fox G P, Kelly A M, Cakir M, Bloustein G, Poulsen D M E, Inkerman P A, Henry R J (2006b) Impact of the Husk on Barley Grain Quality. J Inst Brew, 112: 101-107

Fox G P, Nguyen D L, Bowman J G P, Poulsen D M E, Inkerman P A, Henry R J (2007a) Relationship between Hardness Genes and Quality in barley (Hordeum Vulgare). J Inst Brew, 113: 87-95

Fox G P, Osborne B G, Bowman J G P, Kelly A M, Cakir M, Poulsen D M E, Inkerman P A, Henry R J (2007b) Measurement of Genetic and Environmental Variation in Barley (*Hordeum Vulgare*) Grain Hardness. J Cereal Sci, 46: 82-92

Fox G P, Bowman J, Kelly A, Inkerman P A, Poulsen D M E, Henry R J (2008a) Assessing for Genetic and Environmental Effects on Ruminant Feed Quality in Barley (*Hordeum Vulgare*). Euphytica, 163: 249-257

Fox G P, Kelly A, Bowman J G, Poulsen D M E, Inkerman P A, Henry R J (2008b) Relationship between Malt and Feed Quality in Barley (*Hordeum Vulgare*). Euphytica, Submitted

Furtado A, Henry R J, Scott K J (2003) the Promoter of the asi Gene in Transgenic Barley. Plant Mol Biol, 52: 787-799

Gamlath J, Aldred G P, Panozzo J F (2008) Barley (1→3; 1→4)-β-glucan and Arabinoxylan Content Are Related to Kernel Hardness and Water Uptake. J Cereal Sci, 47: 365-371

Gibson C E, Evans D E, Proudlove M O (1996) Protein Z4 and Beer Foam. Ferment, 9: 81-84

Glennie-Holmes M (1995) Studies on Barley and Malt With the Rapid Viscoanalyser I: the Effect of Variations in Physical and Chemical Parameters. J Inst Brew, 101: 11-18

Glennie-Holmes M (1995) Studies on Barley and Malt With the Rapid Viscoanalyser II: the Effects of Modification on Viscograms. J Inst Brew, 101: 19-28

Glennie-Holmes M (1995) Studies on Barley and Malt With the Rapid Viscoanalyser III: the Prediction of Malting Potential from Viscograms. J Inst Brew, 101: 29-32

Glennie-Holmes M (1995) Studies on Barley and Malt With the Rapid Viscoanalyser IV: Amyloclasis, Amylolysis and Cytolysis of Malt and Malting Quality. J Inst Brew, 101: 33-38

Goblirsch C A, Horsley R D, Schwarz P B (1996) A Strategy to Breed Low-protein Barley With Acceptable Kernel Color and Diastatic Power. Crop Sci, 36: 41-44

Grimes K H, Briggs D E (1995) Release and Activation of Barley Amylase. J Inst Brew, 101: 337-343

Grimes K H, Briggs D E (1996) the Release of Bound amylase By Macromolecules. J Inst Brew, 102: 261-270

Guerin J R, Lance R C M, Wallace W (1992) Release and activation of Barley Beta-amylase By Malt Endopeptidases. J Cereal Sci, 15: 5-14

Holopainen U, Wilhelmson A, Salmenkallio-Marttila M, Peltonen-Sainio P, Rajala A, Reinikainen P, Kotaviita E, Simolin H, Home S (2005) Endosperm Structure Affects the Malting Quality of Barley (Hordeum Vulgare L.). J Agric Food Chem, 53: 7279-7287

Han F, Ullrich S E, Clancey J A, Romagosa I (1999) Inheritance and Fine Mapping of a Major Barley Seed Dormancy QTL. Plant Sci, 143: 113-118

Han F, Ullrich SE, Romagosa I, Clancy JA, Froesth JA & Wesenberg DM (2003) Quantitative Genetic Analysis of Acid Detergent Fibre Content in Barley Grain. J Cereal Sci, 38: 167-172.

Han J Y, Schwartz P B (1996) Arabinoxylan Composition in barley, Malt and Beer. J Am Soc Brew Chem, 54: 216-220

Harju M A, De Souzza S, Brodl M R (2003) the Heat-shock Response of Developing Barley Aleurone Layers. Seed Sci Res, 13: 229-238

Hartl L, Kipp A, Baumer M, Schweizer G, Friedt W (2000) QTL-Mapping of Malting Quality Parameters in Spring Barley. In: Logue S (eds.) Proceedings of the 8th International Barley Genetics Symposium, Volume 3, 246-248

Hejgaard J, Kaesgaard P (1983) Purification and Properties of the Major Antigenic Beer Protein of Barley Origin. J Inst Brew, 89: 402-410

Henry R J (1984) A Rapid Method for the Determination of Diastatic Power. J Inst Brew, 90: 37-39

Henry R J (1986) Genetic and Environmental Variation in the pentosan and β-glucan Contents of Barley and Their Relation to malting Quality. J Cereal Sci, 4: 269-277

Henry R J (1987) Pentosans and (1-3) (1-4)-β-glucan Concentrations in Endopserm and Whole Grain of Wheat, Barley, Oats and rye. J Cereal Sci, 6: 253-258

Henry R J, Cowe I (1990) Factors Influencing the Hardness (Milling Energy) and Malting Quality of Barley. J Inst Brew, 96: 135-136

Hirota H, Kuroda H, Takoi K, Kaneko, T, Kaneda H, Yoshida I, Takashio M, Ito K, Takeda K (2006) Brewing Performance of Malted Lipoxygenase-1 Null Barley and Effect on the Flavour Stability of Beer. Cereal Chem, 83: 250-254

Hockett E A, White L M (1984) Simultaneous Breeding for feed and Malting Quality. Proceedings of the 4th International Barley Genetics Symposium, 234-241

Holtman W L, Vredenbregt-Heistek J C, Schmitt N F, Feussner I (1997) Lipoxygenase-2 Oxygenates Storage Lipids in Embryos of Germinating Barley. Euro J Biochem, 248: 452-458

Hori K, Sato K, Takeda K (2007) Detection of Seed Dormancy QTL in Multiple Mapping Populations Derived from Crosses Involving Novel Barley Germplasm. Theor Appl Genet, 115: 869-876

Hoseney R C (1986) Minor Constituents of Cereals. In: Principles of Cereal Science and Technology. American Society of Cereal Chemists, St Paul Minnesota, 89-110

Howard K A, Gayler K R, Eagles H A, Halloran G M (1996) the Relationship between D-Hordein and Malting Quality in Barley. J Cereal Sci, 24: 47-53

Hough J S (1985) the Biotechnology of Malting and Brewing. In: Cambridge Studies in Biotechnology 1. Cambridge University Press, Cambridge

Hussein A H (2004) Genetic and Mapping of Quantitative trait Loci of Feed Quality-related Traits in Barley (*Hordeum Vulgare* L.). Ph D Thesis, Montana State University

Izawa M, Kano Y, Koshino S (1993) Relationship between Structure and Solubitity of (1-3) (1-4)-β-glucan from Barley. J Am Soc Brew Chem, 51: 123-127

Izydorczyk M S, Macgregor A W, Billiaderis C G (2001) Effects of Malting on Phase Transition Behaviour of Starch in Barley Cultivars With Varying Amylose Content. J Inst Brew, 107: 119-128

Janes P W, Skerritt J H (1993) High Performance Liquid Chromatography of Barley Proteins: Relative Quantities of Hordeins Fractions Correlate With Malt Extract. J Inst Brew, 99: 77-84

Jones B L (1991) Partial Purification and Characterisation of Two Barley Fractions That Inhibit Malt Endoproteinases. J Am Soc Brew Chem, 49: 158-161

Jones B L (1999) Malt Endoproteinases and How They Affect Soluble Protein Levels. in Proceedings of the 9th Australian Barley Technical Symposium, Melbourne

Jones B L (2001) Interactions of Malt and Barley (*Hordeum Vulgare* L.) Endoproteinases With Their Endogenous Inhibitors. J Agric Food Chem, 49: 5975-5981

Jones B L (2005a) Endoproteases of Barley and Malt. J Cereal Sci, 42: 139-156

Jones B L (2005b) the Endogenous Endoprotease Inhibitors of Barley and Malt and Their Roles in Malting and Brewing. J Cereal Sci, 43: 271-280

Jones B L, Marinac L A (1995a) A Comparison of Barley and Malt Polypeptides That Inhibit Green Malt Endoproteinases. J Am Soc Brew Chem, 53: 160-166

Jones B L, Marinac L A (1995b) Barley LTP1 (PAPI) and LTP2: Inhibitors of Green Malt Cysteine Endoproteinases. J Am Soc Brew Chem, 53: 194-195

Jones B L, Marinac L A (2000) Purification and Partial Characterisation of a Second Cysteine Proteinase Inhibitor from Ungerminated Barley (*Hordeum Vulgare* L.). J Agric Food Chem, 48: 257-264

Jones B L, Marinac L A (2002) the Effect of Mashing on Malt Endoproteolytic Activities. J Agric Food Chem, 50: 858-864

Juge N, Svensson B (2006) Proteinaceous Inhibitors of Carbohydrate Active Enzymes in Cereals: Implication in Agriculture, Cereal Processing and Nutrition. J Sci Food Agric, 86: 1573-1586.

Kaiser A G (1999) Increasing the Utilisation of Grain When Fed Whole to Ruminants. Aust J Agric Res, 50: 737-756

Kanauchi M, Bamforth C W (2002) Enzymic Digestion of Walls Purified from the Starchy Endosperm of Barley. J Inst Brew, 108: 73-77

Kenn D A, Dagg A H S, Stuart I M (1993) Effect of Environment and Cultivar on the Fermentability of Malt Produced from Four Australia Barley Cultivars. J Am Soc Brew Chem, 51: 119-222

Koehler S M, Ho T H D (1990) A Major Gibberellic Acid-induced Barley Aleurone Cystein Proteinase Which Digest Hordein. Plant Physiol, 94: 251-258

Koliatsou M, Palmer G H (2003) A New Method to Assess Mealiness and Steeliness of Barley Cultivars and Relationship of Mealiness with Malting Parameters. J Am Soc Brew Chem, 61: 114-118

Kühbeck F, Dickel T, Krottenthaler M, Back W, Mitzscherling M, Delgado A, Becker T (2005) Effects of Mashing Parameters on Mash β-glucan, FAN and Soluble Extract Levels. J Inst Brew, 111: 316-327

Lee W J, Pyler R E (1984). Barley Malt Limit Dextrinase: Varietal, Environmental and Malting Effects. J Am Soc Brew Chem, 42, 11-17

Lenior P, MacGregor A W, Moll M, Duassant J (1984) Identification of Debranching Enzymes from Barley and Malt By Isoelectric Focusing. C.R. Academy of Science Series, 298: 243-248

Li J, Baga M, Rossnagel B G, Legge W G, Chibbar R N (2008) Identification of Quantitative Trait Loci for β-glucan Concentration in Barley Grain. J Cereal Sci, 48: 647-655

Lin R, Horsley R D, Schwartz P B (2008) Associations between Caryopsis Dormancy, α-amylase Activity, and Pre-harvest Sprouting in Barley. J Cereal Sci, in Press 48: 446-456

Longstaff M A, Bryce J H (1993) Development of Limit Dextrinase in Germinated Barley (*Hordeum Vulgare* L.). Evidence of Proteolytic Activation. Plant Physiol, 101: 881-889

Lusk L T, Duncombe G R, Kay S B, Navarro A, Ryder D (2001) Barley β-glucan and Beer Foam Stability. J Am Soc Brew Chem, 59: 183-186

MacNicol P K, Jacobsen J V, Keys M M, Stuart I M (1993) Effects of Heat and Water Stress on Malting Quality and Grain Parameters of Schooner Barley Grown in Cabinets. J Cereal Sci, 18: 61-68

MacGregor A W (1987) α-amylase, Limit Dextrinase and β-glucosidase Enzymes in Barley and Malt. CRC Critical Reviews in Biotechnology 5: 117-128

MacGregor A W (1996) Malting and Brewing Science: Challenges and Opportunities. J Inst Brew, 102: 97-102

MacGregor A W, Balance D L (1980) Hydrolysis of Large and Small Starch Granules from Normal and Waxy Barley Cultivars By β-amylases from Barley Malt. Cereal Chem, 57: 397-402

MacGregor A W, Bazin S L,Izydorczyk M S (2002) Gelatinisation Characteristics and Enzyme Susceptibility of Different Types of Barley Starch in the Temperature Range 48°C-72°C. J Inst Brew, 108: 43-47

MacGregor A W, Bazin S L, Schroeder S W (2002) Effect of Starch Hydrolysis Products on the Determination of Limit Dextrinase Inhibitors in Barley and Malt. J Cereal Sci, 35: 17-28

Macewicz J, Orzechowski S, Dobrzynska U, Haebel S (2006) Is Quantity of Protein in Barley Forms Determined By Proteins Localized in the Subaleurone Layer? Acta Physiol Plant, 2: 409-416

Maghirang E B, Dowell F E (2003) Hardness Measurement of Bulk Wheat By Single-kernel Visible and Near-infrared Reflectance Spectroscopy. Cereal Chem, 80: 316-322

Maguire T, Collins G, Sedgley M (1994) A Modified CTAB DNA Extraction Procedure for Plants Belonging to the Family Proteaceae. Plant Mol Biol Rep, 12: 106-109

Manners D J, Yellowlees D (1973) Studies on Debraching Enzymes. I. the Limit Dextrinase Activity of Extracts of Certain Higher Plants and Commercial Malts. J Inst Brew, 79: 377-385

Marchylo B A, Kruger J E, Hatcher D (1986) High-performance Liquid Chromatographic and Electrophoretic Analysis of Hordein during Malting for Two Barley Cultivars of Contrasting Malting Quality. Cereal Chem, 63: 219-231

May L H, Buttrose M S (1959) Physiology of Cereal Grain. II. Starch Granule Formation in the Developing Barley Kernel. Aust J Biol Sci, 12: 146-159

McLeary B V (1992) Measurement of the Content of Limit Dextrinase in Cereal Flours. Carbohyd Res, 227: 257-268

McLeod L C, Duffus C M (1988) Reduced Starch Content and Sucrose Synthase Activity in Developing Endosperm of Barley Plants Grown at Elevated Temperatures. Aust J Plant Physiol, 15: 367-75

Meredith W O S, Anderson J A, Hudson L E (1962) Evaluation of Malting Barley. In: Cook A H (eds.) Barley and Malt. Biology, Biochemistry and Technology. Academic Press, NewYork, 207-270

Molina-Cano J L, Ramo T, Ellis R P, Swanston J S, Bain H, Uribe-Echieverria T, Perez-Vendrell A M (1995) Effect of Grain Composition on Water Uptake By Malting Barley: A Genetic and Environmental Study. J Inst Brew, 101: 79-83

Molina-Cano J L, Francesch M, Perez-Vendrell A M, Ramo T, Voltas J, Brufau J (1997) Genetic and Environmental Variation in Malting ad Feed Quality of Barley. J Cereal Sci, 25: 37-47.

Molina-Cano J L, Rubio A, Igartua E, Gracia P, Montoya J L (2000a) Mechanisms of Malt Extract Development in Barleys from Different European Regions I. Effect of Environment and Grain Protein Content on Malt Extract Yield. J Inst Brew, 106: 111-115

Molina-Cano J L, Polo J P, Sopena A, Voltas J, Perez-Vendrell A M, Romagosa I (2000b) Mechanisms of Malt Extract Development in Barleys from Different European Regions II. Effect of Barley Hordein Fractions on Malt Extract Yield. J Inst Brew, 106: 117-123

Molina-Cano J L, Polo J P, Romera E, Araust J L, Zarco J, Swanston J S (2001) Relationships between Barley Hordeins and Malting Quality in a Mutant of Cv. Triumph I. Cultivar By Environment Interaction of Hordein Content. J Cereal Sci, 34: 285-294

Moralejo M A, Garcia-Casado G, Sanchez-Monge R, Lopez-Otin C, Romagosa I, Molina-Cano J L, Salcedo G (1993) Genetic Variants of the trypsin Inhibitor from Barley Endosperm Show Different Inhibitory Activities. Plant Sci, 89: 23-29

Moralejo M, Swanston J S, Munoz P, Prada D, Elia M, Russel J R, Ramsey L, Cistue L, Codesal P, Casas A M, Romagosa I, Powell W, Molina-Cano J L (2004) Use of New EST Markers to Elucidate the Genetic Differences in Grain Protein Content between European and North American Two-rowed Malting Barleys. Theor Appl Genet, 110: 116-125

Morell M K, Kosar-Hashemi B, Cmiel M, Samuel M S, Chandler P, Rahman S, Buleon A, Batey I L, Li Z Y (2003) Barley Sex6 Mutants Lack Starch Synthase IIa Activity and Contain a Starch With Novel Properties. Plant J, 34: 172-184

Morgan A G, Riggs T J (1981) Effects of Drought on yields and on Grain and Malt Characters in Spring Barley. J Sci Food Agric, 22: 339-346

Morris C (2002) Puroindolines: the Molecular Genetic Basis of Wheat Grain Hardness. Plant Mol Bio, 48: 633-647

Morris C F, Masa A N (2003) Puroindoline Cultivars of the U.S. National Institute of Standards & Technology Reference Material 8441, Wheat Hardness. Cereal Chem, 80: 674-678

Murray F, Mathews P, Jacobsen J, Grubler F (2006) Increased Expression of HvGAMYB in Transgenic Barley Increases Hydrolytic Enzyme Production By Aleurone Cells in Response to Gibberellin. J Cereal Sci, 44: 317-322

Mutisya J, Sathish P, Sun C X, Andersson L, Ahlandsberg S, Baguma Y, Palmqvist S, Odhiambo B, Aman P, Jansson C (2003) Starch Branching Enzymes in Sorghum (*Sorghum Bicolor*) and Barley (*Hordeum Vulgare*): Comparative Analyses of Enzyme Structure and Gene Expression. J Plant Physiol, 60: 921-930

National Association Commodity Marketing Authorities, 2001. National Barley Standards

Oliveira A B, Rasmusson D C, Fulcher R G (1994) Genetic Aspects of Starch Granule Traits in Barley. Crop Sci, 34: 1176-1180

Olkku J, Kotaviita E, Salmenkallio-Marttila M, Sweins H, Home S (2005) Connection between Structure and Quality of Barley Husk. J Am Soc Brew Chem, 63: 17-22

Osborne B G, Anderssen R S (2003) Single-kernel Characterization Principles and Applications. Cereal Chem, 80: 613-622

Osborne B G, Fox G P, Kelly A M, Henry R J (2007) Measurement of Barley Grain Rheology for the Prediction of Malting Quality. J Inst Brew, 113: 135-141

Osman A M, De Jersey J, Inkerman P A (1996a) A Novel Approach to a Differential Assay of Barley Malt α-glucosidases, Maltase and Malt Oligosaccharide α-glucosidase. In: Wrigley C W (eds.) Proc. of the 46th Australian Cereal Chemistry Conference. Royal Australian Chemical Institute, Melbourne, 172-175

Osman A M, De Jersey J, Inkerman P A (1996b) the Specific Measurement of Starch Degrading Enzymes in a Common Extract from Barley Malt. In: Wrigley C W (eds.) Proc of the 46th Australian Cereal Chemistry Conference. Royal Australian Chemical Institute, Melbournea, 208-212

Osman A M, Coverdale S M, Cole N, Hamilton S E, De Jersey J, Inkerman P A (2002) Characterisation and Assessment of the Role of Barley Malt Proteases during Malting and Mashing. J Inst Brew, 108: 62-67

Palmer G H (1983) Malting and Mashing. In: An Introduction to Brewing and Science Technology. the Institute of Brewing, London, 10-27

Palmer G H, Agu R C (1999) Effect of Mashing Temperature and Endo-β-glucanase on β-glucan Content of Malt Worts. J Inst Brew, 105: 233-235

Passarella V S, Savin R, Slafer G A (2005) Breeding Effects on Sensitivity of Barley Grain Weight and Quality to Events of High Temperature during Grain Filling. Euphytica, 141: 41-48

Perez-Vendrell A M, Brufau J, Molina-Cano J L, Francesch M, Guasch J (1996). Effects of Cultivar and Environment on β-(1,3)-(1,4)-D-glucan Content and Acid Extract Viscosity of Spanish Barleys. J Cereal Sci, 23: 285-292

Poulle M, Jones B L (1988) A Proteinase from Germinating Barley. I. Purification and Some Physical Properties of a 30kD Cysteine Endoproteinase from Green Malt. Plant Physiol, 88: 1454-1460

Psota V, Vejraka K, Famera O, Hrcka M (2007) Relationship between Grain Hardness and Malting Quality of Barley (*Hordeum Vulgare* L.). J Inst Brew, 113: 80-86

Robinson L, Juttner J, Milligan A, Lahnstein J, Eglinton J K, Evans D E (2007) the Identification of a Barley Haze Active Protein that Influences Beer Haze Stability: Cloning and Characterisation of the Barley SE Protein as a Barley Trypsin Inhibitor of the Chloroform/methanol Type. J Cereal Sci, 45: 348-352

Roumeliotis S, Collins H M, Logue S J, Willsmore K L, Jefferies S P, Barr A R (1999) Implication of Thin Husk. In: Proceedings of the 9th Australian Barley Technical Symposium, Melbourne, http://www.cdesign.com.au/proceedings%5Fabts1999/papers/S_Roumeliotis.pdf

Salcedo G, Fra-Mon P, Molina-Cano J L, Aragoncillo C, Garcia-Olmedo F (1984) Genetics of CM-proteins (A-hordeins) in Barley. Theor Appl Genet, 68: 53-59

Sardana V, Zhang G P (2005) Genotypic Variation in Some Quality Traits of Malt Barley (*Hordeum Vulgare* L.) Caused By Time of Nitrogen Application and Kernel Position Within the Spike. Cereal Res Commun, 33: 817-823

Savin R, Nicolas M E (1996) Effects of Short Periods of Drought and High Temperature on Grain Growth and Starch Accumulation of Two Malting Barley Cultivars. Aust J Plant Physiol, 23: 201-210

Savin R, Stone P J, Nicolas M E, Wardlaw I F (1997) Grain Growth and Malting Quality of Barley 1. Effects of Heat Stress and Moderately High Temperature. Aust J Agric Res, 48: 615-624

Savin R, Nicolas M E (1999) Effects of Timing of Heat Stress and Drought on Growth and Quality of Barley Grains. Aust J Agric Res, 50: 357-364

Schmitt H F, Van Mechelen J R (1997) Expression of Lipoxygenase Isoenzymes in Developing Barley Grains. Plant Sci, 128: 141-150

Schmitt M R, Budde A D (2007) Improved Methods for High-throughput Extraction and Assay of Green Barley Malt Proteinases Activity Facilitating Examination of Proteinases Activity Across large-scale Barley Populations. Cereal Chem, 84: 313-319

Schmitt M R, Marinac L (2008) Beta-amylase Degradation By Serine Endoproteinases from Green Barley Malt. J Cereal Sci, 47: 480-488

See D, Kanazin V, Kephart K, Blake T (2002) Mapping Genes Controlling Variation in Barley Grain Protein Concentration. Crop Sci, 42: 680-685

Sheehan M C, Skerritt J H (1997) Identification and Characterisation of Beer Polypeptides Derived from Barley Hordeins. J Inst Brew, 107: 297-306

Shewry P R (1993) Barley Seed Proteins. In: MacGregor A W, Bhatty R S (eds.) in "Barley: Chemistry and Technology". Am Soc Cereal Chem, ST. Paul MN, USA, 131-197

Shewry P R, Tatham A S, Halford N G (2001) Nutritional Control of Storage Protein Systhesis in Developing Grain of Wheat and Barley. Plant Growth Regul, 34: 105-111

Silva F, Nogueria L C, Goncalves C, Ferreira A A, Ferreira I, Teixeira N (2007) Electrophoretic and HPLC Methods for comparative Study of the Protein Fractions of Malts, Worts and Beer Produced from Scarlett and Prestige Barley (*Hordeum Vulgare* L.) Varieties. Food Chem, 106: 820-829

Sission M J, Lance R C M, Sparrow D H B (1992) Studies on Limit Dextrinase in Barley. I. Purification of Malt Limit Dextrinase and Production of Specific Monospecific Antibodies. J Cereal Sci, 16: 107-116

Skerritt JH & Henry RJ (1988) Hydrolysis of Barley Endosperm Storage Proteins during Malting. II. Quantification By Enzyme- and Radio-immunoassay. J Cereal Sci, 7: 265-281.

Skerritt J H, Janes P W (1992) Disulphide-bonded "gel Protein" Aggregates in Barley: Quality-related Differences in Composition and Reductive Dissociation. J Cereal Sci, 16: 219-235

Smith D B (1990) Barley Seed Protein and its Effects on Malting and Brewing Quality. Plant, Var Seeds, 3: 63-80

Sopanen T, Lauriere C (1989) Release and Activity of Bound Beta-amylase in a Germinating Barley Grain. Plant Physiol, 89: 244-249

Sparrow D H B, Lance R C M, Henry R J (1988) Alternative End Uses of Barley. Royal Australian Chemical Institute, Melbourne

Stahl V, Coates S, Bryce J H, Morris P C (2004) Antisense Downregulation of the Barley Limit Dextrinase Inhibitor Modulates Starch Granule Size Distribution, Starch Composition and Amylopectin Structure. Plant J, 39: 599-611

Stahl Y, Alexandra R D, Coates S, Bryce JH, Jenkinson H R, Morris P C (2007) the Barley Limit Dextrinase Inhibitor: Gene Expression, Protein Location and Interaction With 14-3-3 Protein. Plant Sci, 172: 452-461

Stenholm K, Home S (1999) A New Approach to Limit Dextrinase and its Role in Mashing. J Inst Brew, 105: 205-210

Stevnebø A, Sahlström S, Svihus B (2006) Starch Structure and Degree of Starch Hydrolysis of Small and Large Starch Granules from Barley Cultivars With Varying Amylose Content. Anim Feed Sci Tech, 130: 23-38

Stewart D C, Hawthorne D, Evans D E (1998) Cold Sterile Filtration: A Small Scale Filtration Test and Investigation of Membrane Plugging. J Inst Brew, 104: 321-236

Stewart D C, Freeman G, Evans D E (2000) Development and Assessment of a Small-scale Wort Filtration Test for the Prediction of Beer Filtration Efficiency. J Inst Brew, 106: 361-366

Stuart I M, Loi L, Fincher G B (1988) Varietal and Environmental Variations in (1→3, 1→4)-β-glucan and (1→3, 1→4)-β-glucanase Potential in Barley: Relationships to malting Quality. J Cereal Sci, 7: 61-71

Swanston J S (1995) Effects of Barley Grain Size, Texture and Modification during Malting Associated With Three Genes on Chromosome 1. J Cereal Sci, 22: 157-161

Swanston J S (1996) Associations of the Waxy (wx) Gene With Malting Quality Characteristics in Random Inbred Lines of Barley. J Inst Brew, 102: 355-358

Swanston J S, Taylor K (1988a) the Milling Energy of Malted Barley and its Relationship With Hot Water Extract and Amylase Activity. J Inst Brew, 94: 143-146

Swanston J S, Taylor K (1988b) A Comparison of Some Rapid Tests to Assess Extent of Modification during a Laboratory Scale Malting Procedure. J Inst Brew, 94: 311-314

Swanston J S, Cowe I A (1989) A Rapid Technique to Predict Malting Quality Prior to Harvest. Ann Appl Biol, 115: 529-532

Swanston J S, Ellis R P, Rubio A, Perez-Vendrell A, Molina-Cano J L (1995) Differences in Malting Performance between Barleys Grown in Spain and Scotland. J Inst Brew, 101: 261-265

Swanston J S, Molina-Cano J L (2001) Beta-amylase Activity and Thermostability in Two Mutants Derived from the Malting Barley Cv. Triumph. J Cereal Sci, 33: 155-161

Swanston J S, Ellis R P (2002) Genetics and Breeding of Malt Quality Attributes. In: Slafer GA, Molina-Cano J L, Savin R, Araus J L, Romagosa I (eds.) In, Barley Science, Recent Advances From Molecular Biology to Agronomy of Yield and Quality. Food Products Press, New York, 85-114

Tatham A A, Shewery P R (1995) Mini Review: the S-poor Prolamins of Wheat, Barley and Rye. J Cereal Sci, 22: 1-6

Turnbull K M, Gaborit T, Marion D, Rahman S (2000) Variation in Puroindoline Polypeptides in Australian Wheat Cultivars in Relation to Grain Hardness. Aust J Plant Physiol, 27: 153-158

Ullrich S E, Clancy J A, Lee H, Han F, Matsui K, Del Blanco I A (2005) Genetic Analysis of Preharvest Sprouting in Barley. Proceedings of the 18th North American Barley Researchers Workshop, and 4th Canadian Barley Symposium, Red Deer: Canada.

Vickers J E, Hamilton S E, De Jersey J, Henry R J, Marschke R J, Inkerman P A (1996) Assessment of Bacillus Licheniformis β-amylase as a Candidate Enzyme for Genetic Engineering of Malting Barley. J Inst Brew, 102: 75-78

Vietor R J, Angelino S A G F, Voragen A G J (1991) Arabinoxylans in Barley, Malt and Wort. Proceedings of the 23rd EBC Congress, Lisbon, 139-146

Walker C, Dickie K, Biawa J P, Ueda T, Muller R E (2001) Prediction of Extract Potential in New Barley Cultivars By Measuring Cell-wall Breakdown. J Inst Brew, 107: 167-174

Wallwork M A B, Logue S J, MacLeod L C, Jenner CF (1998a) Effects of a Period of High Temperature during Grain Filling on the Grain Growth Characteristics and Malting Quality of Three Australian Malting Barleys. Aust J Agric Res, 49: 1287-1296

Wallwork M A B, Logue S J, MacLeod L C, Jenner C F (1998b). Effect of High Temperature during Grain Filling on Starch Synthesis in the Developing Barley Grain. Aust J Plant Physiol, 25: 173-181

Wang M, Heimovaara-Dijstra S, Van Duijn B (1995) Modulation of Germination of Embryos Isolated from Dormant and Non-dormant Barley Grains By Manipulation of Endogenous Abscisic Acid. Planta, 195: 586-592

Weidner S, Paprocka J, Kamieniecki B, Zadernowski R (1993) In: Walker-Simmons M K, Reid J L (eds.) the Role of Phenolic Acids in Dormancy of Barley Caryopses. Proceedings of the 6th International Conference on Pre-Harvest Sprouting, American Society of Cereal Chemists, USA, 200-211

Woodward J R, Fincher G B (1982) Substrate Specificities and Kinetic Properties of Two (1-3) (1-4) -D-Glucan Endohydrolyase from Germinating Barley. Carbohyd Res, 106: 111-122

Worbel R, Jones B L (1992a) Appearance of Endoproteolytic Enzymes during the Germination of Barley. Plant Physiol, 100: 1508-1516

Worbel R, Jones B L (1992b) Electrophoretic Study of Substrate and PH Dependence of Endoproteolytic Enzymes in Green Malt. J Inst Brew, 98: 471-478

Worbel R, Jones B L (1993) Identification and Partial Characterisation of *high Mr Neutral Proteinases* from 4-day Germinated Barley Seeds. J Cereal Sci, 18: 225-237

Wu Y, Schwartz P B, Doehlert D C, Dahleen L S, Horsley R D (1997) Rapid Separation and Genotypic Variability of Barley (*Hordeum Vulgare* L.) Lipoxygenase Isoenzymes. J Cereal Sci, 25: 49-56

Yamada, J (1981) Purification of Oat Debranching Enzyme and Occurrence of Inactive Debranching Enzyme in Cereals. Agric Biol Chem, 45: 1013-1015

Yan Y, Zhu J, Xu S, Xu Y (1999) Genetic Effects of Embryo and Endosperm for Four Malting Quality Traits of Barley. Euphytica, 106: 27-34

Zhang G, Chen J, Wang J, Ding S (2001) Cultivar and Environmental Effects on $(1\rightarrow3, 1\rightarrow4)$-$\beta$-D-glucan and Protein Content in Malting Barley. J Cereal Sci, 34: 295-301

Zhang N, Jones B L (1995a) Characterisation of Germinated Barley Endoproteolytic Enzymes By Two-dimensional Gel Electrophoresis. J Cereal Sci, 21: 145-153

Zhang N, Jones B L (1995b) Development of Proteolytic Activities during Barley Malting and Their Localisation in the Green Malt Kernel. J Cereal Sci, 22: 147-155

Zhang N, Jones B L (1999) Polymorphism of Aspartic Proteinases in Resting and Germinating Barley Seeds. Cereal Chem, 76: 134-138

Zhang Y, Darlington H, Jones H D, Halford N G, Napier J A, Davey M R, Lazzeri P A, Shewery P R (2003) Expression of the Gamma-zein Protein of Maize in Seeds of Transgenic Barley: Effects on Grain Composition and Properties. Theor Appl Genet, 106: 1139-1146

Zhao F J, Fortune S, Barbosa V L, McGrath S P, Stobart R, Bilsborrow P E, Booth E J, Brown A, Robson P (2006) Effects of Sulphur on Yield and Malting Quality of Barley. J Cereal Sci, 43: 369-377

4

Identification of Barley Varieties by Protein Profiling

M. N. Scobie and M. G. K. Jones
WA State Agricultural Biotechnology Centre
School of Biological Sciences and BiotechnologyMurdoch University
Perth, Western Australia, 6150
Corresponding author: Michael G. K. Jones, M.Jones@murdoch.edu.au
Telephone: +61 8 9360 2424, Fax: +61 893606303

4.1 Introduction

There is an increasing demand in the grain handling chain to control purity and segregation of different classes and varieties of grains. Accurate segregation is needed to ensure that varieties with specific qualities are delivered at high purity and, for example, in the future to segregate GM and non-GM material. Since the grain handling chain stretches from farmers, to transporters, to storage silos, to end-users in the world market, low-cost rapid tests are needed. In addition, with the advent of end-point royalties, rapid variety identification is needed to monitor compliance in delivering new varieties subject to plant breeders' rights (PBR).

At barley receival points in Australia there is routine testing for moisture, weight, damage, colour, protein content, screenings, foreign seeds or material, insects and sand or stones. Varieties are currently identified mainly by physical appearance. In Canada, the Kernel Visual Distinguishability (KVD) system is used to identify varieties. Under KVD, different classes of barley must be bred to have distinguishing physical features that make them easily identifiable with the human eye, providing the basis by which different classes are kept separate in the grain handling system. However, KVD can lead to cases where higher-yielding varieties are not accepted because they do not meet KVD standards (http://www.westerngrains.com).

Various methods for cereal variety identification have been developed which include: combined aleurone colour, starch gel electrophoresis and isoelectric focusing (McCausland and Wrigley, 1977), AFLP fingerprints (Schut *et al.*, 1997), hordein electrophoregrams (Marchylo and LaBerge, 1980), hordein patterns on SDS-PAGE gels (Howard *et al.*, 1996; Echart-Almeida and Cavalli-Molina, 2001), RAPD and ISSR markers (Fernandez *et al.*, 2002), SSR markers (Macaulay *et al.*, 2001; Karakousis *et al.*, 2003), and capillary electrophoresis (Siriamornpun *et al.*, 2001; Uthayakumaran *et al.*, 2005). These methods tend to be time and labour expensive. Bloch *et al.* (2001) developed a method to identify barley varieties based on protein profiles of storage proteins extracted from the grain and analysed by mass spectroscopy. Seed proteins are classified into three groups: storage proteins, structural and metabolic proteins, and protective proteins (Shewry and Halford, 2002). The major alcohol soluble storage proteins of barley are prolamins (hordeins) and some quantitatively minor globulins. One reason for using hordeins as markers is that their patterns, when run on SDS-PAGE gels, are stable under a range of growing conditions, although the proportions of different proteins may vary. Some structural and metabolic proteins, such as amylases, are not produced under poor growing conditions (Shewry, 1993).

In this Chapter we present the approach of protein profiling by mass spectroscopy as a rapid method to identify barley varieties. This approach is widely applicable, but the examples presented are based on varieties grown mainly in Western Australia. Protein profiling can also be used to study the relatedness of different varieties, and as markers to score specific proteins in breeding or mapping populations, to map their chromosomal location or as a basis for selection of genotypes for specific protein quality.

4.2 Extraction of Proteins for Protein Profiling

Seed samples can be extracted simply in 70% ethanol in a partially automated process by extracting multiple samples in sets of up to 48 samples each of 30 seeds using ball bearings and a robust shaker. To generate protein profiles using MALDI-TOF mass spectroscopy, an ionization matrix (e.g. sinapinic acid in an acetonitrile/methanol solvent) is required. A 1-μL aliquot of protein extract supernatant is mixed with 24 μL of matrix, and then 1 μL of this is spotted onto a MALDI-TOF well plate (up to 400 samples per plate) and allowed to air-dry (if required, additional 1 L spots can be layered on top allowing each spot to dry before adding the next). The process of mixing samples with matrix and spotting into the MALDI-TOF plate can be carried out automatically using a robotic liquid handling system.

Ethanolic barley protein extracts can be analysed by MALDI-TOF Mass Spectrometry (e.g. using a PerSeptive Biosystems Voyager - DE PRO Biospectrometry Workstation), using the linear rather then reflectron mode.

4.3 Characteristics of Barley Protein Profiles

Protein profiles have been generated by MALDI-TOF mass spectroscopy from the barley varieties listed in Table 4.1.

Table 4.1. Names and characteristics of the barley varieties studied

Variety	Sample Numbers	Quality	Origin	Pedigree
Arapiles	1	M/F	V	Noyep/Proctor//CI3576/Union/4/Kenia/3/Research/2/Noyep/Proctor/5/Domen
Barque	1	F	S	Triumph/Galleon
Baudin	3	M	W	Stirling/Franklin
Beecher	2	F	A	[1] Atlas/Vaughn
Bomi	1	-	E	[1] Bonus/Minerva
Carlsberg II	1	M	E	[1] Carlsberg selection, Carlsberg = Prentice/Maja
Clipper	2	F	S	[1] Proctor/PriorA
Dampier	1	F	W	[1] Research/Olli
Doolup	3	F	W	(XBVT210)/3/(B6729)Prior/Lenta//Noyep/Lenta(75S:323)/(MndS,74S:314)
				Dampier//(A14)Prior/Ymer/3/Kristina(70S20-20)/4/(73S13)Clipper/Tenn65-117
Fitzgerald	4	F	W	Onslow/Tas85-466
Forrest	3	F	W	[2] Atlas57/2/A16Prior/Ymer
Franklin	4	M	T	Shannon/Triumph
Gairdner	7	M	W	Onslow/Tas85-466 (CVSG2001 and 2002) or Tas83-587(CVSG2005)
Galleon	1	F	S	[3] Clipper/Hiproly//3/Proctor/CI3576
Grimmett	1	M	Q	[4] Bussell/Zephor
Hamelin	3	M	W	Stirling/Harrington
Harrington	6	M/F	C	Canada Klages/3/Gazelle/Betzes//Centennial
Igri	1	M	E	[1] 820/1427/Ingrid
Jupiter	1	F	UK	[1] Betina/Midas
Keel	1	F	S	CPI18197/Clipper//WI2645
Molloy	3	F	W	Golden Promise/WI2395(WARI2-38)/4/(72S:267)XBVT210/3/
				(66S08-4)Atlas57//(A14)Prior/Ymer(82S837)/O扒onnor
Moondyne	2	F	W	Dampier//(A14)Prior/Ymer/3/Kristina(70S20-20)/4/(73S13)Clipper/Tenn65-117
Morrell	2	F	W	WUM221/P23822(81S806)/5/(81S719)Forrest/4/(80S564)
				Psaknon/Dampier//M19(76T111)/3/Zephyr
Mundah	2	F	W	O'Connor/Yagan
O'Connor	2	F	W	Proctor/CI3576(WI2231)/3/(XBVT212)Atlas 57//(A14)Prior/Ymer
Onslow	2	F	W	Forrest/Aapo
Schooner	3	M	S	Proctor/PriorA/Proctor/CI-3576
Skiff	2	F	S	Abed Deba/3//Proctor/CI-3576//CPI-18197/Beka/4/Clipper/Diamant//Proctor/CI-3576
Stirling	14	M	W	Dampier//Prior/Ymer/3/Piroline
Tallon	1	M	Q	[5] Triumph/Grimmett
Triumph	1	M	E	[1] Diamant/St1402964/9
Unicorn	6	M/F	J	pedigree unknown, introduced as Kinukei 21
Windich	4	F	W	Atlas 57//(A16)Prior/Ymer(68S17-75)/3/(B6729)Prior/Lenta//Noyep/Lenta
Yagan	2	F	M	pedigree unknown

F = feed quality; Q = Queensland; W = Western Australia; E = Europe; UK = United Kingdom
M = malting quality; V = Victoria; S = South Australia; C = Canada; J = Japan
- = data missing; T = Tasmania; M = Mexico; A = USA

[1] John Innes Centre Barley Germplasm selection30/07/2002, www.jic.bbsrc.ac.uk/germplas/bbsrc_ce/Pedb.txt
[2] http://barley.ipk-gatersleben.de/ebdbdetail.php3 acc.-no5923
[3] Williams *et al.* (2003)
[4] http://barley.ipk-gatersleben.de/ebdbdetail.php3 acc.-no6108
[5] Karakousis *et al.* (2003)
All other pedigree information from Crop Variety Sowing Guides 1997-2005

The six varieties with the highest percentage of area sown in Western Australia in the 2004/2005 season, Stirling (33.00%), Gairdner (25.01%), Baudin

(13.43%), Mundah (9.71%), Yagan (5.46%), and Hamelin (4.05%)(CVSG 2005), can be differentiated readily from each other based on their protein profiles (Fig. 4.1(a)). Although only one peak differentiates Stirling from Baudin, this is consistent and enough to separate the two varieties. When there is a mixture of varieties, this can be identified, and also "contaminating peaks" can be identified: for example, Fig. 4.1(b) shows samples of the variety Stirling with extra peaks of low intensity, but these did not cause such samples to be mis-identified.

4.3.1 Peak Size Calling

Due to minor variations in the size calling of peaks, a range of approximately 25 Da is allowable for each peak. For example, peak 21.0 includes peaks between 21,043 Da and 21,070 Da; peak 33.2/33.3, includes peaks between 33,280 Da and 33,335 Da. In this work, samples were analysed over a three year time period, and so "standard" samples were run in each analysis to ensure correct peak calling.

The presence/absence of peaks was taken from a consensus of samples and sub-samples of each variety to give a standard profile for each variety. The standard profiles for each of the varieties are presented in Fig. 4.2.

4.3.2 Location and Year

Fourteen samples of Stirling and 6 samples of Unicorn were analysed for possible variation in profile resulting from growth in different locations and years. The geographical origin and year of sample collection did not affect the diagnostic protein profiles (Fig. 4.3). However, occasionally there was slight variation in peak height.

4.3.3 Phylogeny

The relatedness of the protein profiles can be analysed by developing phylogenetic trees. To create the phylogenetic tree for the barley varieties listed in Table 4.1, a Phyllip PARS module (Discrete Character Parsimony Algorithm, Version 3.65) was used, using the default parameters, except for: "randomise trees = 1000 times with a seed value of 433", and "number of trees to save = 100".

Of the 34 barley varieties studied, 30 could be distinguished from each other. The phylogenetic tree (Fig. 4.4) shows the relationship of the varieties based on their protein peaks. Schooner and Gairdner could not be separated because they generated the same profiles, as did varieties Windich and Doolup.

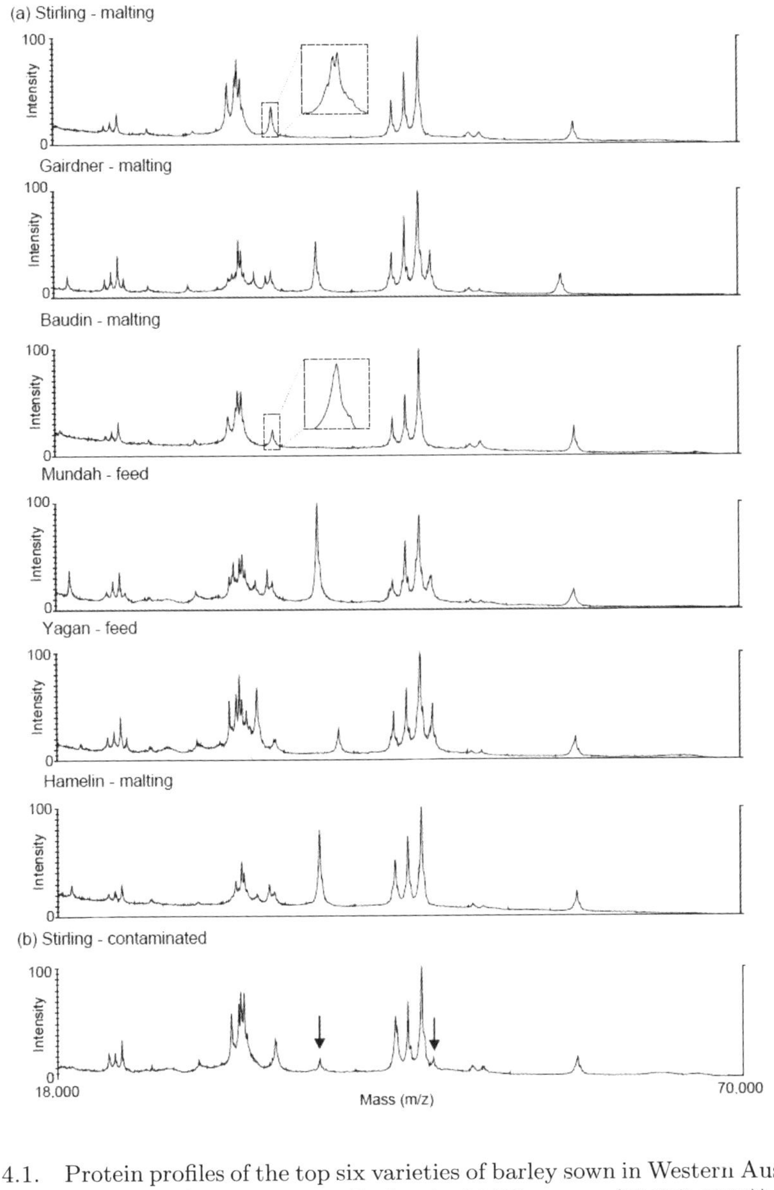

Fig. 4.1. Protein profiles of the top six varieties of barley sown in Western Australia (highest percentage of area sown to Barley in 04/05 season (CVSG 2005)). Each variety has a distinct pattern identifying it from the others (a). Stirling and Baudin are very similar except for one peak present in Stirling, peak 34.5/34.6, which is not present in Baudin, they both have peak 34.6/34.7 (enlargement). Gairdner, Mundah, Yagan and Hamelin are different from each other. (b) Contamination peaks (arrowed) were observed in a Stirling sample, but at reduced intensities than they would normally appear if they were legitimate peaks

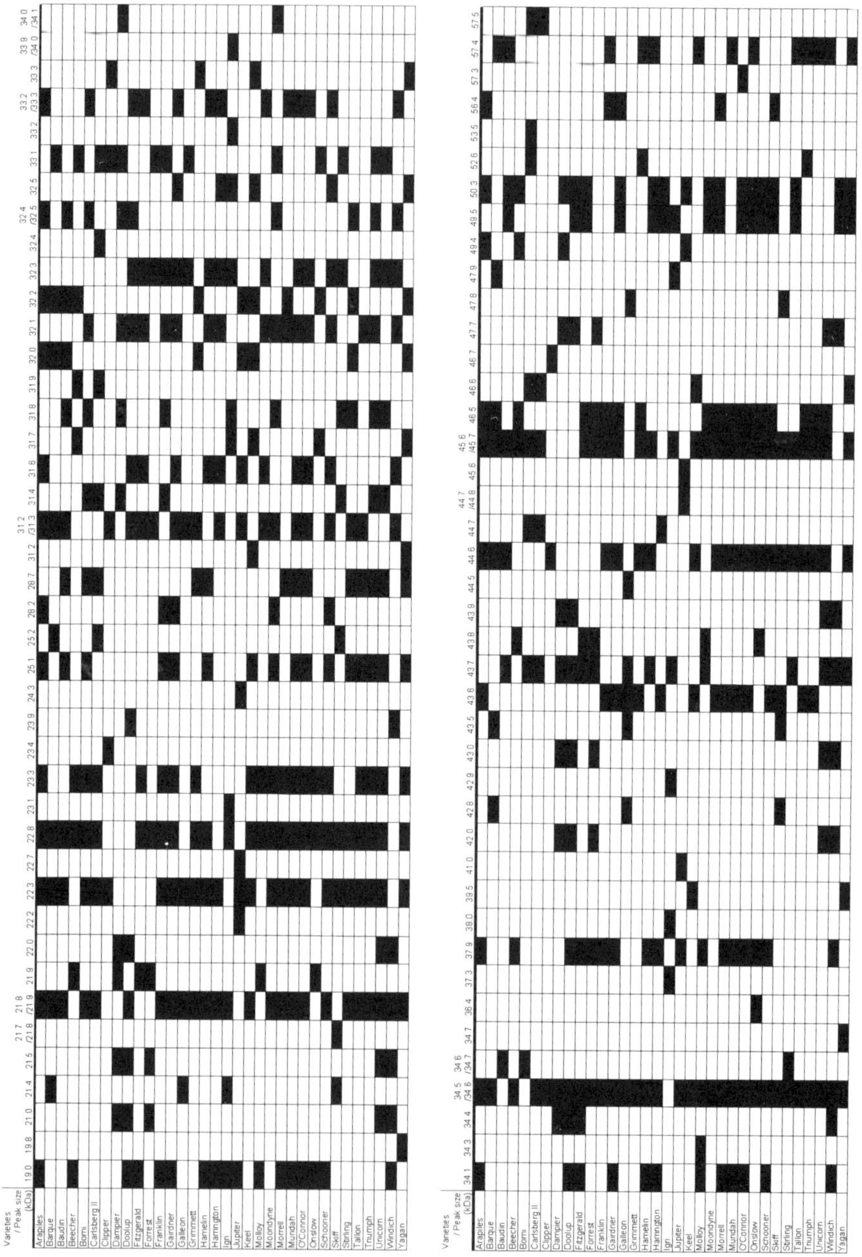

Fig. 4.2. Peak Table for barley varieties. Peak presence (filled block) or absence is taken from a consensus of samples and sub-samples of each variety. Peak sizes are given in kDa

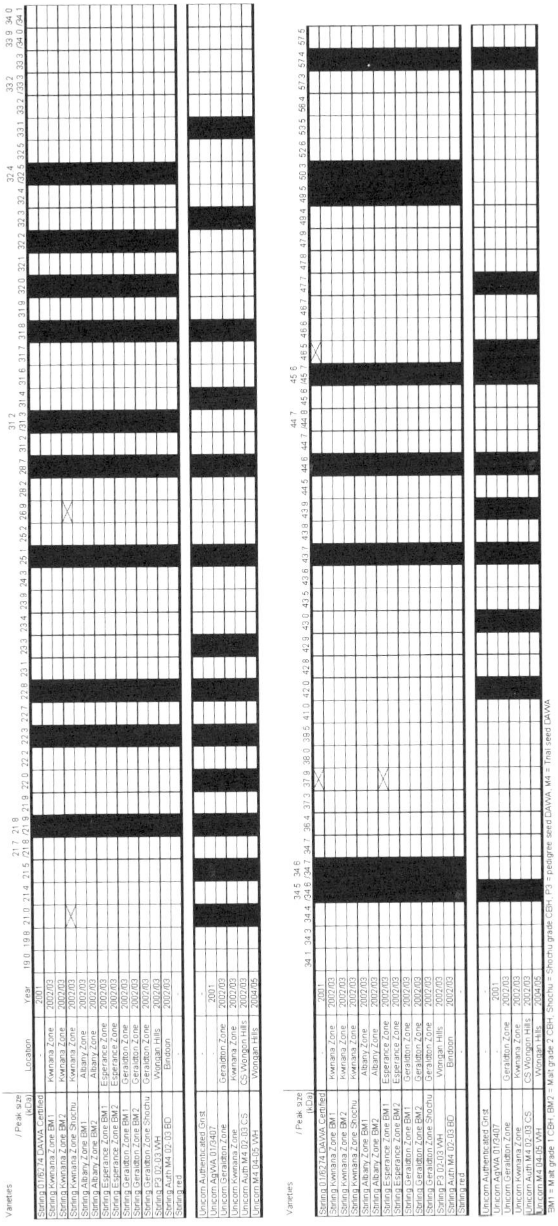

Fig. 4.3. Peak stability in Stirling and Unicorn grown at different locations and in different years. Peaks present in each sample are shown in black. Apart from a few contamination peaks in three Stirling samples, indicated by a cross, there was no difference in profile patterns within a variety when grown at different locations or in different years. Peak sizes are given in kDa

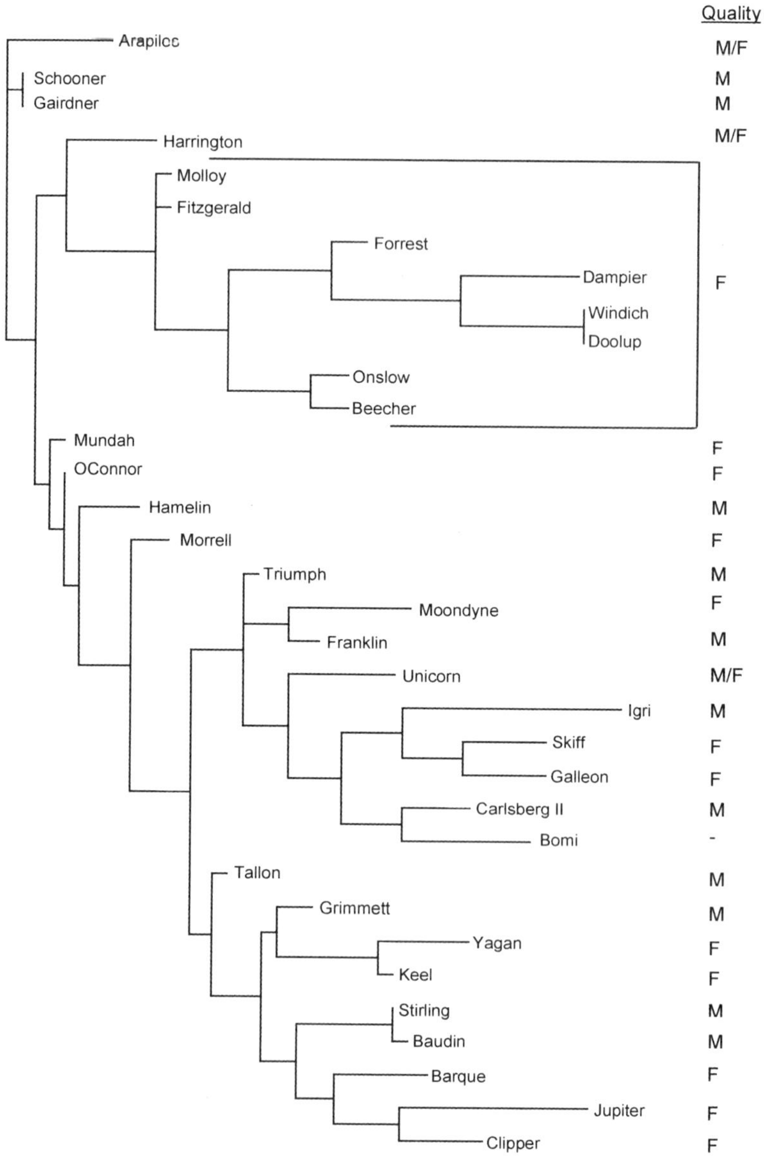

Fig. 4.4. Phylogenetic tree (Phyllip PARS module, using the default parameters, except for: "randomise trees = 1000 times with a seed value of 4331", and "number of trees to save = 1000") based on protein profiles of barley varieties: 30 of the 34 varieties can be distinguished from each other. Schooner and Gairdner, and Windich and Doolup make two groups that cannot be separated. Feed and malting varieties do not segregate together except for one block of feed varieties, indicated in the boxed section

4.4 Discussion

Although the barley varieties chosen were grown predominantly in Western Australia, the principle of variety identification by protein profiling using mass spectroscopy has generic applicability.

The mass spectra range studied was 18,000–70,000 Da. Using this range avoids background signal presence for proteins with lower masses, which could interfere with peak calling, and allows the use of informative peaks in the higher molecular weight range. Although some of the 34 barley varieties studied had common peaks, there were enough polymorphic peaks to allow 30 of these to be distinguished on a routine basis.

4.4.1 Specificity of Variety Identification

In general, the identification of varieties is based on combining both the need to distinguish particular varieties, such as those designated as malting quality from feed varieties, or new varieties subject to PBR or end-point royalties, and knowledge of the proportions of the varieties being grown. Thus, although using the protein profiling method Schooner and Gairdner could not be separated, for practical purposes their discrimination is not a problem. Variety Gairdner was registered in 1997 (still under PBR) and variety Schooner, registered in 1983, is not. However, because the area of Schooner being grown in Western Australia is reducing each year (from 5.90% of the area sown to barley in 00/01 to 3.44% in 04/05), and it is only being delivered as a malting barley at one receival zone, Esperance, this reduces the problem of identification of Schooner and Gairdner by protein profiles. If required, further testing using DNA Simple Sequence Repeat (SSR) markers can be used to distinguish between the two varieties (Macaulay *et al.*, 2001). Alternatively, profiles of proteins extracted using a different method could provide diagnostic data, or the extracted proteins could be digested with trypsin, which can yield different, diagnostic, peptide fragments. The trypsin digest approach has been used to differentiate otherwise identical protein profiles of seeds of different varieties of subterranean clover (Dr Chris Florides, Saturn Biotech, personal communication).

Similarly, varieties Windich and Doolup are also not separated using this method, but since both are feed varieties, their correct identification is not generally required (Doolup, registered in 1998, is protected by PBR, Windich, registered in 1989, is not).

4.4.2 Location and Year

The protein profiles obtained over a number of seasons from barley crops grown in widely differing locations are remarkably constant for each variety. These results are supported, for example, by the work of Marchylo and LaBerge (1980) who found that hordein electrophoregrams for barley cultivars were not affected qualitatively by location or year of growth or grain protein. These data suggest that this method of variety identification is not unduly affected by the location in which a variety is grown or by the environmental conditions under which it was grown.

4.4.3 Malting versus Feed

In analysing the relationships presented by the phylogenetic tree, there is no clear segregation of malting and feed varieties, although one block is made up of only feed varieties (Fig. 4.4). Echart-Almeida and Cavalli-Molina (2001) found an association between hordein patterns on SDS-PAGE gels and high malting quality, in 12 Brazilian barley varieties. Their phenograms were based on the relationship between hordein bands and malting quality measurements. They suggest that the clustering may have been a result of the closer genetic relationship of the varieties, independent of malting quality. Marchylo and LaBerge (1981) distinguished 25 out of 68 Canadian varieties, using SDS-PAGE, with the others divided into 17 groups. They also found that malting and feed varieties segregated but, again, it was thought that the primary cause of the relationship seen was the closely related genetic backgrounds of the varieties they had studied.

With the standard extraction method no significant relationship was found to link a profile peak specifically with malting or feed quality. This is consistent with hordeins being extracted using aqueous alcohol and not being linked to malting quality (Shewry, 1993). Howard *et al.* (1996) found D hordein measurement provided an estimate of malting quality. We were not able to detect the high molecular weight D hordeins (100,000~120,000 Da), even when using an expanded range of 18,000~120,000 Da. B and C hordeins are 70%~80% and 10%~20% respectively of the total hordein fraction, but D hordeins are less than 5% of the total (Shewry, 1993), and thus may be below the limit of detection.

4.4.4 Peak Inheritance

Some of the relationship patterns observed in the phylogenetic tree stem from a common pedigree. For example, Windich and Doolup have common ancestors, and Dampier is included in the progenitors of Doolup. These three groups are together in one block. Gairdner and Schooner have a common ancestor, Proctor, and could not be differentiated based on the profiles obtained.

In general, the position of the variety in the phylogenetic tree based on MALDI-TOF protein profiling patterns did not reflect the pedigree of the variety. That is, many varieties are not positioned adjacent to their ancestors. For example, Hamelin is derived from Stirling and Harrington (CVSG 2005) and all three are widely separated on the tree. Baudin, however, is derived from Stirling and Franklin (CVSG 2005) and appears very close to Stirling.

In looking at the peak present in each variety and comparing these to the peaks in their parents, it can be seen that they acquire their peaks from their parents but the proportions of acquisition from different parents can vary. Hamelin has peaks common to Stirling and Harrington and there are no peaks from either parent that are missing (Fig. 4.5). There are some peaks in Harrington only, and in Stirling only, and one peak that is common to both parents, that Hamelin does not have. This mixing of peaks indicates that there is a certain amount of recombination occurring in developing a variety that ultimately affects the protein content in the seed. Baudin has peaks common to Stirling, and to both parents, but no peaks common only to Franklin. Overall, Baudin is very similar to Stirling but for 1 peak that Stirling has and Baudin has not (Figs. 4.1; 4.5). This similarity may be a reflection of the way Baudin was bred. If there were multiple backcrossing events to Stirling then it is likely that Baudin would have more peaks in common with its parents. The acquisition and loss of peaks may be explained by genetic recombination during breeding, causing changes in the protein expression and thus the profiles.

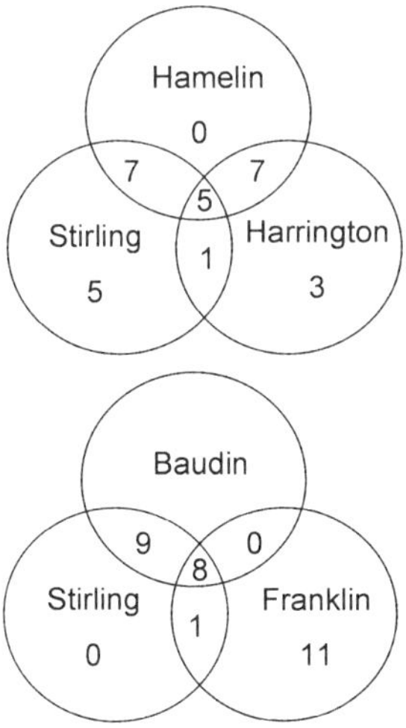

Fig. 4.5. The relationship of protein peaks in Hamelin and Baudin with those of their parental varieties. Hamelin is derived from Stirling and Harrington and has an equal number of peaks from both parents. Baudin is derived from Stirling and Franklin, but its peaks are either common to both parents or only from Stirling

4.4.5 Contamination

Routine barley grain samples can have mixtures of varieties, and they can also be contaminated with weed seeds or been stored with insect pests. Of commercial samples tested, only three Stirling samples showed contamination peaks of a low intensity, below 20%, nevertheless the primary variety could still be identified. If an accurate measure of seed contamination is required, 300 individual seed samples can be analysed. This in fact does not take much time to prepare if the process is semi-automated, and is routinely undertaken for commercial samples when this method is used by a commercial testing service. Identification of single seed in this way enables both identification of the contaminating seed and an accurate estimate of the degree of contamination.

Depending on the degree of contamination, presence of weed seeds may generate additional 'unknown' peaks. In one case, contamination was thought to be present, caused either by insect damage or presence of weed seeds, because no other barley variety studied had a peak at 26.9 kDa. However, the presence of weevils has not been shown to generate peaks under the conditions tested, although it can reduce the intensity of expected peaks.

Another issue is that barley varieties are mainly in-breed and have about 1% outcrossing per generation (von Bothmer, 1992), so that commercial varieties are usually not 100% pure genetically, and variant genotypes can sometimes be found.

An additional source of low level contamination in segregation of varieties for quality control is the possible contamination of storage and transport containers due to inadequate cleaning between harvesting or handling different varieties.

4.4.6 Automation

Using the automation of multi-sample grinding, robotic liquid handling and automatic sample analysis, we routinely processed 192 samples a day, with a 99.5% success rate in correctly identified varieties. This throughput could readily be increased if required, both by increasing the throughput of sample extraction and preparation, and with newer models of MALDI-TOF mass spectrometers, which have much faster lasers and electronics. Uthayakumaran *et al.* (2005) used a "Lab-on-a-Chip" to analyse ten wheat samples at a time. This is a relatively slower approach and requires the sample plate changed approximately every thirty minutes. The method described here requires manual involvement for the extraction process only. Once extracted, the handling and analysis can be automated.

Bloch *et al.* (2001) used an artificial neural network system to train and identify varieties from protein profiles, and we have also investigated the use of an artificial neural network to identify wheat varieties by MALDI-TOF protein profiling (Schibeci *et al*; 2001). However, our experience is that the profiles of the varieties of interest differ significantly, and with experience, users are able to identify each variety readily from the mass spectrometer profiles. This eliminates the need to develop and train an artificial neural network to identify varieties.

4.5 Conclusion

The method presented provides a quick, reliable, large throughput, automated method for barley variety identification that is not influenced by seasonal variation. This approach has been used routinely for commercial identification of barley varieties in Western Australia (http://www. saturnbiotech.com.au), with a throughput of about 50,000 samples per annum.

Acknowledgement

We thank W. Kenworthy, for assistance with the phylogenetic tree analysis, B. Flanigan and C. Li of DAFWA, and C. Grime of UWA, for barley samples. This project was funded by an ARC Linkage Grant, Number LP0455493, and the Centre for High-throughput Agricultural Genetic Analysis.

References

Bloch H A, Petersen M, Sperotto M M, Kesmir C, Radzikowski L, Jacobsen S, Sondergaard I (2001) Identification of Barley and Rye Varieties Using Matrix-assisted Laser Desorption/Ionisation Time-of-flight Mass Spectrometry with Neural Networks. Rapid Commun in Mass Spectrometry, 15: 440-445

von Bothmer R (1992) The Wild Species of *Hordeum*: Relationships and Potential Use for Improvement of Cultivated Barley. In: Shewry, P.R. (Ed), Barley: Genetics, Biochemistry, Molecular Biology and Biotechnology. Oxford, CAB International, UK, 3-18

Crop Variety Sowing Guides (CVSG) 1997-2005, Department of Agriculture, Western Australia, South Perth, WA

Echart-Almeida C, Cavalli-Molina S (2001) Hordein Polypeptide Patterns in Relation to Malting Quality in Brazilian Barley Varieties. Brazilian J Agri Res, 36: 211-217

Fernandez M E, Figueiras A M, Benito C (2002) The Use of ISSR and RAPD Markers for Detecting DNA Polymorphism, Genotype Identification and Genetic Diversity Among Barley Cultivars with Known Origin. Theor Appl Genet, 104: 845-851

Harvest Handbook 2003/04 Season, Co-operative Bulk Handling Ltd, West Perth, WA 6005

Howard K A, Gayler K R, Eagles H A, Halloran G M (1996) The Relationship Between D Hordein and Malting Quality in Barley. J Cereal Sci, 24: 47-53

Karakousis A, Barr A R, Chalmers K J, Ablett G A, Holton T A, Henry R J, Lim P, Langridge P, (2003) Potential of SSR Markers for Plant Breeding and Variety Identification in Australian Barley Germplasm. Aust J Agri Res, 54: 1197-1210

Macaulay M, Ramsay L, Powell W, Waugh R (2001) A Representative, Highly Informative "Genotyping Set" of Barleys SSRs. Theor Appl Genet, 102: 801-809

Marchylo B A, LaBerge D E (1980) Barley Cultivar Identification by Electrophoretic Analysis of Hordein Proteins I. Extraction and Separation of Hordein

Proteins and Environmental Effects on the Hordein Electropherogram. Can J Plant Sci, 60: 1343-1350

Marchylo B A, LaBerge D E (1981) Barley Cultivar Identification by Electrophoretic Analysis of Hordein Proteins II. Catalogue of Electrophoregram Formulae for Canadian-Grown Barley Cultivars. Can J Plant Sci, 61: 859-870

McCausland J, Wrigley C W (1977) Identification of Australian Barley Cultivars by Laboratory Methods: Gel Electrophoresis and Gel Isoelectric Focusing of the Endosperm Proteins. Aust J Exp Agri & Animal Husbandry, 17: 1020-1027

Schibeci D, Potter R, Wathen-Dunn K, Jones M G K, Bellgard M (2001) Applying Artificial Neural Networks to Classification of Wheat Varieties Processed by MALDI-TOF Mass Spectrometry. Proceedings of the Neural Networks Applications (NNA'01), Tenerife, Canary Islands

Schut J W, Qi X, Stam P (1997) Association between Relationship Measures Based on AFLP Markers, Pedigree Data and Morphological Traits in Barley. Theor Appl Genet, 95: 1161-1168

Shewry P R (1993) Barley Seed Proteins In: MacGregor A W, Batty R S (Eds.), Barley: Chemistry and Technology. American Association of Cereal Chemists Inc., St Paul Minnesota, USA, 131-183

Shewry P R, Halford N G (2002) Cereal Seed Storage Proteins: Structures, Properties and Role in Grain Utilization. J Exp Bio, 53: 947-958

Siriamornpun S, Wootton M, Cox J M, Bekes F, Wrigley C M (2001) Capillary Electrophoresis of Wheat Gliadin Proteins and its Potential for Wheat Varietal Identification Using Pattern Matching Software. Aust J Agri Res, 52: 839-843

Uthayakumaran S, Batey I L, Wrigley C W (2005) On-the-spot Identification of Grain Variety and Wheat Quality Type by Lab-on-a-chip Capillary Electrophoresis. J Cereal Sci, 41: 371-374

5

β-glucans and Arabinoxylans

J. M. Wang[1] and G. P. Zhang[2]
[1]Zhejiang Academy of Agricultural Sciences, Hangzhou 310021, China.
wangwangjm@yahoo.com.cn
[2]Agronomy Department, Zhejiang University, Hangzhjou, 310029, China.
zhanggp@zju.edu.cn

The major targets in malting and brewing are to obtain maximum extract yield, enough nutrients for yeast growth, fermentable sugars for alcohol production and a balanced combination of high molecular weight compounds to achieve a beautiful and stable foam and smooth mouth feel in beer and to avoid haze formation during storage. Wort and beer filtration difficulties frequently happen in beer processing because of insufficient hydrolysis of cell wall components. It is well documented that insufficient degradation of cell walls hinders the diffusion of germination enzymes, the mobilization of kernel reserves and hence reduces malt extract. β-glucans and arabinoxylans are the major components of endosperm cell walls. In general, insufficient degradation of cell walls is closely related to more residues of these sugars, which may lead to highly viscous wort and beer, giving rise to filtration problems in the brewery. In malt barley breeding, low β-glucan and arabinoxylans contents, or high activities of β-glucanase and xylosidase, which degrade β-glucan and arabinoxylans respectively, have been considered as an important target.

5.1 β-glucan

Barley (1$\rightarrow$3), (1$\rightarrow$4)-β-glucan, in short referred to as β-glucan thereafter, is a major structural component of the endosperm cell wall of cereal crops, in particular barley and oah. It can be considered as a linear chain of (1$\rightarrow$3)-linked cellotriosyl and cellotetraosyl units, arranged randomly. β-glucan represents only a minor fraction of the total carbohydrate content in barley kernels, but it accounts for about 75% of total carbohydrates in the cell walls of the endosperm (Bamforth, 1982), the remainders being 20%~25% arabinoxylan and minor amounts of cellulose, glucomannan and protein (Fincher, 1992).

In recent two decades, intensive investigation had been done on biosynthesis, genetic control and environmental variation of barley β-glucan as it is considered as a very important quality trait of barley used as malting, feed and food.

5.1.1 Structure of β-glucan in Barley Grain and its Influence on Malting Quality

The distinct influence of β-glucan on malt quality is attributed to its particular chemical structure.

5.1.1.1 Structure and molecular weight of β-glucan

Barley β-glucan is linear polysaccharides, which can be viewed as a cellulose chain (70% 4-O-linked β-D-glucopyranosyl units) interrupted by 3-O-linked β-D-glucopyranosyl units (30%) (Woodward *et al.*, 1983). The (1→3) linkages occur singly, leading to a structure of predominantly β-(1→3)-linked cellotriosyl and cellotetraosyl units. This structure, established many years ago for oat β-glucan (Parrish *et al.*, 1960), has not required substantial modification as newer techniques for evaluation, such as High-Performance Anion-Exchange Chromatography (HPAEC) analysis, have been developed. The molar ratio of (1→3)-linked cellotriosyl to (1→3)-linked cellotetraosyl units, which constitute 85%~90% by weight of the cereal β-glucan (Wood *et al.*, 1994; Skendi *et al.*, 2003), is relatively higher in barley (2.8-3.3), in comparison with that in oats (2.1-2.4) (Wood *et al.*, 1994; Cui *et al.*, 2000). Cellulose-like sequences containing more than three consecutive β - (1→4) - linked D-glucopyranosyl units are present in less than 15% of the polysaccharide, and seem to be conserved amongst all the cereal β-glucans. Considering the generalized β-glucan structure as shown in Fig.5.1, assignment of several resonances to carbons of individual β-glucose residues (A, B, C, D) is feasible based on literature data (Cui *et al.*, 2000). Gómez *et al.* (1997) gave the scheme of the (l→3), (1→4) - β-D-glucan molecule.

Storsley *et al.* (2003) studied the molecular characteristics of β-glucans and arabinoxylans from selected hull-less barleys using sequential extraction with water (45°C and 95°C), $Ba(OH)_2$, $Ba(OH)_2/H_2O$, and NaOH, which resulted in five distinct fractions designated as WE45, WE95, $Ba(OH)_2$, $Ba(OH)_2/H_2O$, and NaOH, respectively. Both water-extractable fractions consisted mainly of β-glucans, but up to 20% of arabinoxylans were found

A B C A B C D

$$\rightarrow {}^{4}G_{1} \rightarrow [\ {}^{3}G_{1} \rightarrow {}^{4}G_{1} \rightarrow {}^{4}G_{1} \rightarrow\ ]_{m}[\ {}^{3}G_{1} \rightarrow {}^{4}G_{1} \rightarrow {}^{4}G_{1}$$

A B C

$$\rightarrow {}^{4}G_{1} \rightarrow\ ]_{n} \rightarrow {}^{3}G_{1} \rightarrow {}^{4}G_{1} \rightarrow$$

Fig. 5.1. The structure module of barley β-glucan (Skendi *et al.*, 2003)

in WE45 of normal and high amylose barley varieties. Analysis of the water-extractable β-glucans, indicated differences in molecular features (weight average molecular weight, β-(1→4) to β-(1→3) linkage ratios, and amounts of cellulosic regions along the β-glucan chain) between WE45 and WE95, and among β-glucans from different cultivars. The variation observed in the viscoelastic behaviour of WE45 and WE95 fractions confirmed differences in the molecular weight among water-extractable β-glucans derived from various barley cultivars.

There is general disagreement about the molecular weight data of β-glucans, as presented in various literature reports, as well as substantial differences in the viscosity values of their aqueous solutions. These differences could be attributed to the sample source, extraction conditions, partial depolymerization by endogenous enzymes during the isolation step and the nature of analytical techniques employed for molecular weight determination (Izydorczyk and Biliaderis, 2000). The magnitude of molecular weights that has been reported for barley varies between 9 and 4×10^{7} Da (Fincher and Stone, 1986; Varum and Smidsrod, 1988; Saulnier *et al.*, 1994; Grimm *et al.*, 1995; Beer *et al.*, 1997; Gómez *et al.*, 1997). Many of the lower values reported for oat and barley β-glucan may be isolated that have been depolymerized by endogenous enzymes or by enzymes (such as amylases) introduced during isolation. On the other hand, some of the higher values reported might be overestimates caused by use of inappropriate standards (such as pullulan) or an overestimation of the molecular weight of the β-glucan standards used (Wood *et al.*, 1991).

5.1.1.2 Effect of β-glucan on malt quality

β-glucans extracted from malted barley or from non-germinated grain are generally regarded as undesirable in the malting and brewing processes (Woodward and Fincher, 1983; Bamforth, 1985). Barley used for brewing should be firstly malted. During the malting process, the storage reserves of grain endosperm, mainly starch and protein, have to be depolymerized by a series of hydrolytic enzymes synthesized in the aleurone layer and in the scutellum of the germinated grains (Mc Fadden *et al.*, 1988; Henry, 1989; MacGregor *et al.*, 1994). Intact cereal walls form effective barriers to the movement of degraded enzymes within the grain and that they must be modified before the digestion of protein and starch can occur during germination. Of the components of barley starchy endosperm cell walls, (1→3),(1→4)-β-D-glucans form an important constituent (Fincher, 1975). The degradation of endosperm cell walls and subsequent changes in β-glucan levels during malting have been linked to successful malt production and the availability of hydrolytic enzymes (Bourne and Wheeler, 1984; Etokakpan, 1993). Incomplete degradation of the cell walls during malting leads to poor modification of the endosperm, which results in low wort extract levels and the presence of high levels of β-glucan in the wort (Brennan *et al.*, 1997). This not only results in a substantial decrease in brewhouse yield but also contributes to filtration problems and haze formation (Bamforth, 1982). Thus, good quality barley cultivars tend to have a more homogeneous pattern of cell wall breakdown than poor malting cultivars do (Fenwick, 1993; Brennan *et al.*, 1997).

During the malting process, insufficient degradation of β-glucan can also lead to viscous initial extracts. The viscosity, which is attributed to the high molecular weight and extreme asymmetry of the polysaccharides, can lead to filtration difficulties at various points (Bamforth and Barclay, 1993). Because the rate of beer filtration often limits the speed and hence the volume of production (Scott, 1972; Fincher, 1975; Prentice *et al.*, 1980; Bathgate, 1983), the decreased rate of filtration and increased time required for filter maintenance are major problems in the brewery. The undegraded β-glucans also have a detrimental effect on the stability of beer during storage. Undegraded β-glucan in beer may precipitate at elevated ethanol concentrations, at low temperatures or after industrial centrifuges, thus causing chill haze and gelations precipitates both during storage and in bottled beer (Barrett *et al.*, 1973; Bamforth, 1982; McCleary and Gennie-Holmes, 1985; Yamashita *et al.*, 1989).

5.1.2 Environmental and Genotypic Variation of β-glucan Content in Barley Grain

It has been found that there is a large difference in β-glucan content among barley cultivars. Moreover β-glucan content is also strongly affected by environmental factors.

5.1.2.1 The variation of β-glucan content in barley grain

The levels in dehulled or naked oats and barleys range mostly from about 3% to 7% (Peterson *et al.*, 1991; Wood, 1994; Lee *et al.*, 1997), in rye about 2% (Ragaee *et al.*, 2001) and in wheat lower than 0.5% (Beresford and Stone, 1983). Milling to separate coarse particle fractions is used to increase the levels of β-glucan (Wood *et al.*, 1989; Knuckles *et al.*, 1992; Wood, 1994; Dexter and Wood, 1996; Nilsson *et al.*, 1997). According to Bamforth (1982) and Wood and Weisz (1984), compared with other cereals, barley and oat grains have a significantly higher level of β-glucan. Aman and Graham (1987) determined the β-glucan content in barley and oat genotypes from many countries, and showed that β-glucan content in barley ranged from 3.0% to 6.9% with a mean of 4.5%, and in oat ranged from 2.2% to 4.2% with a mean of 3.2%. Barley was highly enriched in β-glucan content but, unlike the β-glucans from oats, they were mainly unextractable. The extractable β-glucan content in barley was 54%, and in oats 80%. The variation of β-glucan content in barley grain has been reported in many studies (Prentice *et al.*, 1980; Wood and Weisz, 1984; Ahluwalia and Ellis, 1984; Aman, 1986; Amen and Graham, 1987; Henry, 1987; Lehtonen and Aikasalo, 1987; Genc *et al.*, 2001), and there is a great difference between these reports. In the early days, such as in the 1950s, it was shown that the β-glucan content in barley grain was only about 1%. Recently, most reports showed that the level of β-glucan in barley grain ranged from 3% to 6%, and some was higher than 10%. This difference is due to the changed methodologies in β-glucan measurement. Since the development of the enzymatic method by McCleary and Gennie-Holmes (1985), it has been used as a standard method for β-glucan analysis in cereal grains. Grain β-glucan contents among barley cultivars have been reported to range from 3.0% to 8.2% (Aman and Graham, 1987; Henry, 1987; Lehtonen and Aikasalo, 1987; Carr *et al.*, 1989; Fastnaught *et al.*, 1996). Kato *et al.* (1995) determined β-glucan content of 268 barley genotypes in Japan and found that the range was 2.86%-6.35% with a mean of 4.26%. Han *et al.* (1995) evaluated β-glucan content of 150 DH lines in USA, the range was from 3.4% to 5.3% with a mean of 4.3%. Studies in other countries including Denmark, Finland, Britain and Sweden showed that the mean content of barley β-glucan was about 4.1%∼4.4%. We measured β-glucan contents of more than 200 barley genotypes, collected from different areas of China, as well as from Australia and Canada. It was found that the barleys from Tibet and Xinjiang had higher β-glucan content, and barley cultivars planted in the southeast of China had a basically similar content as those from Australia and Canada. In general, the β-glucan content of all measured genotypes ranged from 2.98% to 8.62% with a mean of 4.58% (Wang *et al.*, 1998; Zhang *et al.*, 2002).

5.1.2.2 Genotypic variation of β-glucan content in barley grain

Cultivar and environmental variations in the β-glucan content of barley have been intensively studied (Molina-Cano *et al.*, 1989; Narasimhalu *et al.*, 1994; Perez-Vedrell *et al.*, 1996; Zhang, *et al.*, 2001). Genotypic factors have been found to be more important (Henry, 1986; Molina-Cano *et al.*, 1997) than environmental ones in determining β-glucan content. Molina-Cano *et al.* (1989) reported that there were significant differences in β-glucan content of barley grain among cultivars and environments. Lehtonen and Aikasalo (1987) determined the β-glucan content of barley in Finland and showed that β-glucan content of 68 two-rowed varieties grown in southern Finland varied between 3.5% and 5.3%, and in central Finland it varied from 4.0% to 5.2%.

The corresponding ranges in 50 six-rowed varieties grown in southern, central and northern Finland were 2.8%~4.3%, 3.3%~5.6%, and 3.6%~4.0%, respectively. Stuart *et al.* (1988) found that the total β-glucan content in barley grain was influenced more strongly by genotypic factors than environmental ones. Narasimhalu *et al.* (1994) studied the variation of β-glucan content in 75 barley cultivars grown at six locations in the east and west of Canada. The range in 32 eastern feed barley cultivars was 3.71%~4.41%, and the range in 43 western feed and malt barley cultivars was 3.37%~4.42%. Perez-Vedrell *et al.* (1996) evaluated β-glucan content of five winter barley cultivars (Albacete, Alpha, Barbarrosa, Hatif de Grignon and Plaisant) and five spring cultivars (Beka, Dobla, Kym, Pallas and Zaida) grown at seven locations in Spain with diverse climates in 1989, 1990 and 1991. They found that there was highly significant variation in β-glucan content among cultivars, locations and years. As an average over all cultivars, β-glucan content ranged from 1.9% to 5.5% with a mean of 3.5%, and winter barley cultivars was higher than spring barley cultivars. We investigated genotypic and environmental effects of β-glucan content by planting 10 barley cultivars in eight locations differing in climatic conditions in the southeast China (Zhang *et al.*, 2001; Wang *et al.*, 2004). The results showed that there was a highly significant difference in grain β-glucan content among cultivars, locations and between the two planting years. The average β-glucan content over the two years for the 10 cultivars in the eight locations ranged from 3.91% for Gangpi, 1%~4.95% for Xiumei 3. Holtekjølen *et al.* (2006) recently studied the variation of β-glucan content in 39 barley varieties differing in genetic background and planting area. It was found that the total β-glucan content ranged from 2.4% to 8.3%. The high-amylose sample contained the highest amount of β-glucan. With two exceptions (Olve and Sunnita), all Norwegian samples were found to contain less than 4% β-glucans, while the atypical amylose genotypes had a β-glucan content above 6%, with the exception of SW 2680. The hull-less atypical amylose varieties had a significantly higher concentration of β-glucan than the normal amylose ones. The difference in β-glucan content between the hulled and hull-less varieties of normal starch type was not significant; however, there was a tendency towards lower β-glucan content in the hulled samples. The main soluble non-starch polysaccharides (NSP) component was β-glucan, which was also indicated by a strong correlation between β-glucan and soluble NSP ($r = 0.88$).

5.1.2.3 Environmental variation of β-glucan content in barley grain

β-glucan content in barley grain is also affected strongly by environmental factors (Stuart *et al.*, 1988; Narasimhalu *et al.*, 1994; Fastnaught *et al.*, 1996). It is a common phenomenon that a barley cultivar shows great difference in β-glucan content when planted in different years and locations. For example, barley cultivar Schooner, mainly planted in the southeast of Australia showed poor malt quality in some years with low rainfall during growth and high temperature during the grain-filling stage (Stuart *et al.*, 1988). The poor malt quality is partly associated with higher grain β-glucan content. Fastnaught *et al.* (1996) reported that barley grains contained a higher level of β-glucan in years with low humidity and mild climate factors. Narasihalu *et al.* (1994) indicated that the environment significantly affected total β-glucan content in barley, with the grains of same cultivars harvested in the east of Canada being higher than those in the west. They considered that higher β-glucan content from the east of Canada was attributed to a higher temperature during the grain development. Perez-Vendrell *et al.* (1996) found that barleys grown at locations characterized by mild temperatures and high rainfall had lower β-glucan content, and Spanish barleys had lower β-glucan content than those grown in Australia, Canada and other European countries. The reason was that the climatic conditions for Spainish barleys were very different from those in other countries. The grains of Spainish barleys contained moisture of less than 12% at maturity, while in Northern Europe the grain was sometimes harvested with moistures up to 18%. Moreover, in Spain the growing season was also different, the barley being sown in October-November and harvested in July. Wallwork *et al.* (1998) evaluated the effects of a period of high temperature during grain filling on the malting quality of Australian malting barleys. It was found that a short period of high temperatures (up to 35°C) during mid grain filling appeared to reduce total β-glucan content in barley endosperm, but not to change the pattern of the accumulation of β-glucan content. Macnicol *et al.* (1993) determined the effects of heat and water stresses on malt quality and grain β-glucan content of Schooner barley grown in cabinets. They found that the water stress at the early filling stage of barley grains appeared to greatly reduce β-glucan content, but the effect became quite small when the treatment was conducted at the late filling stage. Coles (1991) reported that there was a negative relationship between moisture stress in soil and β-glucan content in barley grain. Morgan and Riggs (1981) showed that water stress led to the increase of β-glucan content in barley grains, and the decrease in malt extract hence caused the worse malting performance. In an experiment for determining the genotypic and environmental effect on β-glucan content in barley grain, we found that variation of β-glucan content for 10 barley cultivars planted over eight locations ranged from 3.76% for Hangzhou to 4.75% for Taian, and the distinct difference between the two locations is that Hangzhou has a lower temperature and more rainy days than Taian during the grain-filling stage. Six climatic factors, namely the total accumulated temperature, the mean daily accumulated temperature, the total precipitation, the mean daily precipitation, the total sunshine hours and the mean daily sunshine hours, significantly affected β-glucan content during the grain-filling stage. During the seed development, higher mean daily accumulated temperature (>20°C), lower mean daily precipitation and longer sunshine hours resulted in higher β-glucan content (Wang *et al.*, 2001; Chen and Zhang, 2002; Wang *et al.*, 2004). However, Henry (1986) considered that the β-glucan content in barley grains was mainly controlled by genotypic factor, and there was no significant effect of environmental factors on the β-glucan content.

The agronomic managements also have a great effect on β-glucan content in barley grains. Nitrogen levels in soil and nitrogen fertilization are major factors affecting β -glucan content. High nitrogen levels in soil and an increasing N fertilizer application rate significantly increased β -glucan contents in oats and barley (Brunner and Freed, 1994; Baur and Geisler, 1996). Humphreys *et al.* (1994) determined the effect of fertilization on malt quality and found that protein content in barley grains was significantly increased with increasing application of N fertilizer, but no effect was found on β-glucan content. We found that there was a significant effect of the timing of N application on β-glucan content in barley grains, and more application of N fertilizer at the late growth stage increased the β-glucan accumulation in grains (Chen and Zhang, 2000; Zhang *et al.*, 2002). Agu and Palmer (2001) compared the malt quality of high and low nitrogen barleys (cultivars Alexis, Optic (A), Regina, Scarlett, Sloop and Optic (B)). The results confirmed that low nitrogen content facilitated the malting rate and extract development. The modification rate was not related to β-glucan content. Bichonski *et al.* (2003) investigated the effects of different N fertilizer doses (0, 20, 40, 60 and 80 kg/ha) and sowing densities (200, 300, 400 and 500 grains/m) on the β-glucan content of winter brewery barley cv. Rudzik in Poland. They found that high N doses resulted in an undesirable increase in β-glucan content of wort, whereas a high sowing density had a positive effect on the content of this component. Gler (2003) reported increased nitrogen levels positively affected grain β-glucan content, and the greatest grain β-glucan content was generally obtained under 80kg ha^{-1} N application. They also found a negative effect of irrigation on β-glucan content in barley grain. Increased levels of irrigation tended to decrease β-glucan content. Non-irrigated treatments commonly had the greatest values for grain β-glucan content in all treatments.

5.1.3 Genetic Improvement of β-glucan Content in Barley Grain

Greenberg (1977) firstly investigated the genetic characteristics of β-glucan content in barley. In his study, the two-rowed spring varieties Golden Promise and Maris Badger, with low β-glucan content in grains, Proctor and Maris Mink with intermediate β-glucan content and Julia and Zephyr with high β-glucan content were used in the diallel cross. It was found that gum (mainly β-glucan) content was controlled by a simple additive-dominance genetic system and low content was strongly dominant. It also showed that gum content was principally controlled by two or three genes showing a high degree of dominance. Genotype by environment interaction was detected through comparison between F_2 and F_3 generations grown in different years and locations. Sanada (1983) measured the β-glucan content in barley grains with the method described by Morgan (1977), and determined its genetic effect. The results showed that β-glucan content was mainly controlled by additive and dominant effects, and the genetic variances of epistasis and dominance were very low. Molina-Cano (1982) analyzed the effects of genetic and environment on β-glucan content in barley grains. It figured out that genetic factors played a predominant role in the variation of β-glucan content, accounting for 89%. Molina-cano (1979) reported that β-glucan content in barley grains was controlled by the cultivar itself, and its broad heritability was 0.54. In 1984 Lance studied the inheritance of β-glucan content using populations derived from crosses between the two parents with different β-glucan content and their F1 populations. He found that additive, dominant, additive $\times$ dominant and dominant $\times$ dominant effects were responsible for the variability of β-glucan content in observed populations. Most variability in the segregating populations was due to additive effects. The two parents differed by four additive genes, and a low but undetermined number of dominant genes for low β-glucan content. Powell *et al.* (1985) evaluated the β-glucan content in random inbred lines produced by doubled haploid (DH) and single seed descent (SSD). It indicated that β-glucan content was controlled by a simple additive genetic system. There were significantly negative genetic correlations between β-glucan content, 1000-grain weight and height in the DH population, and these correlations were much reduced in the SSD samples, suggesting linkage of the characters. Yu and Yang (1987) studied the genetic effects of barley gum content, and showed that barley gum content was controlled by an additive-dominant genetic system, and additive effects played the predominant role. There was no effect of epistasis. It was also shown that dominance was partly responsible for the low β-glucan content. Han *et al.* (1995) mapped β-glucan content loci in barley grain and malt. Three QTL for grain β-glucan, six QTL for malt β-glucan, were detected by interval mapping procedures. The QTL with the largest effect on grain and malt β-glucan content were identified in chromosomes 2 and 1, respectively. Mather *et al.* (1997) reported that there were two-four QTLs for barley β-glucan content.

Horsley (1995) suggested that further improvement in grain β-glucan content could be expected by intercrossing closely related genotypes. Greenberg (1977) reported that the gum content was highly heritable, and F_3 generations were higher than F_2 generations. In breeding for improvement of β-glucan content, it is important to develop a simple and rapid method of determining β-glucan content. On the other hand, the impact of β-glucanase on malt β-glucan and quality has been more and more noted. According to Stuart *et al.* (1988), low β-glucan content in the malt was the most important indicator of malt quality. In contrast to Morgan's (1977) emphasis on selection for low levels of β-glucan content in the grains, they found that it was necessary to combine both relatively low grain β-glucan content and the ability to develop high levels of β-glucanase activity. They concluded that, under Australian conditions, the level of β-glucan content in the grains was not a good indicator of the level of malt β-glucan content and that, as β-glucanase activity showed great environmental variation, selection for stable expression should be advantageous. In a previous study (Zhang *et al.*, 2002), we found that most malt barley cultivars planted in the southern China did not have higher β-glucan content than imported barleys from Australia or Canada, while malting performance of these Chinese barleys was relatively inferior. It was reported that higher levels of β-glucanase activity were produced during malting in the cultivars with high malting quality, indicating the importance of the breakdown of cell walls during malting (Barber *et al.*, 1994). Grain β-glucan content during malting would be greatly changed (Bamforth and Martin, 1983; Stuart *et al.*, 1988; Ellis *et al.*, 1997). Ellis *et al.* (1997) found that the breakdown of β-glucan content in grain happened with the development of β-glucanase activity, and the total grain β-glucan content declined by about 50% after malting. Stuart *et al.* (1988) showed that the β-glucan content was greatly depolymerised by β-glucanase during malting. We found approximately 80% of β-glucan present in grains degraded during malting, using an average of all cultivars and locations. It may be suggested that the ability to develop high levels of β-glucanase activity during malting should be more highlighted than low levels of grain β-glucan content in the breeding programs of malt barley (Wang *et al.*, 2004).

β-glucan content in barley grains can be improved by mutation breeding. A high β-glucan content cultivar Minerva was induced in the Carlsberg Company, Denmark. 400 single plants with high β-glucan content were selected from a M2 population consisting of 40000 plants. All single plants were grown in a glasshouse for one generation, and four lines were selected from 40 lines in M4 populations. Analysis of the endosperm in the four lines showed that all of them had the soft endosperm, and the endosperm cell could be depolymerized rapidly. The endosperm cell walls of these lines were thinner than those of Minerva, with the mutation lines ranging from 3.0 to 3.5 μm, and Minerva being 6.5μm. One of the mutations showed the increase in β-glucanase activity in comparison with the wild parent.

The structure and function of genes encoding β-glucan has been investigated. Aspegren *et al.* (1995) investigated the cDNA of the thermostable β-glucanase (EGI) of *Trichoderma reesei* [T. longibrachiatum] in suspension cells. EGI was used in the brewing process to remove excess β-glucans from beer wort. The cDNA coding for EGI and its signal sequence was placed under the control of the CaMV 35S promoter and the construction was transferred to the cells by particle bombardment. Stably transformed lines were obtained through selecting for a co-transformed antibiotic resistance marker. The expression of EGI cDNA led to accumulation of EGI in the culture medium as shown by analysis of EGI-specific antibodies. Enzymatic assays confirmed that the EGI secreted by the suspension cells retained its activity and thermostable character. Furthermore, it was shown that the enzyme produced by the transgenic suspension culture could be used for degradation of soluble β-glucans during mashing. There have been many reports on the mapping of the QTL controlling β-glucan and β-glucanase (Han *et al.*, 1995; Mather *et al.*, 1997). It was very useful for researchers to manipulate the genes encoding β-glucan. Up to now, the cDNA encoding β-glucanase has been isolated from the barley genomic library, and the nucleotide sequence of the gene and its expression are also basically clarified (Slakeski *et al.*, 1990). Jensen *et al.* (1996) modified a hybrid bacterial gene encoding a thermostable (1,3-1,4)-β-glucanase and compare it with that of the barley (1,3-1,4)-β-glucanase isoenzyme EII gene, and transferred the hybrid gene into barley aleurone protoplasts. It was found that the bacterial gene could express the synthesis of heat stable (1,3-1,4)-β-glucanase. Horvath *et al.* (2001) found that three production levels of recombinant β-glucanase were identified in homozygous transgenic T3 plants, and these remained constant over a three-year period. In micro-malting experiments, the heat-stable enzyme reached levels of up to 1.4 micro g.mg^{-1} protein and survived kiln drying at levels of 70%-100%. In the field trials the transgenic lines had a higher 1000-grain weight than those lacking the transgene. All this information might be very useful in molecular breeding for β-glucan in barley grain.

5.1.4 The Relationships between β-glucan and Malt Quality and Some Agronomic Characters

There is great disagreement about the relationship between β-glucan and protein content and grain weight in the reported literature. Welch *et al.* (1991) reported that β-glucan content in barley grain was positively and significantly correlated with protein content. But Saatamoinen *et al.* (1992) and Miller *et al.* (1993) considered that a negatively significant correlation existed between β-glucan and protein content. Molina-cano *et al.* (1997) indicated that β-glucan content was positively and significantly correlated with the total protein content in barley grain. Swanston (1997) found that β-glucan content was negatively correlated with the nitrogen content in barley grain, and positively correlated with grain weight. Henry (1986) studied genetic and environmental variation in the pentosan and β-glucan content of barley, and their effects on malting quality. It was shown that β-glucan content was positively and significantly correlated with the nitrogen content in barley grain, and positively correlated with grain weight. Powell *et al.* (1985) suggested that β-glucan content was negatively correlated with grain weight. Kalra *et al.* (1998) reported that there was no significant relationship between β-glucan and protein content.

The relationships between β-glucan content and agronomic characters have been also studied. The disagreement in the relationship between β-glucan and row type exists among the current reports. Lehtonen and Aikasalo (1987) found that the two-rowed varieties were somewhat higher in β-glucan content than the six-rowed barleys. But Narasimhhalu *et al.* (1995) reported that β-glucan content in six-rowed cultivars was higher than that in two-rowed cultivars. Fustnaught *et al.* (1996) indicated that there was no significant relationship between β-glucan and row type. Powell *et al.* (1985) suggested that β-glucan content was negatively correlated with plant height. Variation in barley β-glucan content is associated with genetically controlled starch and hull type. Commonly hull-less cultivars contain higher β-glucan content, compared with hulled cultivars (Bengtsson *et al.*, 1990; Bhatty *et al.*, 1991; Yoon *et al.*, 1995; Fustnaught *et al.*, 1995). In addition, the cultivars with 100% amylopectin (waxy) starch of endosperm are higher in β-glucan content (Ullrich, *et al.*, 1986; Yoon *et al.*, 1995; Swanston, 1997).

Close relationships could be found between grain β-glucan content and malt quality (Henry, 1989; Ellis *et al.*, 1997; Knuckles *et al.*, 1999). Henry (1986) found that β-glucan content in grain was negatively and significantly correlated with mat extract. Bhatty (1987) figured out that the total β-glucan content in grain was positively and significantly correlated with viscosity. Stuart *et al.* (1988) also found that β-glucan content in grain was negatively and significantly correlated with mat extract. But Bourne and Wheeler (1984) indicated that there was no significant relationship between β-glucan content in grain and malt extract. Molina-Cano *et al.* (1997) investigated the variation of genetic and environment in β-glucan content and its impact on the qualities of malt and feed barley. It was shown that β-glucan content in grain was positively and significantly correlated with malt viscosity, and negatively and significantly correlated with malt extract and Kolbach index. Mather *et al.* (1997) reported that β-glucan content in grain was positively and significantly correlated with malt viscosity, and negatively and significantly correlated with diastatic power. Ellis *et al.* (1997) found that β-glucan content in grain was negatively correlated with β-glucanase. We found that malt β-glucan content was very significantly correlated with malt extract, Kolbach index, diastatic power and viscosity, while the association between grain β-glucan content and these quality parameters was relatively weak. Malt β-glucan content was negatively and highly significantly correlated with malt β-glucanase activity, and positively and significantly correlated with grain β-glucan content (Wang *et al.*, 2004).

5.1.5 Measurement of β-glucan Content

It is very important to precisely and rapidly determine the β-glucan content for barley researchers in carrying out the relevant investigation of this chemical in the fields of genetics, breeding, physiology and ecology. Before the mid 1980s, the progress of studies on β-glucan in barley had been very slow, mainly because of limitations in its analysis methodology. After that, much progress was made in the methods of β-glucan content measurement. Several standard assay methods have been developed for determining the β-glucan content. Now the enzymatic method, developed by McCleary and Glennie-Holmes (1985), the FIA-Calcofluor method, described by Jorgensen and Aastrup (1986), and HPLC method, developed by Perez-Vedrell *et al.* (1995) have been accepted as official methods by the European Brewery Convention.

5.1.5.1 The enzymatic method

A commercially available enzymatic method for the quantitative measurement of β-glucan content was developed by McCleary *et al.* (1985, 1991, 1997). It was simplified to allow analysis of up to 10 grain samples in 70 min. The assay method has been successfully evaluated by Analytical Committees of the Royal Australian Chemical Institute and the European Brewing Convention. A modified version is AOAC Official Method 995.16, and an AACC (Method 32-23) and ICC (Method No. 168) recommended/standard method. The main principle is as follow: Samples are suspended and hydrated in a buffer solution of pH 6.5 and then incubated with purified lichenase enzyme and filtered. An aliquot of the filtrate is then hydrolysed to completion with purified β-glucosidase. The D-glucose produced is assayed using a glucose oxidase/peroxidase reagent.

The assay procedure can be described in detail as follows: milling barley grain to pass a 0.5 mm screen using a Tecator Cyclotec mill; accurately weighing barley flour samples (about 0.5 g) of known moisture content into polypropylene tubes; adding an aliquot (1.0 ml) of aqueous ethanol (50 % v/v) to each tube to aid in the subsequent dispersion of samples; adding 5.0 ml of sodium phosphate buffer (20 mM, pH 6.5) and stirring the tubes on a vortex mixer; incubating the tubes in a boiling water bath for 2 min; removing the tubes and vigorously stirring them on a vortex mixer; heating the tubes for a further 3 min in the boiling water; cooling the tubes to 40°C and adding 0.2 ml of lichenase (10 U) to each tube; capping the tubes, stirring, and incubating at 40°C for 1 h; adjusting the volume in each tube to 30.0 ml by the addition of distilled water; thoroughly mixing the contents of the tubes and filtering an aliquot from each tube through a Whatman No. 41 filter; carefully and accurately transferring aliquots (0.1 ml) from each filtrate to the bottom of three test tubes; adding an aliquot (0.1 ml) of sodium acetate buffer (50 mM, pH 4.0) to one of these (the reaction blank), while to the other two (the reaction) adding 0.1 ml of β-glucosidase (0.2 U) in 50 mM acetate buffer (pH 4.0); incubating the tubes at 40°C for 15 min; adding GOPOD Reagent (3.0 ml) to each tube and incubating at 40°C for 20 min; measuring the absorbance at 510 nm for each reaction and reaction blank.

5.1.5.2 Calcofluor flow-Injection analysis (FIA)

The Calcofluor-FIA method, based on the shift and increase of the emission maximum in the fluorescence spectrum of the dye Calcofluor when it forms a complex with β-glucan, allows rapid, easy and accurate quantification of β-glucan content in brewing samples. The basis of this methodology is that the yield of the complex formed between Calcofluor and β-glucan depends on the molecular weight of the latter as well as on the ionic strength of the eluant used, the shape of the yield curve being a sharp sigmoid. As a consequence and for a given eluant, the detection level of a given molecule of substrate is a function of its molecular weight.

The procedure can be separated into two parts. One part is extraction of β-glucan in samples described by Manzanares and Sendra (1996): 0.2 g barley flour was weighed into a 25 ml cap-sealed glass tube, and 20 ml 0.05 M HNO_3 was added. The contents were mixed on a vortex mixer, avoiding the formation of lumps, placed in a boiling water bath for 20 min and occasionally shaken gently by hand. The sample was cooled to room temperature and centrifuged at 100 g for 10 min and the supernatant was recovered and filtered through a 0.45 μm membrane filter. The residual pellet was re-extracted under the same conditions, but the volume of 0.05 M HNO_3 was reduced to 5 ml. The other part is flow-injection analysis (FIA). In the typical FIA set-up, a constant flow of Calcofluor solution is delivered to the mixing coil by a peristaltic pump and a fixed volume of the sample solution is injected into the same column. They are mixed in a mixing coil and passes through a fluorescence detector.

5.1.5.3 High performance liquid chromatography method

β-glucan content could be determined by high-performance liquid chromatography (HPLC) (Perez-Vedrell *et al.*, 1995). The principle was as follows: the β-glucan was hydrolysed with lichenase to oligosaccharides, which were analyzed by reversed-phase HPLC using water as the mobile phase at a flow-rate of 0.7 ml/min. The separation of the oligosaccharides was performed in a C18 stainless-steel column (Spherisorb ODS-2) with 5-μm particles in less than 10 min, with refractive index detection.

The procedure was as follows: accurately weighing 0.5 g barley flour samples into polypropylene tubes and adding 1 ml 50% ethanol and 5 ml 20mM sodium phosphate buffer (pH 6.5); incubating the tubes in a boiling water bath for 10 min; removing and cooling the tubes to 40°C; adding 0.2 ml of lichenase to each tube and incubated at 40°C for 1 h; adding 24 ml distilled water and mixing; centrifuging the content at 5,000 g for 15 min; filtering the supernatant through a 0.45 m membrane filter; adding 6ml sample solution to reversed-phase HPLC for analysis.

5.1.5.4 Near-infrared reflectance spectroscopy method

Near infrared reflectance (NIR) spectroscopy has become the method of choice for the analysis of grain as it is received from farmers into marketing systems. It is also widely used in plant breeding programs for the selection of cross-breed with the desired qualities. NIR spectroscopy provides fast, safe and inexpensive analysis. It is, however, a comparative technique that relies on multivariate calibration of sample spectra and accurate reference analysis. It has the potential to be exploited as a rapid analytical method for β-glucan. Over the past 25 years, the number of applications of NIR spectroscopy has increased rapidly, as demonstrated by the emergence of dedicated journals and large international conferences (Henry, 1985; Batten, 1998; Garrido *et al.*, 2002; Blakeney and Peter, 2005).

The basis of this method is as follows: NIR forms parts of the electromagnetic spectrum in the wavelength range 780 to 2500nm. Absorption bands relating to many chemical bonds, such as C-H, N-H, O-H, S-H, C=O and C=C, are found in the NIR region. The NIR spectrum shows overtone and combination bands of these groups. The absorption bands in the NIR region are broad and overlap. Calibration is required to correlate the spectral response of each sample at individual wavelengths to recognize chemical data from laboratory analysis. The operating principle is that β-glucans have molecular bonds that stretch and bend causing absorbance at certain wavelengths of infrared radiation in a linear proportion to the amount of β-glucan present. An equation that relates log 1/R values at selected key wavelengths to chemistry values enables a calibration to be developed, provided sufficient samples have been scanned to represent the full range of β-glucan. NIR spectroscopy method has been gradually accepted by the American Association of Cereal Chemists (AACC), The American Association of Official Analytical Chemists (AOAC), and the American Oil Chemists' Society (AOCS). The procedure is very simple: selection of calibration and validation samples, reference data obtained by routine laboratory analysis, NIR spectral data obtained by scanning samples, selection of optimum equations between spectral data and reference values by calibration, confirmation of optimum equations by validation.

5.2 Arabinoxylans

Arabinoxylans are the other major non-cellulosic polysaccharides in barley cell walls, besides β-glucan. These polysaccharides are not only particularly abundant in the walls of aleurone cells and the starchy endosperm, but also are found in the husk (Aspinall and Ross, 1963). It was assumed that they were present in the wall remnants surrounding the grain (Fincher and Stone, 1986). Arabinoxylans consisted predominantly of the pentose, arabinose and xylose, and therefore often are also referred to as pentosans.

On the whole arabinoxylans has relatively less influence on malting procedure than β-glucan does because of its lower content in barley grains, however it frequently causes negative effect on malting quality.

5.2.1 Structure of Arabinoxylans in Barley Grain and Its Influence on Malt Quality

Like the β-glucan, the negative influence of arabinoxylans on malt quality is closely related to its chemical structure.

5.2.1.1 Structure and molecular weight

The arabinoxylans of barley walls, like the β-glucans, consisted of a family of polysaccharides in which individual members differ in molecular size, monosaccharide composition, structure and solubility (Viëtor *et al.*, 1993). Henry (1988) reported that arabinoxylan contained a backbone structure of D-xylopyranose residues linked by β-(1→4)-glycosidic bonds with units such as L-arabinofuranose attached as branches by β-(1→2)- or β-(1→3)-linkages. Typically, L-arabinofuranose or other side chains were carried on the main chain as non-reducing end groups. Forrest and Wainwright (1977) reported that the pentosans isolated from endosperm cell walls were not covalently linked to the β-glucans, and these had a higher xylose/arabinose ratio than previously reports for barley pentosans. Recently, it was reported that arabinoxylan consisted of a linear back bone of (1→4)-β-D-xylopyranosyl units, to which a-L-arabinofuranosyl substituents were attached through O-2, O-3, or O-2, 3. One of the unique features of arabinoxylan was the presence of ferulic acid covalently linked via an ester linkage to C (O)-5 of the arabinose residue. Furthermore, ferulic acid dimers were detected at a rather high proportion in malt arabinoxylan (Dervilly *et al.*, 2002; Han, 2000).

According to Holtekjølen *et al.* (2006), the degree of branching in the arabinoxylans (A/X) was generally the same for both the total and insoluble fraction. Hull-less, waxy, and high-amylose varieties had the highest degree of substitution in these fractions. In the soluble arabinoxylans, the substitution pattern was different from those in the total and insoluble arabinoxylans. Generally the A/X ratio was higher, except for the hull-less genotypes.

Han (2000) studied three malting barley samples to determine glycosyl linkage composition of highly substituted arabinoxylans by analysis of partially methylated alditol acetate (PMAA) sugar derivatives using gas chromatography (GC) and gas chromatography mass spectrometry (GC-MS). Arabinoxylan, total non-starch polysaccharide (NSP) and β-glucan levels between all samples were not significantly different ($P<0.05$). The NSP fractions decreased significantly ($P<0.05$) during malting and brewing. No apparent differences in arabinoxylan structure were found between malt and beer samples. Arabinoxylans appear to contain randomly branched regions, mostly consisting of an un-substituted and a double arabinofuranosylated xylose residue. The number of xylose residues O-2,3 di-substituted, and O-2 and O-3 monosubstituted were similar in all varieties.

The molecular weights of barley arabinoxylans were generally lower than those of the β-glucans (Balance *et al.*, 1986; Fincher and Stone, 1986). Using sedimentation velocity techniques, Podrazky (1964) obtained a molecular weight of 58,800 for an arabinoxylan fraction extracted from barley flour with water. Extremely high molecular weight values for barley endosperm cell wall arabinoxylans of up to 5×10^6, estimated by gel filtration chromatography was also reported (Bathgate *et al.*, 1974; Forrest and Wainwright, 1977). The molecular weight of arabinoxylan extracted by water was smaller than that extracted by alkali.

5.2.1.2 Effects of arabinoxylan on malt quality

As the major constituents of the walls of the starchy endosperm in barley grain, there are the great effects of arabinoxylans on the end-use quality of barley. In under-modified malts, the fact that arabinoxylans couldn't be degraded sufficiently, might cause many problems such as low extract yield, high wort viscosity, decrease of filtration rate, and haze formation in brewing (Coote and Kirsop, 1976). Water-extractable arabinoxylans might be partially cross-linked with the possibility of causing filtration problems (Figueroa-Espinoza and Rouau, 1998). Historically, reduced beer filtration efficiency had been mainly attributed to β-glucan in the brewing process. Recently, it was found that other large molecules such as arabinoxylans, proteins and polyphenols were also associated with reduced beer filtration, in particular micro-filtration. In fact, it has been reported that the amount of arabinoxylans in commercial beer was approximately 10 times greater than that of β-glucan (Schwarz and Han, 1995).

The enzymes that degrade arabinoxylans are often produced late in the germination process (Banik *et al.*, 1997), and high levels of arabinoxylans can survive through the brewing into the final beer (Coote and Kirsop, 1976). Some arabinoxylans are solubilized from the cell walls but are not extensively degraded by endogenous enzymes during malting (Voragen *et al.*, 1987). Malt extracts can contain high levels of arabinoxylans and attendant difficulties associated with the filtration of viscous extracts may significantly deteriorate the performance of the brewing processes (Bamforth, 1985). These brewing problems were more pronounced when wheat malt was used as adjunct for its higher arabinoxylans content and higher molecular weight.

Sadosky and Schwarz (2002) indicated that the effects of arabinoxylans on viscosity and filterability were equivalently important to those of β-glucan. Stewart *et al.* (1998) found that viscosity and membrane filterability of the pilot-brewed beer were correlated with arabinoxylan content, whereas β-glucan was only correlated with viscosity. Although germination of barley reduced total arabinoxylans content, considerable amounts of water-extractable arabinoxylans (WEAX) were solubilized from cell walls and could survive during processing and fermentation. WEAX could reach levels of 4 g/l in beers (i.e. 0.4% w/v), which was considerably more than the β-glucan contents (maximum 0.2 g/l) (Han, 2000; Han and Schwarz, 1996). At these levels, viscous arabinoxylans solutions contributed to filtration problems and haze formation during brewing (Viëtor *et al.*, 1993). WEAX levels must be influenced by the relative rate of solubilization of arabinoxylans from cell walls and their rate of depolymerisation to oligosaccharides and monosaccharides by the concerted action of endoxylanases, the exo-acting β-D-xylosidases (Debyser *et al.*, 1997) and a-L-arabinofuranosidases (Lee *et al.*, 2001). Considering that cell wall β-glucan and arabinoxylan degradation might be synergistically linked (Kannauchi and Bamforth, 2002), it was possible that arabinoxylan degradation might be important to malting quality.

Sungurtas *et al.* (2004) studied xylan-degrading enzymes and arabinoxylan solubilisation in barley cultivars of differing malting quality. The barley cultivars Chariot, Optic, Static and Hart were grown at the Scottish Crop Research Institute (SCRI) in 1999 and micro-malted under standard conditions. Chariot and Optic are recognised as good quality malting cultivars, Static is a moderate quality malting cultivar and Hart is a feed cultivar with poor malting quality, which could be characterized by the hot water extraction. Chariot and Optic achieved the highest levels of extract after five days whereas in Static, extract levels peaked after four days malting then declined more rapidly with extended malting than the other two cultivars. By contrast, Hart increased in extract level throughout the seven-day germination period without reaching the levels attained by the other cultivars. The levels of xylanolytic activities ((1→4)-β-endoxylanases, β-D-xylosidases and α-L-arabinofuranosidases) and the levels of water extractable arabinoxylans (WEAX) in hot water extracts of malts were compared during malting of four barley cultivars of differing quality. Endoxylanase activity increased from day 1 after steeping showing a slight lag compared with the development of (1→3,1→4)- β-glucanase activity. The levels of endoxylanase activity were different between the cultivars. For example, on day 4, which would be a suitable endpoint for malting of good quality cultivars, total endoxylanase activities of Static and Chariot malts were twice those of Hart and Optic. This variation in level of endoxylanase between cultivars did not correlate with the content of water extractable arabinoxylans in hot water extracts of the malts. However, those cultivars that accumulated most water extractable arabinoxylan had a different set of endoxylanase isoforms. This suggests that expression of specific endoxylanase isoforms may influence the extent of arabinoxylan solubilisation and/or degradation. The β-xylosidases and -arabinosidases increased from low levels after steep to a plateau of activity at day 6 or 7 of germination. Total xylosidase was always higher than the arabinosidase activity. There were differences between cultivars but these were small compared with the variation in endoxylanase activity.

5.2.2 Environmental and Genotypic Variation of Arabinoxylans Content in Barley Grain

There is a large difference in arabinoxylans content among the cultivars and over the environments for a given cultivar.

5.2.2.1 Amount and synthesis of arabinoxylans in barley brain

Arabinoxylans were concentrated in the outer, protective layers of the barley grain (Hashimoto *et al.*, 1987). They represented about 25% and 70% of the cell wall polysaccharides of endosperm and aleurone layer, respectively (Fincher and Stone, 1986). Henry (1987) reported that only 22% of total barley arabinoxylans were found in the barley endosperm. In *Horduem spontaneum*, the progenitor of modern barleys, total arabinoxylan levels range from 4.2% to 9.9% by weight (Henry and Brown, 1987). Fincher (1989) also found that arabinoxylans comprised 4%~10% of the dry weight of barley grain but made up 20%~25% of endosperm cell walls. Most researches showed that arabinoxylans constituted 3%~11% of the barley kernel (Lethonen and Aikasalo, 1987; Han and Schwarz, 1996; Jadhav *et al.*, 1998). Recently, Gibeaut *et al.* (2005) reported that arabinoxylan content remained essentially constant at 25%~30% during growth of the coleoptiles at about five days, although the ratio of substituted to unsubstituted 4-linked xylosyl units decreased from about 4:1 to 1:1.

β-(1→4)-xylosyltransferase (XylTase; EC 2.4.2.24) participating in the synthesis of arabinoxylans was investigated using microsomal membranes prepared from developing barley (Hordeum vulgare L.) endosperms (Urahara *et al.*, 2004). The microsomal fraction transferred Xyl from uridine diphosphoxylose (UDP-Xyl) into exogenous β-(1→4)-xylooligosaccharides derivatized at their reducing ends with 2-aminopyridine. HPLC analysis showed chain elongation of pyridylaminated β-(1→4)-xylotriose (Xyl3-PA) by repeated attachment of one to five single xylosyl residues depending on the reaction time, leading to the formation of Xyl4-8-PA. Methylation analysis and enzymatic digestions with β-xylosidase (EC 3.2.1.37) and endo-β-(1→4)-xylanase (EC 3.2.1.8) confirmed that the transfer of xylosyl residues into the newly synthesized products occurred through β-(1→4)-linkages. The activity of the XylTase was maximal at pH 6.8 and 20 °C and most enhanced in the presence of 0.5% Triton X-100 and 5 mM $MnCl_2$. The enzyme also transferred [14C]Xyl from UDP-[14C]Xyl into higher β-(1→4)-xylooligosaccharides and birchwood xylans through β-(1→4)-linkages. The enzyme activity varied according to the stage of development (7~35 days after flowering) of the endosperms. Maximal activity occurred at 13~16 days; no activity was detectable in mature seeds. A comparison of endosperms from 10 different cultivars of barley harvested 11~22 days after flowering showed no correlation between enzyme activity and the amount of Xyl in the cell walls.

5.2.2.2 Genotypic and environmental variation of arabinoxylans

The level of barley arabinoxylans or pentosans varied somewhat with genotype, but was largely controlled by environmental factors (Henry, 1986). Jadhav *et al.* (1998) reported that arabinoxylan contents of barley grain ranged from 4% to 7%, and small grains contained more pentosans, mainly because they were found in the outer layers of the grain and small grains had relatively less endosperm than larger ones. Han and Schwarz (1996) figured out that arabinoxylans comprised 4%~8% of the barley kernel and were major constituents of both aleurone and endosperm cell walls. Debyser *et al.* (1997)

estimated the water-extractable and total arabinoxylan contents in six barley malts by gas-liquid chromatography. The water-extractable and total arabinoxylan contents of six barley malts varied between 0.49%~0.69%. Centrifuged wort arabinoxylan (0.62%~0.93%) represented 10% to 14% of the total barley malt arabinoxylan. During brewing some monomeric arabinose (0.02%~0.03%) and xylose (0.03%~0.06%) were released. There was a linear relationship between the centrifuged wort arabinoxylan contents and endoxylanase activities in the malts as well as between the levels of arabinoxylan solubilized during brewing and the endoxylanase activities of malts. The levels of monomeric xylose released during brewing were also related in a linear way to the β-D-xylosidase activities. The levels of monomeric arabinose released were not directly related to α-L-arabinofuranosidase activities.

Arabinoxylan is partly solubilized during brewing. Apparently, most of the arabinoxylan in wort originates from the water-extractable part of the malt. Some relations have been found between arabinoxylan degrading enzyme activities of malt and arabinoxylan contents in worts. As endoxylanase activity of malt influences the amount of arabinoxylan in the resulting wort, it may have an impact on the wort and beer characteristics. The malting conditions of barley appear to influence the degradation of arabinoxylan during both malting and brewing. It is unclear at this stage whether endoxylanases have a harmful rather than a beneficial effect during beer production. Although endoxylanases solubilized arabinoxylan which could cause problems such as poor filtration, they may reduce such problems by degrading soluble arabinoxylan to small oligomers.

Han (2000) analyzed the identity and levels of methylated glycosyl components of total non-starch polysaccharide (T-NSP) and arabinoxylan by analysis of partially methylated alditol acetate sugar derivatives using gas chromatography and gas chromatography/mass spectrometry. They found that the arabinoxylan contents in barley and malt flour were 7.1%~8.0% and 3.1%~4.0%, respectively, and there were no statistically significant differences in arabinoxylan and total non-starch polysaccharide levels between all barleys and malt flour and those present in beer ($P<0.05$). Holtekjølen *et al.* (2006) studied the contents of arabinoxylans of 39 barley varieties of diverse genetic origin and from several barley growing areas. It was found that total content of arabinoxylans ranged from 7.4%~15.7%. The variety with the highest amount of T-NSP (Olve) also had the highest amount of total arabinoxylans (T-AX). The lowest amount was found among the hull-less varieties. Both six-rowed and the two-rowed contained more than 12% T-AX, with one exception in each group. The hulled varieties contained a significantly higher amount of T-AX than did the hull-less samples. This can be explained by the presence of the arabinoxylans in the hull, and showed that the amount of hull influences the T-AX level. There was also a significantly higher level of T-AX in the Norwegian hulled two-rowed compared to the foreign hulled two-rowed varieties. Since these were varieties with the same characteristics, but grown in different places, it indicated that growing conditions may also influence the level of T-AX. Also, the amount of insoluble arabinoxylans (I-AX) varied much between the different varieties, with a range of 4.4%~13.7%. Olve had the highest amount, while CDC McGwire had the lowest. Both the Norwegian six- and two-rowed varieties contained high amounts, and showed a small variation in I-AX content. The hull-less had generally little (<7%) I-AX, which was significantly less than the hulled samples. Two varieties, Justina and Otira, contained no or little soluble arabinoxylans (S-AX), while a hull-less sample contained the most (5.0%). The hull-less genotypes also had a wider variation in the content of S-AX. 0%~42.2% were soluble (of the total arabinoxylans) with the hull-less varieties having a significantly higher content of soluble arabinoxylans than the hulled samples.

Choct (1997) indicated that the non-starch polysaccharides in cereal grains were composed mainly of arabinoxylans (pentosans), β-glucans, and cellulose, and the contents of arabinoxylans were 7.4%~15.7%. They also reported that the variety with the highest amount of total non-starch polysaccharide had the highest amount of total arabinoxylans.

Variation in barley arabinoxylan contents was also associated with hull type (Storsley *et al.*, 2003). Andersson *et al.* (1999) found that the hull barley varieties had a significantly higher content of total arabinoxylans than that in the hull-less barleys. Holtekjølen *et al.* (2006) reported that the hulled varieties contained a significantly higher amount of total arabinoxylans than did the hull-less samples. The hulled barleys also had a greatly higher level of insoluble arabinoxylans than did the hull-less barleys (<7%), and the hull-less genotypes had a wider variation in the content of soluble arabinoxylans.

Many factors influence arabinoxylan contents in barley during malting and brewing. Debyser *et al.* (1997) reported that considerable amounts of water-extractable arabinoxylans were solubilized from cell walls and survived during processing and fermentation. Han (2000) reported that germination of barley reduced total arabinoxylan content, and the arabinoxylan content of de-husked barley was reduced by 50% after germination under malting conditions. The malting conditions of barley appeared to influence the degradation of arabinoxylan during both malting and brewing (Debyser *et al.*, 1997). According to Debyser *et al.* (1997), a linear relationship ($R^2 = 0.65$; $n = 6$) existed between the centrifuged wort arabinoxylan contents and endoxylanase activities in the malts as well as on the levels of arabinoxylan solubilized during brewing and the endoxylanase activities of malts ($R^2 = 0.88$; $n = 6$). Li *et al.* (2005) studied water-extractable arabinoxylans (WEAX) of barley during malting and brewing. They found that during the germination period, the content of total arabinoxylan in barley grain partially decreased, while WEAX content increased. A slight change was also observed between total arabinoxylan content and WEAX content after kilning. Most of the arabinoxylan in barley malt was insoluble. WEAX content of 100% malt wort (0.50%~1.01%) was higher than WEAX content of malt (0.42%~0.98%). WEAX contents of beers were lower than the worts. Holtekjølen *et al.* (2006) indicated that the growing conditions might influence the level of total arabinoxylans. For the insoluble and soluble arabinoxylans, the amount of insoluble arabinoxylans varied greatly among varieties, with a range of 4.4%~13.7%.

There are no comprehensive researches on the genetics of arabinoxylans in barley, nor is there any information on the number of genes involved or their location in the genome. Detailed analysis of the genetics of arabinoxylans will be very important for the genetic improvement of arabinoxylans in barley grain.

Acknowledgements

We are deeply indebted to Chinese Natural Science Foundation for its financial support (30630047) in our barley research programs.

References

Agu R C, Palmer G H (2001) The Effect of Nitrogen Level on the Performance of Malting Barley Varieties during Germination. J Inst Brew, 107: 93-98

Ahluwalia B, Ellis E E (1984) A Rapid and Simple Method for the Determination of Starch and β-glucan in Barley and Malt. J Inst Brew, 90: 254-259

Aman P (1986) A Note on the Content of Mixed-linked β-glucans in Swedish Barleys. Swed J Agr Res, 16: 73-75

Aman P, Graham H (1987) Analysis of Total and Insoluble Mixed-linked (1-3),(1-4)- β-D-glucans in Barley and Oats. J Agr Food Chem, 35: 704-709

Andersson A A M, Elfverson C, Andersson R, Regner S, Aman P (1999) Chemical and Physical Characteristics of Different Barley Samples. J Sci Food Agr, 79: 979-986

Aspegren K, Mannonen L, Ritala A, Puupponen-Pimiä R, Kurtén U, Salmenkallio-Marttila M, Kauppinen V, Teeri T H (1995) Secretion of a Heat-stable Fungal β-glucanase from Transgenic Suspension-cultured Barley Cells. Mol Breeding, 1: 91-99

Aspinall G O, Ross K M (1963) The Degradation of Two Periodate-oxidised Arabinoxylans. J Chem Soc, 2: 1681-1689

Balance G M, Hall R S, Manners D J (1986) Studies of Some Arabinoxylans from Barley Endosperm Walls. Carbohydr Res, 150: 290-294

Bamforth C W (1982) Barley β-glucans: Their Role in Malting and Brewing. Brewer Dig, 3: 22-35

Bamforth C W (1985) Biochemical Approaches to Beer Quality. J Inst Brew, 91: 154-160

Bamforth C W, Barclay A H P (1993) Malting Technology and the Uses of Malt. In: Macgregor A W, Bhatty R S (eds.) Barley: Chemistry and Technology. American Association of Cereal Chemistry, St. Paul, MN. 297-354

Bamforth C W, Martin H L (1983) The Degradation of β-glucan during Malting and Mashing: the Role of β-glucanase. J Inst Brew, 89: 303-307

Banik M, Li C D, Langridge P, Fincher G B (1997) Structure, Hormonal Regulation, and Chromosomal Location of Genes Encoding Barley (1→4)-β-xylan Endohydrolases. Mol Gen Genet, 253: 599-608

Barber M G, Jackson E A, Smith D B (1994) Total and Individual Barley (1-3,1-4)- β-D-glucanase Activity in Some Green and Kilned Malts. J Inst Brew, 100: 91-97

Barrett J, Clapperton J F, Divers D M, Rennie H (1973) Factors Affecting Wort Separation. J Inst Brew, 79: 407-411

Bathgate G N (1983) The Relationship between Malt "friability" and Wort Viscosity. J Inst Brew, 89: 416-419

Bathgate G N, Palmer G H, Wilson G (1974) The Action of Endo-β-glucanases on Barley and Malt β-glucans. J Inst Brew, 80: 278-285

Batten G D (1998) Plant Analysis Using Near Infrared Reflectance Spectroscopy: the Potential and the Limitations. Aust J Exp Agr, 38: 697-706

Baur S K, Geisler G (1996) β-Glucan Content in Caryopses of Oat Varieties with Regard to Cultivation Year and Nitrogen Level. J Agron Crop Sci, 176: 5-14

Beer M U, Wood P J, Weisz J (1997) Molecular Weight Distribution and (1$\rightarrow$3)(1$\rightarrow$4)-β-D-glucan Content of Consecutive Extracts of Various Oat and Barley Cultivars. Cereal Chem, 74: 476-480

Bengtsson S, Aman P, Graham H, Newman C W, Newman R K (1990) Chemical Studies on Mixed-linked β-glucans in Hull-less Barley Cultivars Giving Different Hypocholesterolaemic Responses in Chickens. J Sci Food Agr, 52: 435-445

Beresford G, Stone B A (1983) (1-3), (1-4)- β-D-glucan Content of Triticum Grains. J Cereal Sci, 1: 111-114

Bhatty R S (1987) Relationship between Acid Extract Viscosity and Total Soluble and Insoluble β-glucan Contents of Hulled and Hulless Barley. Can J Plant Sci, 67: 997-1008

Bhatty R S, Macgregor A W, Rossnagel B G (1991) Total and Acid-soluble β-glucan Content of Hulless Barley and Its Relationship to Acid-extract Viscosity. Cereal Chem, 68: 221-227

Bichonski A, Pecio A, Radecki-Pawlik A (2003) The Effects of Sowing Density and Nitrogen Fertilization on the Beta-glucan Content of Wort in Barley Cultivar Rudzik. Biuletyn Instytutu Hodowli I Aklimatyzacji Roslin 230: 305-310

Blakeney A B, Flinn P C (2005) Determination of Non-starch Polysaccharides in Cereal Grains with Near-infared Reflectance Spectroscopy. Mol Nutr Food Res, 49: 546-550

Bourne D T, Wheeler R E (1984) Environmental and Varietal Differences in Total β-glucan Contents of Barleys and the Effectiveness of Its Breakdown under Different Malting Conditions. J Inst Brew, 90: 306-310

Brennan C S, Amor M A, Harris N, Smith D, Cantrell I, Griggs D, Shewry P R (1997) Cultivar Differences in Modification Patterns of Protein and Carbohydrate Reserves during Malting of Barley. J Cereal Sci, 26: 83-93

Brunner B R, Freed R D (1994) Oat Grain β-glucan Content as Affected by Nitrogen Level, Location, and Year. Crop Sci, 33, 473-476

Carr J M, Glatter S, Jeraci J L, Lewis B A (1990) Enzymic Determination of β-glucan in Cereal Based-food Products. Cereal Chem, 67: 226-229

Chen J X, Zhang G P (2000) The Effect of Nitrogen and Potassium Fertilizer Application Rates and Timing on Barley Grain Quality. Acta Agriculturae Zhejiangensis, 12: 74-77

Chen J X, Zhang G P (2002) Dynamics of β-glucan Content during Kernel Development and Its Relationship with Meteorological Factors. Chinese Journal of Applied Ecology, 13: 417-420

Choct M (1997) Feed Non-starch Polysaccharides: Chemical Structures and Nutritional Significance. Feed Milling International, 6: 13-26

Coles G D, Jamieson P D, Haslemore R M (1991) Effect of Moisture Stress on Malting Quality in Triumph Barley. J Cereal Sci, 14: 161-177

Coote N, Kirsop B H (1976) A Haze Consisting Largely of Pentosas. J Inst Brew, 82: 34

Cui W, Wood P J, Blackwell B, Nikiforuk J (2000) Physicochemical Properties and Structural Characterization by Two Dimensional NMR Spectroscopy of Wheat-β-D-glucan-comparison with Other Cereal β-D-glucans. Carbohydr Polym, 41: 249-258

Debyser W, Derdelinckx G, Delcour J A (1997) Arabinoxylan and Arabinoxylan Hydrolysing Activities in Barley Malts and Worts Derived from Them. J Cereal Sci, 26: 67-74

Dervilly G, Leclercq C, Zimmermann D, Roue C, Thibault J F, Saulnier L (2002) Isolation and Characterization of High Molar Mass Water-soluble Arabinoxylans from Barley and Barley Malt. Carbohydr Polym, 47: 143-149

Dexter J E, Wood P J (1996) Recent Applications of Debranning of Wheat Before Milling. Trends Food Sci Technol, 7: 35-41

Ellis R P, Swanston J S, Rubio A, Perez-vendrell, Romagosa I, Molina-cano J L (1997) The Development of β-glucanase and Degradation of β-glucan in Barley Grown in Scotland and Spain. J Cereal Sci, 26: 75-82

EtokAkpan O U (1993) Enzymic Degradation and Nature of the Endosperm Cell Walls of Germinating Sorghums and Barley. J Sci Food Agr, 61: 389-393

Fastnaught C E, Berglund P T, Holm E T, Fox G J (1996) Genetic and Environmental Variation in β-glucan Content and Quality Parameters of Barley for Food. Crop Sci, 36: 941-946

Fenwick R (1993) Cereal Cultivar Quality Evaluation in a Changing Agricultural Environment. Aspects Appl Biol, 36: 169-178

Figueroa-Espinoza M C, Rouau X (1998) Oxidative Cross-linking of Pentosans by a Fungal Laccase and Horseradish Peroxidase: Mechanism of Linkage between Feruloylated Arabinoxylans. Cereal Chem, 75: 259-265

Fincher G B (1975) Morphology and Chemical Composition of Barley Emdosperm Cell Walls. J Inst Brew, 81: 116-122

Fincher G B (1989) Molecular and Cellular Biology Associated with Endosperm Mobilisation in Germinating Cereal Grains. Annu Rev Plant Physiol Plant Mol Biol, 40: 305-346

Fincher G B (1992) Cell Wall Metabolism in Barley. In: Shewry P R (ed.) Barley: Genetics, Biochemistry, Molecular Biology and Biotechnology, Biotechnology in Agriculture Series No. 5, C.A.B. International, Wallingford, 413-437

Fincher G B, Stone B A (1986) Cell Walls and Their Components in Cereal Grain Technology. In: Pomeranz Y (ed.) Advances in Cereal Science and Technology, Vol 8. American Association of Cereal Chemistry, St. Paul, MN, 207-295

Forrest I S, Wainwright T (1977) The Mode of Binding of β-glucans and Pentosans in Barley Endosperm Cell Walls. J Inst Brew, 83: 279-286

Garrido A, Pérez-Marín M D, Guerrero J E*et Al* (2002) Near Infrared Reflectance Spectroscopy as an Essential Tool in Food Safety Programmes: Predicting Ingredients in Commercial Compound Feed. In: Davies A M C, Cho R K (eds.) Proceedings of the 10th International Conference on Near Infrared Spectroscopy. NIR Publications, Chichester, West Sussex, UK, 145-150

Genc H, Ozdemir M, Demirbas A (2001) Analysis of Mixed-linked (1→3),(1→4)-β-D-glucans in Cereal Grains from Turkey. Food Chem, 73: 221-224

Gibeaut D M, Pauly M, Bacic A, Fincher G B (2005) Changes in Cell Wall Polysaccharides in Developing Barley (*Hordeum Vulgare*) Coleoptiles. Planta, 221: 729-738

Gómez C, Navarroa A, Manzanare P, Horta A, Carbonell J V (1997) Physical and Structural Properties of Barley (1→3),(1→4)-β-glucan-D-glucan. Part I. Determination of Molecular Weight and Macromolecular Radius by Light Scattering. Carbohydr Polym, 32: 7-15

Greenberg D C (1977) A Dillel Cross Analysis of Gum Content in Barley (*Hordeum Vulgare*). Theor Appl Genet, 50: 41-46

Grimm A, Krüger E, Burchard W (1995) Solution Properties of β-D-(1→3) (1→4)-glucan Isolated from Beer. Carbohydr Polym, 27: 205-214

Gler M (2003) Barley Grain Beta-glucan Content as Affected by Nitrogen and Irrigation. Field Crop Res, 84: 335-340

Han F, Ullrich S E, Chirat S, Menteur S, Jestin L, Sarrafi A, Hayes P M, Jones B, Blake T, Wesenberg D, Kleinhofs A, Kilian A (1995) Mapping of β-glucan Content and β-glucanase Activity Loci in Barley Grain and Malt. Theor Appl Genet, 91: 921-927

Han J P (2000) Structural Characteristics of Arabinoxylan in Barley, Malt, and Beer. Food Chem, 70: 131-138

Han J Y, Schwarz P B (1996) Arabinoxylan Composition in Barley, Malt, and Beer. J Amer Soc Brew Chem, 54: 216-220

Hashimoto S, Shogren M D, Bolte L C, Pomeranz Y (1987) Cereal Pentosans: Their Estimation and Significance. III. Pentosans in Abraded Grains and Milling By-products. Cereal Chem, 64: 39-41

Henry R J (1985) Near-infrared Reflectance Analysis of Carbohydrates and Its Application to the Determination of (1-3), (1-4)-β-D-glucan in Barley. Carbohydr Res, 141: 13-19

Henry R J (1986) Genetic and Environmental Variation in the Pentosan and β-glucan Contents of Barley, and Their Relation to Malting Quality. J Cereal Sci, 4: 269-277

Henry R J (1987) Pentosan and (1-3), (1-4)-β-glucan Concentrations in Endosperm and Wholegrain of Wheat, Barley, Oats and Rye. J Cereal Sci, 6: 253-258

Henry R J (1988) The Carbohydrates of Barley Grains - A Review. J Inst Brew, 94: 71-78

Henry R J (1989) Factors Influencing the Rate of Modification of Barleys during Malting. J Cereal Sci, 10: 51-59

Henry R J, Brown A H D (1987) Variation in the Carbohydrate Composition of Wild Barley (*Hordeum Spontaneum*) Grain. Plant Breed, 98: 97-103

Holtekjølen A K, Uhlen A K, Brøthen E, Sahlstr?m S, Knutsen S H (2006) Contents of Starch and Non-starch Polysaccharides in Barley Varieties of Different Origin. Food Chem, 94: 348-358

Horsley R D (1995) Genetic Diversity in Malt Quality of Northern American Six-rowed Spring Barley. Crop Sci, 35: 113-118

Horvath H, Jensen L G, Wong O T, Kohl E, Uilrich S E, Cochran J, Kannangara C G, Von Wettstein D (2001) Stability of Transgene Expression, Field Performance and Recombination Breeding of Transformed Barley Lines. Theor Appl Genet, 102: 1-11

Humphreys D G, Smith D L, Mather D E (1994) Nitrogen Fertilizer and Seedling Date Induced Changes in Protein, Oil and β-glucan Contents of Four Oat Cultivars. J Cereal Sci, 20: 283-290

Izydorczyk M S, Biliaderis C G (2000) Structural and Functional Aspects of Cereal Arabinoxylans and β-glucans. In: Doxastakis G, Kiosseoglou V (eds) Novel Macromolecules in Food Systems, Elsevier, Amsterdam, pp 361-384

Jadhav S J, Lutz S E, Ghorpade V M, Salunkhe D K (1998) Barley: Chemistry and Value-added Processing. Crit Rev Food Sci Nutr, 38: 123-171

Jensen L G, Olsen O, Kops O, Wolf N, Thomsen K K, Von Wettstein D (1996) Transgenis Barley Expressing a Protein-engineered, Thermostable (1-3,1-4)-β-glucanase during Germination. Appl Biol Sci, 93: 3487-3491

Jorgensen K G, Aastrup S (1986) Analysis of β-glucan in Wort. Monograph-European Brewery Convention, 11: 262-273

Kalra S, Jood S (1998) Biological Evaluation of Protein Quality of Barley. Food Chem, 61: 35-39

Kannauchi M, Bamforth C W (2002) Enzymic Digestion of Walls Purified from the Starchy Endosperm of Barley. J Inst Brew, 108: 73-77

Kato T, Sasali A, Taleda G (1995) Genetic Variation of β-glucan Contents and β-glucanase Activities in Barley, and Their Relationships to Malting Quality. Breed Sci, 45: 476-477

Knuckles B E, Chiu M C M (1999) β-glucanase Activity and Molecular Weight of β-glucans in Barley after Various Treatments. Cereal Chem, 76: 92-95

Knuckles B E, Chiu M C M, Betschart A A (1992) β-glucan-enriched Fractions from Laboratory-scale Dry Milling and Sieving of Barley and Oats. Cereal Chem, 69: 198-202

Lance R C M (1984) Genetic Studies of the Beta-glucan Content of Barley (*Hordeum Vulgare* L.). Dissertation Abstracts International, B (Sciences and Engineering), 45: 742

Lee C J, Horsley R D, Manthey F A, Schwarz P B (1997) Comparisons of β-glucan Content of Barley and Oats. Cereal Chem, 74: 571-575

Lee R C, Burton R A, Hrmova M, Fincher G B (2001) Barley Arabinoxylan Arabinofuranohydrolases: Purification, Characterisation and Determination of Primary Structures from CDNA Clones. Biochem J, 356: 181-189

Lehtonen M, Aikasalo R (1987) β-glucan in Two- and Six-rowed Barley. Cereal Chem, 64: 191-193

Li Y, Lu J, Gu G X, Shi Z P, Mao Z G (2005) Studies on Water-extractable Arabinoxylans during Malting and Brewing. Food Chem, 93: 33-38

MacGregor A M, Dushnicky L G, Schroeder S W, Balance G M (1994) Chances in Barley Endosperms during Early Stages of Germination. J Inst Brew, 100: 85-90

Macnicol P K, Jacobsen J V, Keys M M, Stuart I M (1993) Effects of Heat and Water Stress on Malt Quality and Grain Parameters of Schooner Barley Grown in Cabinets. J Cereal Sci, 18: 61-68

Manzanares P, Sendra M (1996) Determination of Total (1-3), (1-4)-β-D-glucan in Barley and Malt Flour Samples. J Cereal Sci, 23: 293-296

Mather D E, Tinler N A, LaBerge D E, Edney M, Jones B L, Rossnagel B G, Legge W C, Briggs K G, Irvine R B, Falk D E, Kasha K J (1997) Regions of the Genome that Affect Grain and Malt Quality in a North American Two-row Barley Cross. Crop Sci, 37: 544- 554

McCleary B V, Codd R (1991) Measurement of (1-3),(1-4)-D-glucan in Barley and Oats: A Streamlined Enzymic Procedure. J Sci Food Agr, 55: 303-312

McCleary B V, Glennie-Holmes M (1985) Enzymic Quantification of (1-3)-(1-4)-β-D-glucan in Barley and Malt. J Inst Brew, 91: 285-295

McCleary B V, Mugford D C (1997) Determination of Beta-glucan in Barley and Oats by Streamlined Enzymatic Method: Summary of Collaborative Study. J AOAC Int, 80: 580-583

Mcfadden G I, Ahluwalia B, Clarke A E (1988) Expression Sites and Developmental Regulation of Genes Encoding (1-3),(1-4)-β-glucanase in Germinated Barley. Planta, 173: 500-508

Miller S S, Wood P J, Pietrazak L N, Fulcher R G (1993) Mixed Linkage β-glucan, Protein Content and Kernel Weight in Avena Species. Cereal Chem, 70: 231-233

Molina-Cano J L (1979) Effect of Aeration during Sodium Azide Treatment of Barley Seeds on the Physiological Damage in M1 Seedlings. Barley Genet Newsl, 9: 62-65

Molina-Cano J L (1982) Genetic, Agronomic and Malting Characteristics of a New Erectoides Mutant Induced in Barley. Z Pflanzenzuecht, 88: 34-42

Molina-cano J L, Francesch M, Perez-Vendrell A M, Ramo T, Voltas J, Brufau J (1997) Genetic and Environmental Variation in Malting and Feed Quality of Barley. J Cereal Sci, 25: 37-47

Molina-Cano J L, Roca De Togores F, Royo C, Perez A (1989) Fast Germinating Low Beta-glucan Mutants Induced in Barley with Improved Malting Quality and Yield. Theor Appl Genet, 78: 748-754

Morgan A G (1977) The Relationship between Barley Extract Viscosity Curves and Malting Ability. J Inst Brew, 83: 231-234

Morgan A G, Riggs T J (1981) Effects of Drought on Yield and on Grain and Malt Characters in Spring Barley. J Sci Food Agr, 32: 339-346

Narasimhalu P, Kong D, Choo T M, Ferguson T, Therrien M C, Ho K M, May K W, Jiu P (1994) Effects of Environment and Cultivar on Total Mixed-linkage β-glucan Content in Eastern and Western Canadian Barleys (*Hordeum Vulgare* L.). Can J Plant Sci, 75: 371-376

Nilsson M, Aman P, Harkonen H, Hallmans G, Knudsen K, Mazur W, Adlercreutz H (1997) Content of Nutrients and Lignans in Roller Milled Fractions of Rye. J Sci Food Agr, 73: 143-148

Parrish F W, Perlin A S, Reese E T (1960) Selective Enzymolysis of Poly-β-D-glucans, and the Structure of the Polymers. Can J Chem, 38: 2094-2104

Perez-Vendrell A M, Brufau J , Molina-Cano J L, Molina-Cano, Francesch F, Guasch J (1996) Effects of Cultivar and Environment on β- (1-3), (1-4)-D-glucan Content and Acid Extract Viscosity of Spanish Barleys. J Cereal Sci, 23: 285-292

Perez-Vendrell A M, Guasch J, Francesch M, Molina-Canoc J L, Brufau J(1995) Determination of Beta-(1-3),(1-4)-D-glucans in Barley by Reversed-phase High-performance Liquid Chromatography. J Chromatogr A, 718: 291-297

Peterson D M, Wesenberg D M, Burrup J (1991) Genotype and Environment Effects on Oat β-glucan Concentration. Crop Sci, 31: 1517-1520

Podrazky V (ed.) (1964) Some Characteristics of Cereal Gums. Chem. Ind. (Land.), 712-713

Powell W, Caligari P D S, Swanston J S, Jinks J L (1985) Genetical Investigations into β-glucan Content in Barley. Theor Appl Genet, 71: 461-466

Prentice N, Babler S, Faber S (1980) Enzymic Analysis of β-D-glucans in Cereal Grains. Cereal Chem, 57: 198-202

Ragaee S M, Campbell G L, Scoles G J, McLeod J G, Tyler R T (2001) Studies on Rye (Secale Cereale L.) Lines Exhibiting a Range of Extract Viscosities. 1. Composition, Molecular Weight Distribution of Water Extracts, and Biochemical Characteristics of Purified Water-extractable Arabinoxylan. J Agr Food Chem, 49: 2437-2445

Saastamoinen M, Plaami S, Kumpulainen J (1992) Genetic and Environmental Variation in β- Glucan Content of Oats Cultivated or Tested in Finland. J Cereal Sci, 16: 279-290

Sadosky P, Schwarz P B (2002) Effect of Arabinoxylans, Beta-glucans, and Dextrins on the Viscosity and Membrane Filterability of a Beer Model Solution. J Amer Soc Brew Chemist, 60: 153-162

Sanada G (1983) Genetic Analysis of Components if Malting Quality in Barley. Japan J Breed, 33: 283-295

Saulnier L, Gévaudan S, Tribault J F (1994) Extraction and Partial Characterisation of β-glucan from the Endosperm of Two Barley Cultivars. J Cereal Sci, 19: 171-178

Schwarz P B, Han J Y (1995) Arabinoxylan Content of Commercial Beer. J Amer Soc Brewing Chemist, 53: 157-159

Scott R W (1972) The Viscosity of Worts in Relation to Their Content of β-glucans. J Inst Brew, 78: 411-412

Skendi A, Biliaderis C G, Lazaridou A, Izydorczyk M S (2003) Structure and Rheological Properties of Water Soluble β-glucans from Oat Cultivars of Avena Sativa and Avena Bysantina. J Cereal Sci, 38: 15-31

Slakeski N, Baulcombe D C, Devos K M, Ahluwalia1 B, Doan D N P, Fincher G B (1990) Structure and Tissue-specific Regulation of Genes Encoding Barley (1-3),(1-4)-β-glucan Endohydrolases. Mol Gene Genet, 224: 437-449

Stewart D C, Hawthorne D, Evans D E (1998) Cold Sterile Filtration: A Small-scale Filtration Test and Investigation of Membrane Plugging. J Inst Brew, 104: 321-326

Storsley J M, Izydorczyk M S, You S, Biliaderis C G, Rossnagel B (2003) Structure and Physicochemical Properties of β-glucans and Arabinoxylans Isolated from Hull-less Barley. Food Hydrocolloid, 17: 831-844

Stuart I M, Loi L, Fincher G B (1988) Varietal and Environmental Variations in (1-3),(1-4)- β-glucan Levels and (1-3), (1-4) - β-glucanase Potential in Barley: Relationships to Malting Quality. J Cereal Sci, 7: 61-71

Sungurtas J, Swanston J S, Davies H V, McDougall G J (2004) Xylan-degrading Enzymes and Arabinoxylan Solubilisation in Barley Cultivars of Differing Malting Quality. J Cereal Sci, 39: 273-281

Swanston J S (1997) Waxy Starch Barley Genotypes with Reduced β-glucan Contents. Cereal Chem, 74: 452-455

Ullirich S E, Clancy J A, Eslick R F, Lance R C M (1986) β-glucan Content and Viscosity of Extracts from Waxy Barley. J Cereal Sci, 4: 279-285

Urahara T, Tsuchiya K, Kotake T, Tohnooka T, Komae K, Kawada N, Tsumuraya Y (2004) A β-(1→4)-xylosyltransferase Involved in the Synthesis of Arabinoxylans in Developing Barley Endosperms. Physiol Plant, 122: 169-180

Varum K M, Smidsrod O (1988) Partial Chemical and Physical Characterization of (1→3)(1→4)-beta-glucans from Oat (Avena Sativa L) Aleurone. Carbohydr Polym, 9: 103-117

Viëtor R J, Angelino S A G F, Voragen A G J (1993) Structural Features of Arabinoxylans from Barley and Malt Cell Wall Material. J Cereal Sci, 15: 213-222

Voragen A G J, Schols H A, Marijs J, Rombouts F M, Angelino S A G F (1987) Non-starch Polysaccharides from Barley: Structural Features and Breakdown during Malting. J Inst Brew, 93: 202-208

Wallwork M A B, Logue S J, MacLeod L C, Jenner C F (1998) Effects of a Period of High Temperature during Grain Filling on the Grain Growth Characteristics and Malting Quality of Three Australian Malting Barleys. Aust J Agr Res, 49: 1287-1296

Wang J M, Zhang G P, Chen J X, Ding S R (1998) The Variation of Genotypes on Beta-glucan Content in Barleyand Its Relationships with Some Characteristics. J Zhejiang Agr Univ, 24: 253-254

Wang J M, Zhang G P, Chen J X, Wu F B (2004) The Changes of β -glucan Content and β-glucanase Activity in Barley Before and after Malting and Their Relationships to Malt Qualities. Food Chem, 86: 223-228

Welch R W, Leggett J M, Lloyd J D (1991) Variation in the Kernel (1→3), (1→4)-β-D-glucan Content of Oat Cultivars and Wild Species and Its Relationship to Other Characteristics. J Cereal Sci, 13: 173-178

Wood P J (1994) Evaluation of Oat Bran as a Soluble Fibre Source. Characterization of Oat β-glucan and Its Effects on Glycaemic Response. Carbohydr Polym, 25: 331-336

Wood P J, Weisz J (1984) Use of Calcoflour in Analysis of Oat Beta-D-glucan. Cereal Chem, 61: 73-75

Wood P J, Weisz J, Blackwell B A (1994) Structural Studies of (1-3),(1-4)-β-D-glucans by 13C-NMR and by Rapid Analysis of Cellulose-like Regions Using High-performance Anion-exchange Chromatography of Oligosaccharides Released by Lichenase. Cereal Chem, 71: 301-307 Item Wood P J, Weisz J, Fedec P, Burrows V D (1989) Large-scale Preparation and Properties of Oat Fractions Enriched in (1-3),(1-4)- β-D-glucan. Cereal Chem, 66: 97-101

Wood P J, Weisz J, Mahn W S (1991) Molecular Characterization of Cereal β-glucans. II. Size-exclusion Chromatography for Comparision of Molecular Weight. Cereal Chem, 68: 530-536

Woodward J R, Fincher G B (1983) Water-soluble β-glucans. Fine Structure, Solution Behaviour and Organization in the Cell Wall. Brew Dig, 58: 28-32

Woodward J R, Phillps D R, Fincher G B (1983) Water-soluble (1-3),(1-4)-β-D-glucans from Barley (Hordeum Vulgare) Endosperm. I. Physicochemical Properties. Carbohydr Polym, 3: 143-156

Yamashita H, Fujino S, Aoyagi S, Tsumura Y, Hayase F, Kato H (1989) Characterization and Mechanisms of Formation of Frozen Beer Precipitates. J Am Soc Brew Chem, 47: 77-81

Yoon S H, Berglund P T, Fastnaught C E (1995) Evaluation of Selected Barley Cultivars and Their Fractions for β-glucan Enrichment and Viscosity. Cereal Chem, 72: 187-190

Yu Z L, Yang Y F (1987) Genetic Analysis of the Gum Content in Barely. Acta Genetics Sinica, 14: 93-199

Zhang G P, Chen J X, Wang J M, Ding S R(2001) Cultivar and Environmental Effects on β-glucan and Protein Content in Malting Barley. J Cereal Sci, 34: 295-301

Zhang G P, Wang J M, Yin C, Ding S R (2002) The Variation of Barley β-glucan and Protein Content as Affected by Production Practices. Agricultural Sciences in China, 10: 1162-1167

6

The Properties and Genetics of Barley Malt Starch Degrading Enzymes

D. E. Evans[1], C. Li[2], J. K. Eglinton[3]
[1]TIAR, University of Tasmania, Private Bag 54, Hobart,Tasmania, 7001, Australia
[2]Dept of Agriculture & Food Western Australia, 3 Baron-Hay Court, South Perth, WA 6151, Australia
[3]School of Agriculture and Wine, University of Adelaide, Glen Osmond, SA, 5064, Australia
Corresponding author: D.E.Evans, eevans@postoffice.usta.edu.au

6.1 Introduction

The properties and quality of barley malt starch degrading enzymes are of primary importance to the efficiency and profitability of brewing (beer and whiskey), and the bio-fuel (bio-ethanol) industries. The barley starch degrading enzymes hydrolyse starch into fermentable sugars that yeast converts into alcohol. This process is key for the alcohol producing industries as the starch substrate makes up approximately 60% of grain weight (Holtekjolen *et al.*, 2006). Malted barley is the main source of the diastase or diastatic power (DP) enzymes that hydrolyse starch. The DP enzymes comprise the combined activity of α-amylase, β-amylase, α-glucosidase and limit dextrinase whose concerted action hydrolyse the α-(1,4) and α-(1,6) glucosyl linkages in starch (Fig. 6.1) into fermentable sugars (i.e., glucose, maltose, etc.), dextrins and limit dextrins. The actions of the DP enzymes are summarised as follows:

(1) α-amylase cleaves α-(1,4)-linkages internally (endo-acting) to primarily produce oligosaccharides, limit dextrins and some fermentable sugars;
(2) β-amylase cleaves α-(1,4)-linkages from the non-reducing ends (exo-acting) to produce maltose;
(3) Limit dextrinase hydrolyses internal α-(1,6)-linkages (endo), to remove branch-points in amylopectin or α-limit dextrins;

(4) α-Glucosidase primarily cleaves α-(1,4)-linkages from the non-reducing ends to produce glucose.

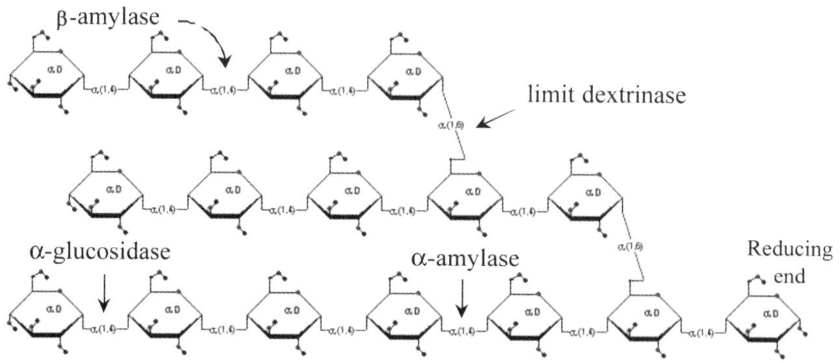

Fig. 6.1. Schematic representation of starch hydrolysis

Generally it is considered that α-amylase, β-amylase and limit dextrinase are of most importance in determining starch degrading potential while in the absence of contrary evidence it is considered that the contribution of α-glucosidase is minimal (Bamforth, 2003).

The efficiency of the alcohol production process in brewing is governed by the amount of soluble material extracted during the mashing stage of brewing (termed extract) and the fermentability of that extract. Brewers extract includes fermentable sugars, dextrins, amino acids, proteins, lipids, non-starch polysaccharides, minerals and other substances. The extent to which yeast can metabolise the extract to produce alcohol, in turn determines the fermentability or attenuation. The fermentability of the wort is not only determined by the availability of fermentable sugars in the wort but also the content of amino acids and other nutritional components required for yeast vigour and biomass production. For the production of whiskey the ideal outcome is that all starch is converted into fermentable sugars that can be converted into alcohol by yeast as the "mash" is subsequently distilled to extract the alcohol. The yield of alcohol, or "spirit yield", is a critical malt characteristic of the production efficiency of whiskey. In brewing beer, however, the extent of starch hydrolysis and the extent of wort fermentability are typically optimised to the style of beer being produced. This is because the residual extract contains components such as dextrins and limit dextrins in part determine consumer perception of beer body or mouth-feel (Langstaff and Lewis, 1993; Bamforth, 2001).

Barley malt is a relatively expensive ingredient in the brewing process as it is the product of controlled barley germination and kilning (heat drying) that typically takes 6~8 days and consumes substantial amounts of energy and water. As such, many conventional brewing styles (e.g. Asia and north America) use unmalted starch adjuncts such as barley, wheat, maize or rice that are added to the mash at levels up to 60% or even > 75% (of malt charge) for the Happoshu style in Japan (Brandee *et al.*, 1999; Haywood, 1996). These are added not only as a cheap source of extract, but also give brewers control over attenuation, mouth-feel, flavour, foam head retention or are the result of constraints in local availability. As these adjuncts contain starch but very limited levels of the starch degrading enzymes, it is incumbent on the malt to supply extra levels of the DP enzymes to efficiently degrade this extra starch. Consequently, malt with high to very high levels of starch degrading capacity is typically required for brewing styles using unmalted starch adjuncts. In other regions, such as Australia, liquid sugar adjunct, typically 15%~25% of the malt charge, may be used by the brewer. As liquid sugar (sucrose) is yeast fermentable, lower levels of malt DP enzymes are required by the brewers using these brewing practices (e.g. Australia) so that the desired level of attenuation and beer body are produced.

It is the objective of this chapter to provide an overview of the quality parameters of barley malt that determine the potential for starch hydrolysis and the subsequent fermentability of the wort produced. Particular focus will be directed to the genetic determinants of the level and attributes of the DP enzymes.

6.2 The Substrate: Starch

The quality and characteristics of the starch substrate is fundamental to understanding how efficiently it may be hydrolysed into fermentable sugars by the DP enzymes. Perspectives on starch composition and structure have been reviewed by Tester *et al.* (2004), while a more brewing orientated outlook is discussed by Bamforth (2003). As such the discussion of starch characteristics in this chapter will be limited to a short overview that is a necessary basis for an understanding of its hydrolysis to fermentable sugars.

6.2.1 Starch Structure

Starch consists of two large polysaccharides, amylose and amylopectin. Amylose is an essentially linear molecule comprising α(1,4) glucosyl linkages. Its molecular weight is typically in the range 10^5 to 10^6 (i.e. 500~5,000 glucose residues) with a small number (9~20 per molecule) of α(1,6) branches. The structure of amylopectin is much more complex than that of amylose in that it contains three types of α(1,4) glucosyl linked chains; A, B and C, that are linked via α(1,6) branches to form a molecule of weight in the order of 10^7 to 10^8 (Bamforth, 2003). The C chain contains approximately 15~45 glucose residues and contains the sole reducing end. The B chains branch from the C chain or from each other and are also of 15~45 glucose residues in length.

Finally, the unbranched A chains terminally branch from the B chains, in chains for approximately 15 residues. Bamforth (2003) likens this structure to the appearance of a "weeping willow". Barley breeders have selected lines that have either low (waxy) or high proportions (high amylose) of amylose with respect to amylopectin. Typically in normal varieties that are used for brewing, the proportion of amylose is 25%~30% of total starch (Bamforth, 2003).

A properly malted barley endosperm essentially consists of a "bag" of starch granules that are composed of amylose and amylopectin. There is a bimodal distribution of large lenticular A granules (diameter 15~25 μm) and small spherical B granules (diameter <10 μm). The small granules account for 80%~90% of the total but only 10%~15% of the total weight of starch (Oliveira *et al.*, 1994). Sun and Henson (1991) observed that α-amylase, and to a lesser extent α-glucosidase, were able to slowly hydrolyse the surface of the intact and un-gelatinised starch granules. For brewing, starch hydrolysis typically approaches completion in less than an hour during the mashing phase where ground malt (grist) is mixed with hot water (typically 65°C) conventionally in a ratio of approximately 1:3. This quantum leap in the efficiency of starch hydrolysis over that which occurs during normal seedling growth, is the direct result of starch gelatinisation in which the molecular order in the starch granule collapses when heated beyond a critical temperature so that the granules absorb water and swell to many times their original size. As a consequence, the starch granule structure is "opened up" to allow the efficient access of the DP enzymes to undertake starch hydrolysis. This process is normally referred to as "pasting" or "gelatinisation".

6.2.2 Starch Gelatinisation

Starch gelatinisation temperature can be determined by a number of methods, all of which have some impact on the gelatinisation temperature recorded. Traditional methods include observance of the loss of granule birefringence or Congo red absorption during heating (Watson, 1964). Currently the most widespread and accepted method is differential scanning calorimetry (DSC) (Lund, 1984). Typically, the peak temperature of gelatinization or T_p is reported. As DSC analysis is relatively expensive, some investigators have recently assessed the viability of using the relatively common rapid viscoanalyser (RVA) to estimate the likely mashing starch gelatinisation temperature (i.e., Kessler *et al.*, 2005).

In general, most barley malt starch gelatinises between 59°C and 63°C (Table 6.1). This observation is consistent with the general perception of most brewers that *"mash conversion temperatures of around* 65°C *will comfortably ensure the gelatinisation of starch"* (Bamforth, 2003). It is perhaps not surprising that when starch gelatinisation temperatures exceed 64°C~65°C, sub-optimal levels of wort fermentability begin to be noted by brewers (Stenholm *et al.*, 1996, 1998; Bekkers *et al.*, 2007). The growing environment of the barley crop, in particular higher temperatures during grain fill, have been observed to lead to higher starch gelatinisation temperatures (Myllarinen *et al.*, 1998; Tester *et al.*, 1991). More recently, Ao and Jane (2007) reported that this observation may be the result of different proportions of A and B granules that were found to contain varying levels of amylose and amylopectin.

Table 6.1. Reported starch gelatinisation temperatures as summarised by Bamforth (2003)

Starch	Gelatinization Temperature (°C)	
Normal	Lowest	52
	Highest	60~68
	Majority	59~63
High amylose		64~66
Waxy	Lowest	54.1
	Highest	67.5
	Majority	61~65

6.3 The Relationship between Malt DP Enzymes and Fermentability

It has been well-known that malt DP enzymes have a great impact on fermentability of barley grains when they are malted.

6.3.1 Measurement of DP Enzymes

In defining the trading specifications for brewer malt, brewers have conventionally relied on the DP quality parameter to enable them to gauge a malts starch degrading capacity. Brewers have used the diastase measure for over 100 years. It was one of the five methods included in the very first Institute of Brewing's (IoB) methods of analysis (Anon, 1906). Before this official recognition, the term was coined by two French chemists in 1833 (Payen and Persoz, 1833). Later Lintner (1886), described a method for estimating the "diastatic properties" of malt which was based on observations made by Kjeldahl (1879). Not surprisingly, equivalent methods for determining DP feature prominently in the official methods of analysis of each of the three main international brewing organizations, American Society of Brewing Chemists (ASBC, 1992), European Brewery Convention (EBC, 1998) and IoB (1998). These procedures are relatively inexpensive and easy to conduct in malt quality analysis laboratories. Although the DP specification generally gives an indication of potential malt starch degrading capacity, brewers are increasingly loosing confidence in the value of the DP parameter. To illustrate brewers concerns, a series of commercial case studies were presented in Evans *et al.* (2007) which provides examples of where the DP specification was of questionable value or in some cases downright misleading.

6.3.2 Measurement of Fermentability/AAL

As such, brewers have been increasingly requesting extra measures to predict brewery fermentation performance such as fermentability and α-amylase activity over the past 20 years. The determination of α-amylase activity is relatively straight forward, while the measurement of fermentability, or apparent attenuation limit (AAL), seeks to emulate the brewing process on the small scale. That is a rapid fermentation of a small sample of wort (100~200 mL) in less than 24 h at 20°C with agitation, using excess yeast. The mash temperature program for the widely used Congress mashing protocol to produce wort for fermentation is presented in Fig. 6.2. The "Congress" mash is an example of a temperature controlled stepped mashing program that was developed in Europe in the late 19th century to emulate brewing practices of the time. This typically entailed a decoction-mashing regime that used relatively poorly modified malts. Mashing in at 45°C, often termed the "protease or β-glucanase rest", ensures the completion of the modification process of malting. In contrast, the "modified 65°C" or 65°C "infusion" mash is an example of a mashing program more typical of modern brewing practices that use better modified malt.

A desirable attribute of the methods for determination of AAL are that they entail a practical trade off between the mash temperature needed to gelatinise starch and the intrinsic thermostability or mash durability of the DP enzymes. There have been many early reports that the thermostability of purified solutions of β-amylase and limit dextrinase are in the range 40°C~60°C (for summary see Bamforth, 2003). Given that the gelatinisation temperature of starch is between 59°C~63°C (Table 6.1), it is initially difficult to understand how these enzymes could have a substantive role in starch hydrolysis in brewing. However, the dilute concentration of the purified solutions does not imitate the high solute concentrations in the brewers mash where the ratio of the malt grist to water is typically between 1:3 to 1:4. It is well known that protein thermostability is increased by the presence of sugars and other solute components (Back *et al.*, 1979; Arakawa and Timasheff, 1982). Thus high solute concentrations during mashing provide some thermal protection to limit dextrinase and β-amylase to enable these enzymes to have a substantive role during mashing. More recently Henson *et al.* (2008) provided evidence that higher solute concentrations in the mash improve fermentability.

A more realistic evaluation of the thermostability of α-amylase, limit dextinase and β-amylase is depicted in Fig. 6.3. Under simulated mashing conditions at 65°C, α-amylase activity remains stable and high through 60 min, while limit dextrinase and β-amylase activity are relatively thermolabile. Both limit dextrinase and β-amylase are rapidly inactivated within the first 30 min

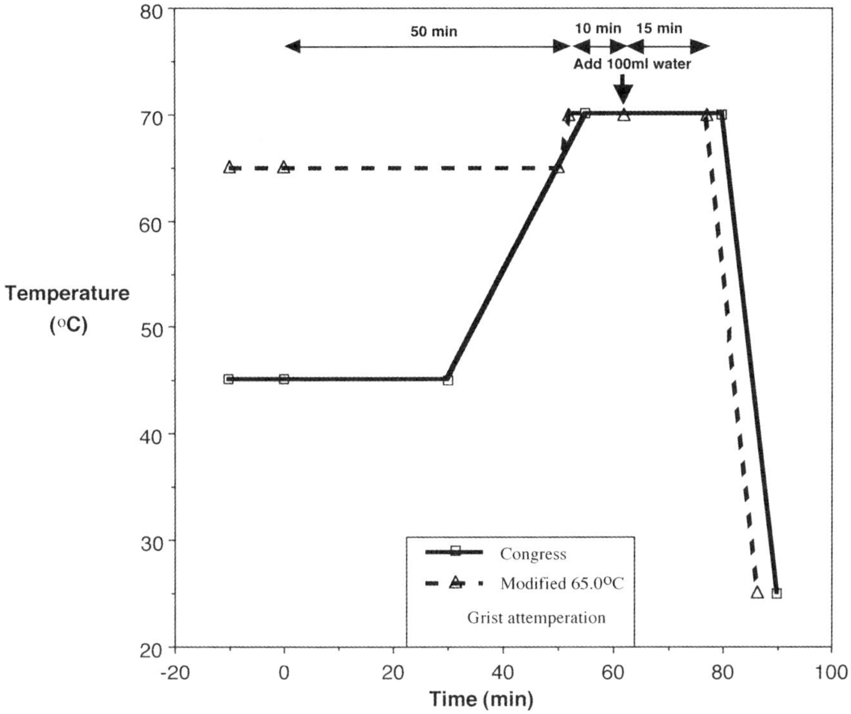

Fig. 6.2. Comparison of the laboratory scale temperature programs for the EBC "Congress" (method 4.5.1, EBC, 1998), equivalent ASBC (method Malt/4, ASBC, 1992), with the modified infusion at 65°C program described by Evans *et al.* (2005). The malt was milled at 0.2mm in a disc mill and mashed in at water ratios, 1:4 initial and 1:6 final

of mashing. It appears that the level of limit dextrinase activity is higher than for β-amylase but this is most likely an over estimate as it was later shown that the maltodextrin hydrolysis products from its action and other DP enzymes boost apparent activity (MacGregor *et al.*, 2002b) as measured by the Limit DextriZyme method (dyed pullulan substrate) used by Sjöholm *et al.* (1995). Similarly, with the Ceralpha assay, high concentrations of small carbohydrates or substrate could potentially lower the activity measured by this assay (Baks *et al.*, 2006a, 2006b). As such the level of activity measured during mashing is likely to be underestimated in Fig. 6.3.

It follows that for starch hydrolysis, mash temperature is a compromise between the requirement for starch gelatinization and the loss of DP enzyme activity, particularly β-amylase and limit dextinase. Evans *et al.* (2005) showed that the optimum compromise temperature for mashing is approximately 65°C (Fig. 6.4) which is consistent with brewers received wisdom that mash conversion temperatures are around 65°C (Bamforth, 2003) and earlier studies

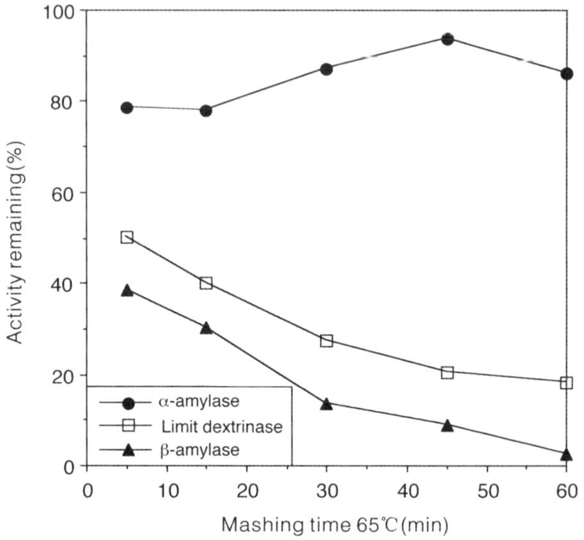

Fig. 6.3. Percentage of α-amylase, β-amylase and limit dextrinase remaining in the liquid phase of the mash during isothermal mashing at 65°C, Congress coarse grist, 1:4 grist to water ratio and pH 5.5. Activity remaining calculated as a percentage of the maximal activity detected in a similar 55°C isothermal mash. Data extracted from Sjøholm *et al.*(1995).

that suggested that this optimum was in the vicinity of 60°C~65°C (Hamilton and Lewis, 1974; Moll *et al.*, 1981; Piendl, 1973; Stenholm and Home, 1999). Interestingly, Fig. 6.4 also shows that as a brewer changes mashing in temperature from 45°C~76°C, perhaps as an option to control attenuation, the relative proportions of the fermentable sugars change quite drastically. Such changes in the sugar spectrum, particularly the proportion of glucose and fructose to maltose, have been shown to influence yeast metabolism and the subsequent ester content and flavour of the beer (Verstrepen *et al.*, 2003).

Despite the attractions of the methods determining fermentability, they are relatively time consuming and suffer from the use of mashing, lautering and fermentation conditions that are not typical of commercial production (Schwarz *et al.*, 2007). In addition, it is also well known that different yeast strains attenuate wort differently (Dowhanick, 1999; Lewis and Young, 2001; Evans and Hamet, 2005) and that the way the malt is milled (particle size) also impacts on starch hydrolysis (Mousia *et al.*, 2004; Edney *et al.*, 2004). Similarly, wort oxygenation, cellar temperature programs and tank pressure also differ. Although more predictive than DP, Goldsmith (2007) recently observed that even the determination of AAL may not always provide an accurate prediction of malt fermentability performance.

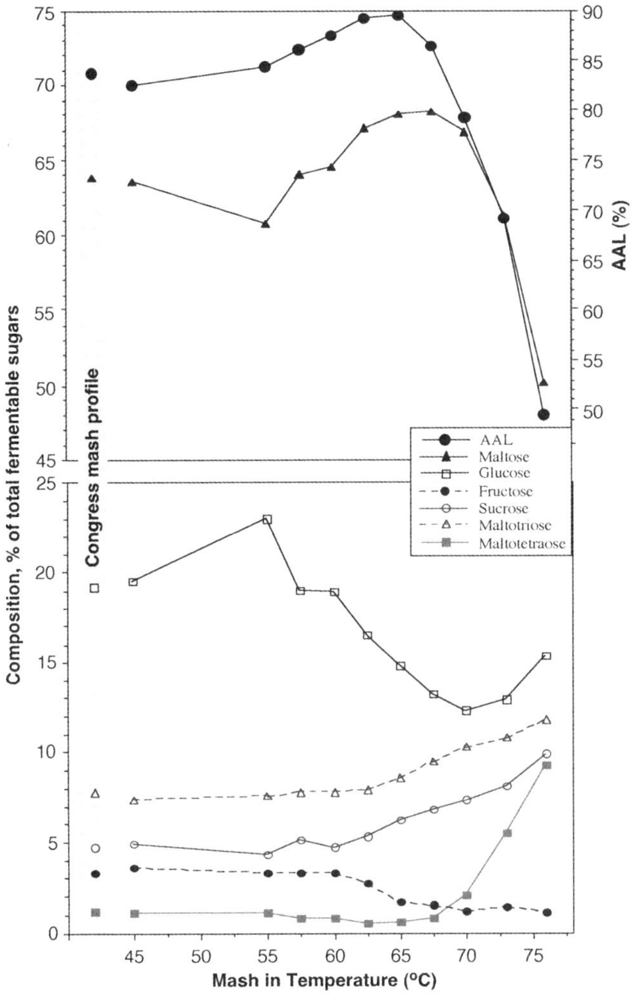

Fig. 6.4. Fermentable sugar profiles mashed in at 10 different temperatures and compared with the Congress mash and apparent attenuation limit (AAL). (Data extracted from Evans *et al.*, 2005, and as compiled by Palmer 2006)

6.3.3 Prediction of Fermentability

These understandings have turned brewers and malt quality researcher's attention to predict potential malt fermentability from the activities and attributes of the starch degrading DP enzymes. Although DP is considered an overall measurement of potential starch degrading activity, most investigations have observed that DP is essentially correlated to β-amylase activity (Arends *et al.*, 1995; Delcour and Verschaeve, 1987; Erdal *et al.*, 1993; Gibson *et al.*, 1995; Santos and Riis, 1996). Other studies have found significant correlations with limit dextrinase (Arends *et al.*, 1995) or all four DP enzymes (Evans *et al.*, 2005). It has also been observed that small increases in the thermostability of the β-amylase enzyme are associated with an increase in fermentability such as with the Sd2H or A type β-amylase (Eglinton *et al.*, 1998; Kihara *et al.*, 1998). More recently, Evans *et al.* (2005) estimated that the impact of malt containing the Sd2H enzyme resulted in a commercially important 2% point increase in AAL. However, malt that contains very high levels of β-amylase activity can also compensate for the absence of a more thermostable β-amylase type (Edney *et al.*, 2007a).

The principal fermentable sugar of wort is maltose which typically comprises >60% of all fermentable sugars. Maltose is the product of β-amylase action, which along with the correlation with DP and the impact of β-amylase thermostability provides a tempting *prima facie* case for the central importance of this enzyme in determining wort fermentability (Eglinton *et al.*, 1998; Gunkel *et al.*, 2002; Kihara *et al.*, 1998). It is acknowledged that both α-amylase (MacGregor *et al.*, 1994) and limit dextrinase (Enevoldsen and Schmidt, 1973) are also capable of producing maltose as part of their action, albeit at lower efficiency. As indicated by a β-amylase-free barley line from Tibet (Kihara *et al.*, 1999), wort produced from malt of this β-amylase-free line (extract = 81.8, AAL = 58.3%, DP = 36°WK) by the Congress AAL protocol contains 48% maltose compared to 73% for Haruna Nijo (Sd2H, extract = 83.3, AAL = 83.9%, DP = 351°WK) when comparably mashed (Kaneko *et al.*, 2000). This mashing experiment confirms that α-amylase and limit dextrinase are capable of producing maltose under mashing conditions.

The other school of thought is that limit dextrinase, not β-amylase, is the more important enzyme in determining resultant wort fermentability (MacGregor *et al.*, 1999; Sjöholm *et al.*, 1995; Stenholm *et al.*, 1996; Stenholm and Home, 1999). The basis for this hypothesis is the estimation that during mashing only 75%~80% of starch is converted into fermentable sugars, while 20%~25% of the remaining hydrolysis products are small α-(1,6) branched dextrins that are carried through into the finished beer (Enevoldsen and Bathgate, 1969). This concept was supported by experiments modeling the contribution of the DP enzymes (MacGregor *et al.*, 1999) and observations of the progression of starch hydrolysis during mashing with respect to enzyme temperature optimums, activity and pH (Sjöholm *et al.*, 1995; Stenholm *et al.*, 1996; Stenholm and Home, 1999). More recently, Evans *et al.* (2005) confirmed in a series of small scale mashing experiments that increased levels of limit dextrinase can potentially result in a 2%~4% point increase in fermentability.

It would thus appear that the remaining two DP enzymes may not have an important impact on determining final wort fermentability. Bamforth (2003) contends that "*there is no evidence to contradict the long-held appreciation that α-amylase is not limiting in its activity*". Dextrin profiles in wort and beer suggest that α-amylolysis of starch is usually complete by the end of mashing (MacGregor, 1995). It has also been demonstrated that the level of

α-amylase in most malts is sufficient to enable maximum production of soluble dextrins not only from the starch of the malt itself, but also from the starch of relatively large proportions of adjuncts (Norris and Lewis, 1985). To date, the impact of α-glucosidase on mashing is thought to be minimal in the absence of information to the contrary (Bamforth, 2003). However, Muslin *et al.* (2002, 2003) suggested that the inclusion of thermostable, recombinant α-glucosidase could increase the yield of fermentable sugars and fermentability. However, in brewing this change would presumably change the proportion of glucose in the wort, thus influence yeast metabolism and the subsequent ester content and flavour of the beer (Verstrepen *et al.*, 2003).

The relative importance of the starch degrading enzymes in producing fermentable extract has recently been satisfactorily reconciled. In a series of holistic experiments, Evans *et al.* (2005), showed that the prediction of malt AAL could be improved from around 50% of the variation with DP to around 90% by using an algorithm (Eq.1), that combined the activities of limit dextrinase, α-amylase, β-amylase, and β-amylase thermostability which also included malt KI as a parameter. The levels of the α-amylase, β-amylase and limit dextrinase were measured by the specific Megazyme Ceralpha, Betamyl and Limit DextriZyme assay substrates, respectively (McCleary and Sheehan, 1987; McCleary and Codd, 1989; McCleary, 1992). The measurement of total β-amylase and limit dextrinase was favoured in this investigation as the levels of the free forms of these enzymes were slightly less predictive. The measurement of the free and total forms of these enzymes will be defined later in this chapter.

$$\begin{aligned} AAL &= 69.9 + 0.017 \times a + 0.0096 \times b + 0.195 \times c \\ &\quad + 0.007 \times d + 0.538 \times e - 0.001 \times d \times e \\ r^2 &= 0.91 \end{aligned}$$

Eq.1 The parameters for the DP enzyme multi linear regression equation (MLR) are:

$$\begin{aligned} a &= \alpha - amylase\ activity\ (\mathrm{U/g}) \\ b &= Total\ limit\ dextrinase\ activity\ (\mathrm{U/kg}) \\ c &= KI\ (\%) \\ d &= Total\ \beta - amylase\ activity\ (\mathrm{U/g}) \\ e &= \beta - amylase\ thermostability\ (\%) \end{aligned}$$

This algorithm conformed to the projected biochemical roles of these parameters in starch hydrolysis during mashing. These roles may be summarised as follows:

- α-amylase: Primary starch attack, cleaving internal α-(1,6)-glycosidic bonds to produce substrates for β-amylase and limit dextrinase.
- Limit dextrinase: α-(1,6)-glycosidic bond cleavage to produce fermentable sugars and substrates for β-amylase.

- β-amylase activity: cleavage of α-(1,6)-glycosidic bonds to produce the primary fermentable sugar maltose.
- β-amylase thermostability: maintain β-amylase activity during mashing at temperatures above 60-65°C. Temperatures that are required for starch gelatinisation to achieve rapid starch hydrolysis.
- KI (%): access of DP enzymes to starch or solutes and proteins in the mash liquor that protect the thermolabile enzymes (Henson *et al.*, 2008).

This holistic concept that the activities of α-amylase, β-amylase and limit dextrinase has been further supported by the analysis of pilot brewing trials conducted by Pilot Brewing Australia and case studies of the commercial brewing performance of a series of malts (Evans *et al.*, 2007). This investigation also emphasized that these DP enzymes were equally important and to some extent mutually substitutable, so that it is the balance of these enzymes that is the key to determining final wort fermentability. Although starch gelatinisation temperature was not found to be an important factor in these studies, it was noted that starch T_p in excess of 64°C could result in reduced levels of wort fermentability (Bekkers *et al.*, 2007; Stenholm *et al.*, 1996, 1998) as a result of higher temperatures during grain fill (Myllarinen *et al.*, 1998; Tester *et al.*, 1991). Finally, recent research has shown that the barley endosperm needs to be suitably modified to enable efficient starch release and gelatinization, probable release of glucose from β-glucan and in some cases, suitable levels of FAN for yeast nutrition (Edney *et al.*, 2004, 2007b). These factors have also been linked with the efficiency with which fermentable extract is converted into alcohol (Edney *et al.*, 2007b; Evans *et al.*, 2005), a small but important factor in brewery profitability.

Overall, these insights into the determinants of wort fermentability provide encouragement and specific targets for barley breeders seeking to optimise the target fermentability of the varieties they are developing. Instead of relatively ill-defined, multi gene characters such as DP and fermentability, characters such as the activity and thermostability of β-amylase, α-amylase and limit dextrinase that are presumably controlled by a limited number of highly heritable genes are provided as selection targets. As such, the remainder of this chapter will be devoted to the further discussion of the biochemical attributes and their genetic controls of these enzymes.

6.4 The DP Enzymes

Several hydrolytic enzymes contribute to DP, including α-amylase, β-amylase, limit dextrinase and -glucosidase. Among these, β-amylase is considered the principal enzyme responsible for diastatic power.

6.4.1 β-amylase

β-amylase (α-1,4-glucan maltohydrolase; EC 3.2.1.2) is a key enzyme in the hydrolysis of starch in germinating barley grains. β-amylase catalyses the liberation of β-maltose from the non-reducing ends of starch and related 1,4-α-glucans. Of the DP enzymes, the genetics and biochemistry of β-amylase is best understood. β-amylase is synthesised during the development of the barley grain (Hardie, 1975) and responds positively to nitrogen nutrition in parallel to the hordein storage proteins (Giese and Hopp, 1984; Giese and Hejgaard, 1984). β-amylase is one of the major proteins present in mature grains accounting for 1%~2% of total protein in the starchy endosperm (Hejgaard and Boisen 1980; Evans *et al.*, 1997a). The very low level of β-amylase in the mutant barley Riso 1508 and the Tibetan β-amylase-less lines suggests that little β-amylase is required for successful germination and seedling growth under standard conditions (Kreis *et al.*, 1987; Kihara *et al.*, 1999).

β-amylase is present in three fractions in the mature grain. One fraction, the active or "free" enzyme is water soluble, the 'bound' form can be extracted in the presence of reducing agents or proteolytic enzymes (Sallans and Anderson, 1940; Sandegrene and Klang, 1950; Bendelow, 1964), and the "latent" fraction can be extracted with detergent (SDS) in combination with a reducing agent (Evans *et al.*, 1997b). The bound form of β-amylase has been referred to as inactive (Bendelow, 1964; Hejgaard, 1976), however it does possess limited activity against soluble starch (Hara-Nishimura *et al.*, 1986; Sopanen and Lauriere, 1989). Total β-amylase is the sum of the free, latent and bound fractions. After germination or malting, little of the latent form can be detected with most (~90%) β-amylase in the free form (Evans *et al.*, 1997b). Under relatively standard malting conditions between 10%~50% of the β-amylase activity in green malt is destroyed during kilning (Runkel, 1975; Bamforth, 1986; Sjöholm *et al.*, 1995; Evans *et al.*, 1997b). In commercial malt typically 78%~99% of β-amylase is in the free form (Evans *et al.*, 2005). Immunocytochemical investigations have localised the enzyme to the endosperm and mid-region aleurone, as major and minor sites respectively (Shen-Miller *et al.*, 1991), and biochemical analysis has shown the periphery of starch granules to be the main site of bound β-amylase deposition (Hara-Nishimura *et al.*, 1986).

In the early investigations, the measurement of barley β-amylase activity was problematic as its action needed to be separated from other amylolytic enzymes such as α-amylase that also produce reducing sugars as a product of their action. In un-germinated barley grain this was less of a problem as α-amylase is only expressed after germination. However, with β-amylase activity measurement in malt, there are substantial difficulties in activity assay by traditional methods that are based on the measurement of reducing sugars. Typically, total amylolytic production of reducing sugars was measured and compared with the activity recorded when either the extract had been heated, had been treated with mercuric chloride that "selectively" inhibits β-amylase function (Meredith, 1970; Etokakpan and Palmer, 1990) or ammonium oxalate that selectively inhibits α amylase (Delcour and Verschaeve, 1987). These methods were cumbersome and suffered from doubts about their veracity. The solution was to use the p-nitrophenyl maltopentose, a substrate that is essentially specific to β-amylase and which yields the p-nitrophenol product that can easily be measured spectrophotometrically (McCleary and Codd, 1989). The PNPG5 substrate is poorly cleaved by α-amylase as its kinetics of binding are very poor on glycosyl chains with a degree of polymerisation (dp) of less than five, while the substrate is also resistant to α-glucosidase hydrolysis. Using the Betamyl assay total β-amylase activity (free + bound) is then measured by inclusion of at least 100mM of cysteine or β-mercapto ethanol reducing agent (Erdal *et al.*, 1993; Evans *et al.*, 1997a). The level of free β-amylase is measured by extraction without a reducing agent.

6.4.1.1 Biochemistry of β-amylase

There are two forms of β-amylase expressed in barley, delineated by their spatial and temporal expression. Two loci encoding barley β-amylase have been identified. The endosperm specific form of β-amylase is the dominant form whose presence and high activity appears limited to the *Triticeace* tribe (i.e. barley, wheat, rye, Ziegler *et al.*, 1999). It is encoded by the *Bmy1* locus on the long arm of chromosome 4H (Powling *et al.*, 1981; Nielsen *et al.*, 1983), that gives rise to a 535 amino acid, 59.7 kDa protein (Kreis *et al.*, 1987; Yoshigi *et al.*, 1994). Analysis of the primary structure of the barley endosperm β-amylase shows that the carboxy-terminal region contains four glycine rich repeats and a cysteine residue at position 503 which may be involved in interactions with protein Z and other cellular components (Kreis *et al.*, 1987; Lundgard and Svensson, 1987). The removal of these structures from the C-terminal region of the protein is presumably one of the key processes in converting the bound form of the enzyme into the free form.

The *Bmy2* locus on chromosome 2H encodes the ubiquitous form of the enzyme (Kreis *et al.*, 1988) for a 505 amino acid protein with a molecular mass of 57 kDa, lacking the glycine-rich repeats found in the C-terminal region of the endosperm form (Jung *et al.*, 2001). This form exhibits only a transient presence in the developing kernel, and is found in other seedling tissues including leaves and roots (Shewry *et al.*, 1988; Daussant *et al.*, 1994; Jung *et al.*, 2001). Interestingly, recent research has indicated that a small number of amino acid substitutions in the ubiquitous, compared to the endosperm for results in improvements in the thermostability of the ubiquitous form (Clarke *et al.*, 2005). However, the endosperm form is the only one of consequence to the malting and brewing uses of barley, so the remainder of the discussion will be dedicated to this form.

The literature contains numerous investigations demonstrating a high degree of polymorphism in β-amylase with respect to size and charge. Chromatofocusing and isoelectric focusing have been used to resolve up to eight β-

amylase isoforms with isoelectric points ranging from pH 4.2 to pH 6.4 (La Berge and Marchylo, 1983; Lundgard and Svensson, 1987; Shewry *et al.*, 1988). A size range from 40~400 kDa has been demonstrated by gel filtration, SDS-PAGE and ultra-centrifugation (Visuri and Nummi, 1972; Niku-Paavola *et al.*, 1973; Bilderback, 1974). This heterogeneity may result at least partly from the formation of polymers with itself and other proteins, including hetero-dimers with protein Z (Hejgaard, 1976; Hejgaard and Carlsen, 1977; LaBerge and Marchylo, 1983). The presence of thiol groups that can participate in intermolecular disulphide bond formation are likely to be involved in the formation of such aggregates (Visuri and Nummi, 1972).

Developmental changes in the isoforms of β-amylase also occur during grain maturation and germination. In mature grain the major isoform is a single polypeptide chain of 59.7 kDa, which is converted during germination to an isoform of 56 kDa via a series of intermediate forms (Lundgard and Svensson, 1987; Shewry *et al.*, 1988). The reduction in molecular weight is accompanied by a basic shift in the isoelectric point of the protein (Evans *et al.*, 1997a). The conversion of isoforms is mediated by limited proteolysis in the carboxy-terminal region of β-amylase by malt endopeptidase activity (Lundgard and Svensson, 1986; Guerin *et al.*, 1992).

It has been identified that there are at least four allelic forms of β-amylase in cultivated barley that exhibit subtle but significant differences in thermostability and electrophoretic mobility (Eglinton *et al.*, 1998; Gunkel *et al.*, 2002; Kihara *et al.*, 1998). Increasing thermostability has been linked to increased levels of fermentability of wort produced during mashing by these authors. Unfortunately, these and other investigations have spawned an unnecessarily confusing β-amylase isoform nomenclature. Based on the first description by Allison (1973) these isoform types, characterised by electrophoretic mobility, are best known as Sd1 and Sd2 (starch degrading 1 and 2, Table 6.4). We concur with the plea by Chiapparino *et al.* (2006) that the nomenclature for β-amylase should follow the recommended rules for nomenclature and gene symbolism for barley as reported in the Barley Genetics Newsletter (ANON, 1992). Eglinton and Evans (1997) demonstrated that Sd1 was equivalent to Bmyl1 Ar and β-amy 1b and Sd2 was equivalent to Bmyl1 Br and β-amy 1a. This nomenclature should take precedence. Therefore the correct naming of the Sd1 allele should be *Bmy1-Sd1* due to the locus (*Bmy1*) and the first naming of the allele Sd1 by Allison (1973). Examples of these β-amylase types are presented in Fig. 6.5.

A major advance in the practical understanding of barley β-amylase biochemistry was that the β-amylase types had subtle differences in thermostability (Eglinton *et al.*, 1998; Kihara *et al.*, 1998). This analysis enabled the differentiation of the Sd2 IEF type into low and high thermostability types (Fig. 6.6). In addition, a new β-amylase type, Sd3 was identified in *Hordeum vulgare ssp. spontaneum* that had different IEF mobility and very high thermostability. As is shown in Table 6.2, this thermostability knowledge can be easily combined with the preceding nomenclature where the Sd2 type is split

Table 6.2. Historical summary of β-amylase allele nomenclature

Key citations	β -amylase nomenclature					
Allison (1973)	Sd1		Sd2		-	-
Nielsen & Johansen (1986)	Bmyl1 Ar		Bmyl1 Br		-	-
Forster *et al.*, (1991)	β-amy 1b		β-amy 1a		-	-
Evans *et al.*, (1997a)	*Bmy1*-Sd1		*Bmy1*-Sd2		-	-
Eglinton *et al.*, (1998)	Sd1		Sd2L	Sd2H	Sd3	-
Kihara *et al.*, (1998)	B		C	A	-	-
Chiapparino *et al.*, (2006)	Sd1		Sd2L Sd4	Sd2H	Sd3	-
Kaneko *et al.*, (2001)	B	A-B	C	A	A+	β amy-less
Zhang *et al.*, (2004a, 2004b, 2007)	B-I	A-B	C-II	A-II	A+-II	β amy-less
Extra IEF mobility variants	I, Ia, II	IIb, IIc	I, Ia, IV	Ia, II, IIa	-	-
Thermostability[a]	intermed	intermed	low	high	V. high	No activity

[a]Thermostability determination based on definitions of Eglinton *et al.*(1998) and Kihara *et al.*(1998).

into high (Sd2H) and low (Sd2L) thermostability types. The Sd1 type displays intermediate thermostability. It was also observed that the three endosperm-specific β-amylase alleles are inherited by co-dominant inheritance at the Bmy1 locus on chromosome 4H among cultivated barleys (Eglinton *et al.*, 1998).

In practical brewing terms, both Eglinton *et al.* (1998) and Kihara *et al.* (1998) demonstrated the inherent thermostability of β-amylase could impact on potential wort fermentability (Fig. 6.7). The key observation was that, on a per unit of DP basis, the Sd2H varieties performed the fermentability of the Sd2L and Sd1 types (Fig. 6.7B). Interestingly, the differentiation between Sd1 and Sd2L samples assessed in this investigation was based on the amount of activity rather than thermostability (Eglinton *et al.*, 1998). More recently, Evans *et al.* (2005) concluded that this increase was in the order of 2% points of AAL. To date, only the Sd2H β-amylase type has been shown to increase fermentability, while no experiments have yet been reported on whether the highly thermostable Sd3 type confers any further increase in potential wort fermentability properties. Similarly, Kaneko *et al.* (2001) has also identified another thermostability variant, "A$\sim$B" (Table 6.4) whose thermostability is slightly greater than that for Sd1 but again the practical impact of this variant on fermentability has not been determined. A recent study identified a putative 4th β-amylase thermostability type, Sd? (very low) was identified (Evans *et al.*, 2005). This hulless malt variety (Torrens) had very high levels of extract but produced wort with the lowest wort fermentability, presumably as a result of the low level of thermostability of its β-amylase. It was suggested that such a characteristic could have advantages in the production of low alcohol beers where low alcohol with optimal limit extract provides desirable extra "body" to the beer. This putative new Sd type is now awaiting further characterisation. Finally, a recombinant β-amylase enzyme with 6°C increase in thermostability (as measured by T50) was produced by Sapporo Breweries,

Japan, by combining seven different amino acid substitutions (Okada *et al.*, 1995; Yoshigi *et al.*, 1995). Although fertile transgenic barleys synthesizing this thermostable β-amylase were reported by Kihara *et al.* (1997), we are not aware of the release of a commercial variety with this attribute.

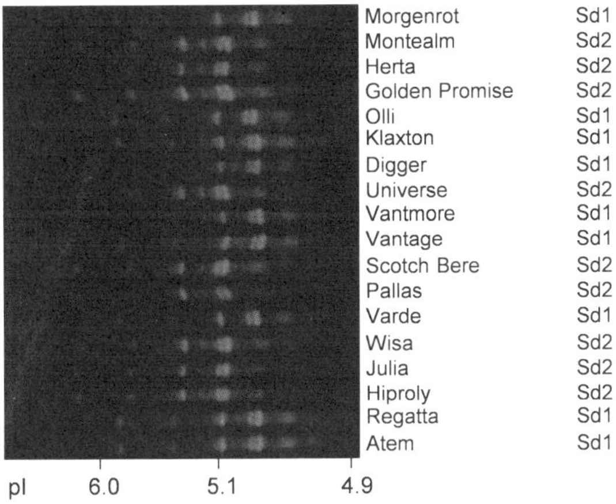

Fig. 6.5. Two types of barley β-amylase resolved by IEF and stained for amylase activity by iodine/starch stain (extracted from Eglinton and Evans, 1997)

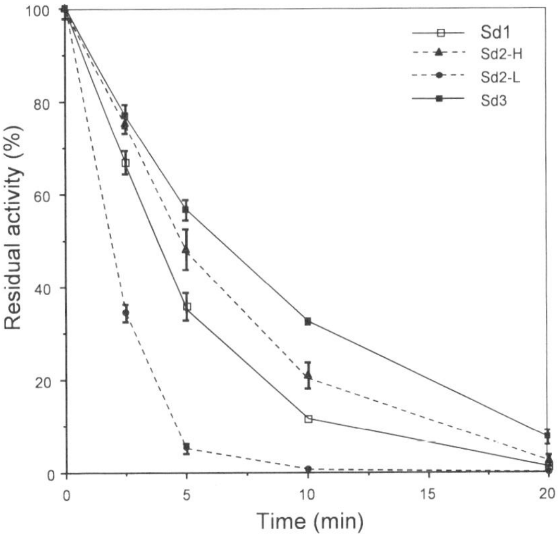

Fig. 6.6. Irreversible thermal inactivation of β-amylase in barley extracts incubated at 60°C. Values are the mean of five barley varieties and the standard error of the mean is shown. (From Eglinton *et al.*, 1998)

Since the original 1998 studies of β-amylase thermostability variation, over 8,000 barley varieties, lines and landrace lines from around the world have been screened (Kaneko *et al.*, 2001; Zhang *et al.*, 2004a, 2004b, 2007). Overall, most barley varieties are of either the Sd2H, Sd1 or Sd2L types with the Sd3 (A+-II) and other types being relatively rare. In two rowed varieties/lines the Sd2L (C-II) type is most common while in the six rowed varieties/lines, the Sd2H (A-II) type is most common (Table 6.5). Among European varieties, the most common types are Sd1 followed by Sd2L, with Sd2H being relatively rare (Malysheva *et al.*, 2004; Ovesna *et al.*, 2006; Polakova *et al.*, 2003). Such analysis agrees with the results of our continuing screening of barley malt varieties (Evans and Eglinton, unpublished) and the observation that most of the highly regarded malting international varieties are of the Sd1 type (i.e., Alexis, Barke, Baudin, Chariot, Gairdner, Harrington, Metcalfe Scarlett). However, following the lead of highly regarded Japanese malting varieties such as Haruna Nijo and Amaji Nijo, the release of Sd2H varieties (i.e., Buloke, Flagship from Australia) to take advantage of increased levels of fermentability, are becoming more prevalent.

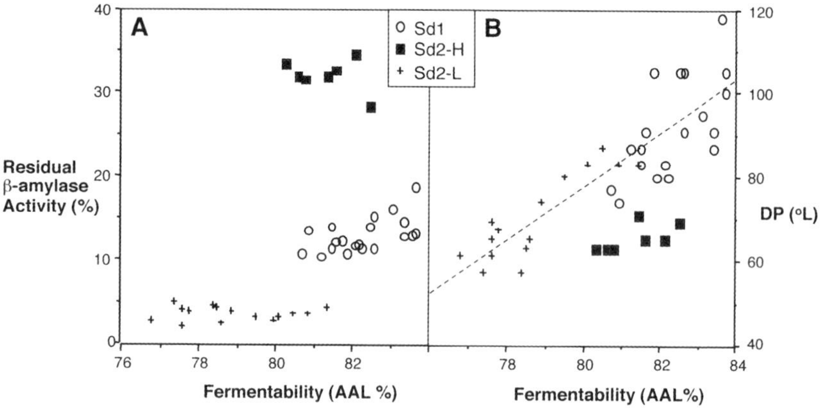

Fig. 6.7. Relationship in 42 commercial malts between fermentability (AAL %) and A. β-amylase thermostability with respect to β-amylase allele; B. diastatic power with respect to β-amylase allele (From Eglinton *et al.*, 1998)

As the β-amylase alleles have been intensively scrutinized further and more lines have been examined further allele variation has come to light. In the main, these have been observed as variation in the IEF mobility patterns within the main Sd types (Kaneko *et al.*, 2001; Zhang *et al.*, 2004a, 2004b, 2007). It is likely that these IEF variations may be similar to those identified by Chalmers *et al.* (1992) who reported 7 β-amylase banding types from 135 accessions of *H. spontaneum* collected from ecologically diverse sites in Israel. To date, the importance of these IEF mobility changes and their biochemical basis, genetic change or post-translational modification, has yet to be ascertained. Polakova *et al.* (2003) identified a second form of Sd1, Sd1b where a nucleotide substitution occurred, Sd1: A702→G702 Sd1b but this change was conservative so that the amino acid specified remained glycine. In addition, a number of putative divergent β-amylase types have been identified based on nucleotide sequence (Clark *et al.* 2003; Eckstein *et al.*, 2004). These may impart further useful variation for manipulation of malt quality. However, comparison of the amino acid substitutions reported by these authors with the established amino acid sequences in the literature (see Table 6.6 and following discussion) indicated several important conflicts that were not reconciled in these studies. Until these divergent changes have been confirmed and validated, they should be considered with caution.

6.4.1.2 Genetics of β-amylase

The genetic control of the level of β-amylase expression has been keenly sought by both barley breeders and malt quality chemists. Two mutant genes on chromosome 7H have been shown to dramatically affect the deposition of β-amylase in the endosperm. The Hiproly high-lysine barley variety contains a mutation in the *lys1* gene and exhibits substantially higher levels of β-amylase in the mature grain (Hejgaard and Boisen, 1980). The Riso mutant 1508 contains an altered *lys3* gene, described as *lys3a* (Ingverson *et al.*, 1989), which contains very low levels of β-amylase (Allison, 1978). This mutation also interferes with the transcription of the B- and C-hordeins whose levels are severely reduced due to hypermethylation of promoters (Sorensen *et al.*, 1996). Results from analysis of both of these mutations are consistent with the established relationship between storage protein accumulation and β-amylase expression.

An important breakthrough was the demonstration that the level of β-amylase activity is influenced by a 126 bp deletion in intron 3 of the *Bmy1* non-coding region of the structural gene (Erkkila *et al.*, 1998). Erkkila *et al.* (1998) suggested that the deletion in intron 3 either increases transcriptional efficiency or post-transcriptional mRNA stability to effectively increase the amount of β-amylase synthesised. The deletion in intron 3 is amenable to detection utilising PCR primers (Fig. 6.8). Due to linkage, selection for the 126 bp deletion (higher levels of β-amylase activity) differentiates the Sd2L from the Sd1, Sd2H and Sd3 alleles. The differences in intron 3 are consistent with the higher activity of the Sd1 samples identified previously in Fig. 6.7. PCR primers have been further refined by Kaneko *et al.* (2000) to allow the separation of the Sd1 and Sd2H alleles. More recently, SNP (single nucleotide polymorphism) markers have been developed to differentiate the Sd1, Sd2L, Sd2H and Sd3 β-amylase alleles (Paris and Eglinton, 2002). It is recognized from population mapping analysis that quantitative trait loci (QTL) in locations other than the structural gene for β-amylase, influence the level of β-amylase activity, DP and fermentability (Hayes and Jones, 2000; Coventry *et al.*, 2003). Recently, Hayden *et al.* (2007) investigated the sequence variation of the *Bmy1* gene, including the 1.5 kb promoter sequence. Numerous SNPs and INDELs were clearly differentiated for each of the *Bmy1* alleles. In total, eight *Bmy1* alleles were distinguishable at the DNA sequence level and seven isozymes were distinguishable at the protein level.

6.4.1.3 Amino acid substitutions and enzyme thermostability

Comparison of the amino acid and nucleotide sequences of the Sd1, Sd2L, Sd2H and Sd3 β-amylase types has revealed that there are eight amino acid substitutions and the removal of 4 kDa from its C-terminal that explains the thermostability, grain free/bound ratio, electrophoretic/IEF mobility and Km differences between the β-amylase types. During malting and mashing, β-amylase is modified by C-terminal proteolysis (4 kDa removed, Evans *et al.*, 1997b; Guerin *et al.*, 1992; Lungard and Svensson, 1987). The removal of the C-terminal has been found to increase the thermostability T_{50} of each β-amylase isoform by approximately 2°C and increase the enzymes affinity only for starch (decrease starch K_m), particularly for the Sd1 type. However, the k_{cat} (reaction velocity) was unchanged, which indicates that the impact on enzyme activity would only be realised when substrate concentrations were limiting (Ma *et al.*, 2000). Given the central role of α-amylase and limit dextrinase providing substrates (low degree of polymerisation) proposed by Evans *et al.* (2005), this kinetic effect would be expected to have little practical impact during mashing.

Ma *et al.* (2001) showed by the assessment of recombinant β-amylase enzymes that only two of these amino acid substitutions, valine 233 (Sd2L, Sd1) to alanine 233 (Sd2H), and leucine 347 (Sd2L) to serine (Sd1, Sd2H), determine thermostability. Further assessment of the recombinant enzymes suggested that the Val233→Ala substitution conferred improved refolding after heating giving a benefit during malt kilning, while the Leu347→Ser substitution conferred the maintenance of activity at higher assay or mashing temperatures Ma *et al.* (2001). This conclusion contradicts the finding both here and by Eglinton *et al.* (1998) that the Sd2H type alone confers improved maintenance of activity at 65°C as demonstrated by the superior fermentability of this Sd2H malt type, which is the only type to contain the Val233→Ala substitution (Fig. 6.8). We suggest that Ma *et al.* (2001) may have inadvertently reversed the Val233→Ala and Leu347→Ser recombinant enzymes in their investigation.

The proportion of bound β-amylase co-segregates with the Sd1 and Sd2 alleles. An examination of 46 cultivars showed those with Sd1 type had less than 50% of β-amylase in the "free" form, whereas Sd2 types had greater than 50% in the "free" form (Allison and Swanston, 1974). Segregation analysis had previously demonstrated that the free/bound ratio was determined by a single locus with incomplete dominance (Bendelow, 1964). As the Sd1 varieties have Cys 115 substituted for Arg 115 in the Sd2 varieties, it is this cysteine residue

Table 6.3. Comparison of amino acid substitutions between the barley β -amylase enzyme types

Impact	**Sd2L**[a]	**Sd1**[b]	**Sd2H**[c]	**Sd3**[d]
IEF, bound/free, K_m[e]	Arg115	Cys115	Arg115	Arg115
	Asp165	Glu165	Asp165	Glu165
Thermostability[e]	Val233	Val233	Ala233	Ala233
	Ser254	Ser254	Ser254	Thr254
Thermostability[e]	Leu347	Ser347	Ser347	Ser347
	Val430	Ala430	Ala430	Ala430
IEF, Thermostability[d]	Gln472	Gln472	Gln472	Lys472
	Ile527	Ile527	Met527	Ile527
C-terminal G489[e]		Thermostability, K_m		

[a]Kreis *et al.* (1987)- EMBL X52321, [b]Ma *et al.* (2001) C Genbank AF300800,[c]Yoshigi *et al.* (1994) C EMBL D21349, [d]Eglinton(2003), [e]Evaluated by production of recombinant proteins after site-directed mutagenesis (Ma *et al.*, 2000, 2001, 2002).

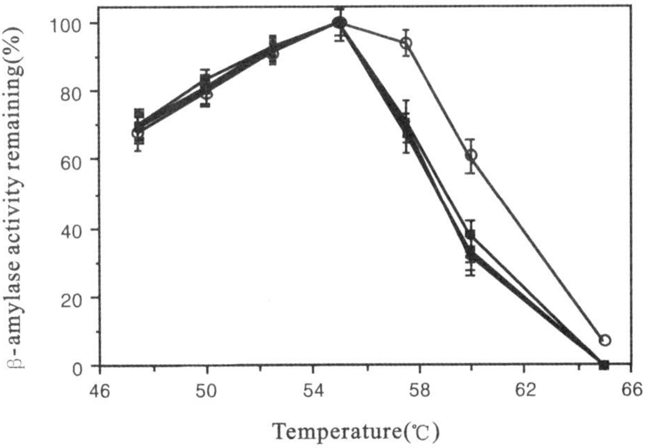

Fig. 6.8. Effect of assay temperature on the activities of mutant and wild type barley β-amylases. ■ Sd2L-L347S, ▲Sd2L-R115C, ♦ Sd2L-D165E, ○ Sd2L-V233A, ◇ Sd2L-V430A. The activities are expressed as percentages of the initial activity. Values are the mean of three independent determinations, with the standard deviation shown as bars (From Ma *et al.* 2001). The figure has been corrected in that the Sd2L-V233A (Sd2H only) was mis-labelled in the original publication

that is capable of forming an extra disulphide bond to increase the proportion of β-amylase bound in Sd1 compared to Sd2 varieties (Li *et al.*, 2002). This amino acid substitution has also been shown to alter the net charge of the protein, thereby making the Sd1 IEF pattern more acidic (Li *et al.*, 2002; Ma *et al.*, 2002).

6.4.1.4 Three dimensional structures of different isoforms

The development of the three dimensional structures for the barley β-amylase forms enables a detailed prediction of what is the likely impact of the amino acid substitutions in terms of the conformation and biochemical properties of this enzyme. Three dimensional structures of the *Bmy1*-Sd2L, Sd1, Sd2H and Sd3 forms of barley β-amylase were developed based on the crystal structure of an engineered form of barley β-amylase (BBA-7) resolved at 2.5$\AA$ resolution (Mikami *et al.*, 1999). BBA-7 was based on a clone of the Sd2H enzyme with five N-terminal amino acids deleted, and methionine added to the N terminus, presumably due to the *E.coli* translation initiation codon (ATG). The amino acid numbering in the model structures was based on the full-length barley sequence. The recombinant β-amylase was also mutated at seven sites to include M185L, S295A, I297V, S350P, S351P, Q352D and A376S (Yoshigi *et al.*, 1995), resulting in homology to the four naturally occurring forms of barley β amylase ranging from 97.3 to 98.1% positional identity.

The crystal structure of the engineered BBA-7 β-amylase does not extend beyond position 504 of the amino acid sequence. This was due to the disordered structure of the extended C terminal loop (Mikami *et al.*, 1999), presumably resulting from the glycine-rich repeats which would be expected to impart high levels of main chain flexibility. For this reason, the barley β-amylase models were developed based on the proteolytically cleaved form of the enzyme present in germinated barley, comprising residues 2–489, as determined by amino acid sequencing of the purified protein (Eglinton, 2003).

Molecular modeling using a template structure that is almost identical in primary sequence is expected to yield models that do not deviate significantly from the original structure. However, analysis of stereochemistry and database related checks were performed to ensure the structures were suitable for detailed molecular analysis. Homology based models were also constructed using the soybean β-amylase structure as a template, and using both soybean and BBA-7 β-amylase simultaneously. The resulting structures were discarded in favor of the models based on only the BBA-7 enzyme.

The RMS Z-scores for all improper dihedrals in the four barley β -amylase structures were within normal ranges. All bond lengths were in agreement with standard bond lengths using a tolerance of 4σ, and bond length deviation was distributed normally from standard bond lengths (values taken from Engh and Huber, 1991). Bond angles were also shown to deviate normally from mean standard bond angles (values taken from Engh and Huber, 1991). The RMS Z-scores of bond angles for the four structures ranged from 0.989 to 0.994, very close to the value of 1.0 expected for normally restrained data sets. Backbone conformation analysis also generated values consistent with well-refined protein structures. Ramachandran plots (Fig. 6.9) of the main chain conformation angles showed that approximately 90% lie within the core region, and 100% within allowed regions (Ramachandran *et al.*, 1963). As expected from the very high level of homology, each of the four structures modeled are almost identical to the experimentally determined structure of the BBA-7 β-amylase. The RMS distance between each of the four models and the original structure was 0.11Å for Cα atoms. Energy minimization of the four barley β-amylase structures resulted in final energies ranging from -27227 KJ mol^{-1} for the Sd2H structure, to -26919 KJ mol-1 for the Sd1 structure. The average B-factor values for all buried protein atoms was outside the normal range in each of the four structures modeled. However the average atomic B-factors for buried atoms were also outside the expected range in the mutant BBA-7 β-amylase, and high in the soybean β-amylase structure resolved at 2.2Å , suggesting this may be an intrinsic property of β-amylase enzymes.

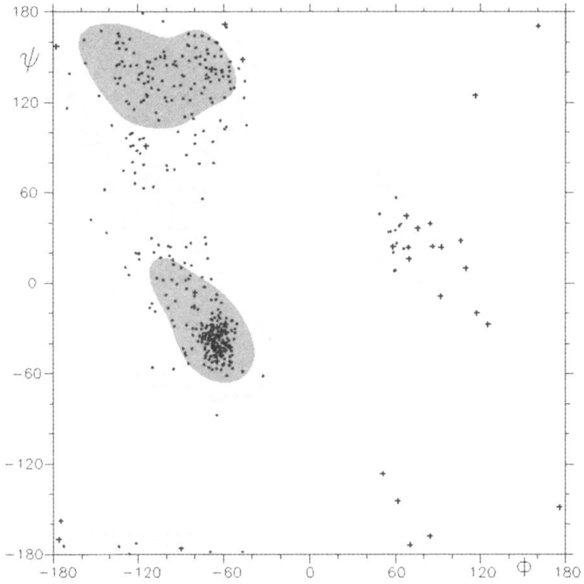

Fig. 6.9. Ramachandran plot (Ramachandran *et al.*, 1963) of the Sd3 β-amylase. Glycine residues are indicated by plus symbols, all other residues are plotted as dots. The ϕ and ψ angles of all residues are within allowed regions of low energy (extracted from Eglinton, 2003)

The active site of β-amylase is comprised of three long loops forming a deep pocket close to the centre of the carboxyl end of the β-barrel. One wall of the pocket is formed by a flexible loop (L3) that extends into the solvent during substrate capture and product release, but closes upon substrate binding, shielding the catalytic groups and the reaction centre from the solvent (Mikami *et al.*, 1994). The active site is shown in the open and closed conformations with loop L3 coloured red in Fig. 6.10(A) and Fig. 6.10(B) respectively.

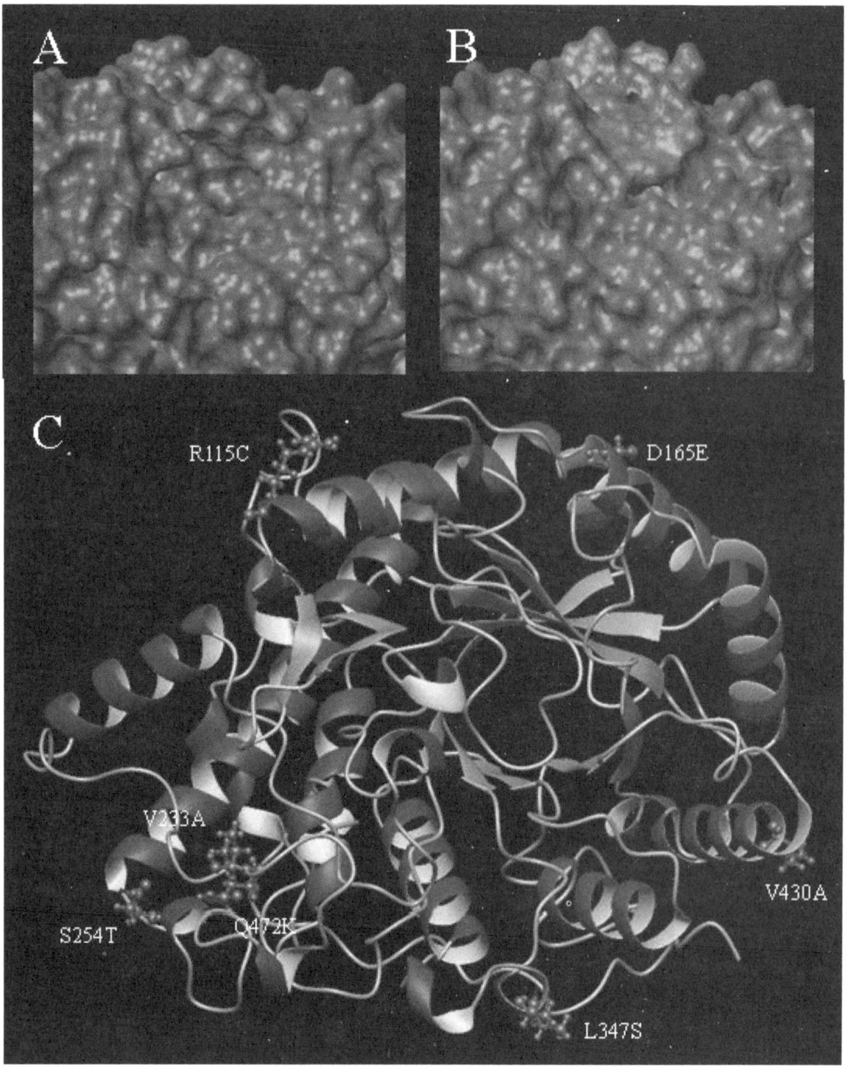

Fig. 6.10. (A) View into the active site pocket of β-amylase showing the van der Waals protein surface. The hinged loop L3 (designated red) is shown in the unliganded open conformation. (B) The hinged loop L3 in the closed conformation showing the van der Waals surface extending over the reaction centre, allowing the methyl groups of Val95 to interact with those of Leu381. (C) Ribbon representation of the Sd2L barley β-amylase, looking towards the carboxyl end of the $(\alpha/\beta)_8$ core with loop L3 centred above the barrel. The seven variant residues in the allelic forms of β-amylase are shown as ball and stick structures. (extracted from Eglinton, 2003)

The global structure of barley β-amylase is represented in Fig. 6.10(C). The protein forms a single domain comprising an $(\alpha/\beta)_8$ barrel core exhibiting extended loops on the C terminal side of the β-barrel, with each loop connecting the C terminus of a β-strand to the N terminus of the next α-helix. The seven amino acid mutations identified between the four allelic forms of the enzyme are shown as ball and stick structures in Fig.6.11c. Each of the mutations occurs close to the surface of the protein and are distal to the active site.

Three of the mutations occur within α-helices or turns. The substitution Arg115→Cys is in an α-helix within loop L3, and Asp165→Glu and Val430→Ala occur in $\alpha3$ and $\alpha8$ respectively. These three mutations do not result in changes to helix dipoles (Sali *et al.*, 1988), helix capping (Richardson and Richardson, 1988), or helix forming propensity (Chou and Fasman, 1978). Therefore the mutations of Arg115→Cys and Val430→Ala in the Sd1 β-amylase, and Asp165→Glu in the Sd1 and Sd3 enzymes are not expected to contribute to the observed changes in thermostability. The substitution of Arg115→Cys in the Sd1 enzyme occurs on the surface of the protein and the respective side chains project into the solvent, consistent with the decreased isoelectric point of the Sd1 enzyme. The side chain of Cys115 is not close enough to any other cysteine residue to allow the formation of an intramolecular disulphide bond, however the location of the side chain does allow the formation of an intermolecular disulphide bond.

The amino acid substitution of Val233→Ala occurs in loop L4 which exhibits higher than average main chain flexibility. In the thermolabile Sd1 and Sd2L enzymes, the side chain of Val233 extends into a cavity delimited by the cyclic side chains of Phe204, Pro230, Phe245 and Phe246 (Fig. 6.11(A)). The interatomic distances between CG1 of V233 and the β-carbon of Pro230, CG of Phe245 and CE2 of Phe204 are 3.2Å , 3.35Å and 3.34Å respectively. The substitution of the smaller side chain of alanine at position 233 in the thermostable Sd2H and Sd3 enzymes yields improved side chain packing into the cavity (Fig. 6.11(B)), with the interatomic distances to the β-carbon of Ala233 increased to 4.01Å , 4.31Å and 3.83Å respectively. The mutation of Val233→Ala is therefore expected to increase the thermostability by decreasing the entropy of refolding in the Sd2H and Sd3 enzymes.

The thermolabile Sd2L β-amylase is the only enzyme to contain leucine at position 347 instead of serine. This position is within loop L6 which also exhibits high main chain flexibility. The side chains of the respective amino acids project directly into the solvent. Although the side chains are not involved in intramolecular hydrogen bonding, serine at this position is expected to stabilise the flexible loop through decreased surface hydrophobicity.

The amino acid substitution of Ser254→Thr is present only in the thermostable Sd3 β-amylase. This mutation also occurs within a flexible loop structure (L4), and is surface exposed. However, this substitution does not significantly alter the main chain conformation or influence hydrogen bonding, and represents a small increase in surface hydrophobicity. Therefore this

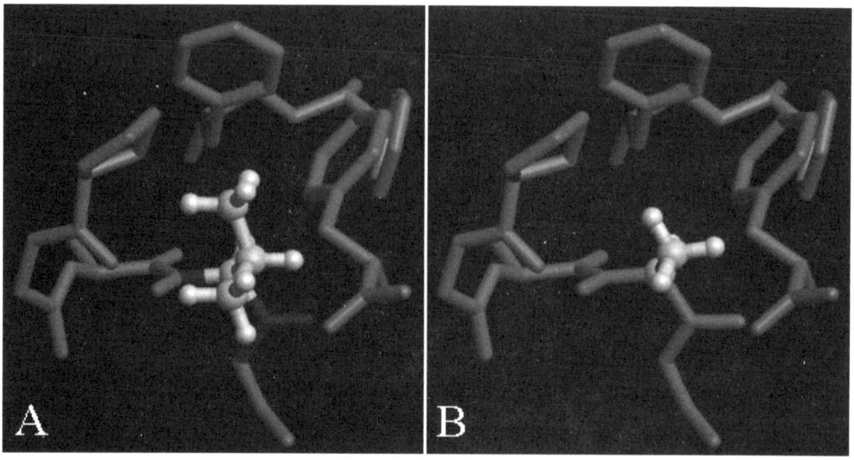

Fig. 6.11. The spatial arrangement of amino acids (clockwise from the left) Pro230, Phe245, Phe246 and Phe204 with respect to (A) valine and (B) alanine at position 233 (extracted from Eglinton, 2003)

mutation is not predicted to be responsible for the observed thermostability difference between the Sd2H and Sd3 enzymes.

The amino acid substitution of *Gln472Lys* occurs in the highly flexible C-terminal loop of the thermostable Sd3 β-amylase. The side chains of the two residues are surface exposed, therefore the presence of Lys472 is consistent with the observed increase in isoelectric point of the Sd3 β-amylase. Fig. 6.12 shows this substitution also results in the formation of an interchain hydrogen bond between NZ of Lys 472 and OD2 of Asp319 in loop L5. This additional hydrogen bond is concluded to be responsible for the increased thermostability of the Sd3 β-amylase by decreasing the flexibility of the C-terminal and L5 loops, and is also the basis of the increased isoelectric point of the Sd3 enzyme.

6.4.2 α-amylase

α-amylase, α-(1,4)-D-glucan glucanohydrolase (EC 3.2.1.1), is synthesised during germination in response to the plant hormone gibberellin (Filner and Varner, 1967) and catalyses the cleavage of internal α-(1,4)-glucosyl linkages of amylose and amylopectin. In contrast to other hydrolytic enzymes of malt, α-amylase is a relatively thermostable enzyme, with only 4%~18% of enzyme activity lost during kilning (Runkel, 1975; Sjöholm *et al.*, 1995). As shown earlier in this chapter, α-amylase is the most stable of the DP enzymes during mashing (see Fig. 6.3). The conventional view that a combination of high α-amylase levels in malt and the thermostability of the enzyme, suggest that it is not normally a rate-limiting enzyme in the degradation of starch during mashing (Bamforth, 2003; MacGregor, 1995), has been shown to be incorrect

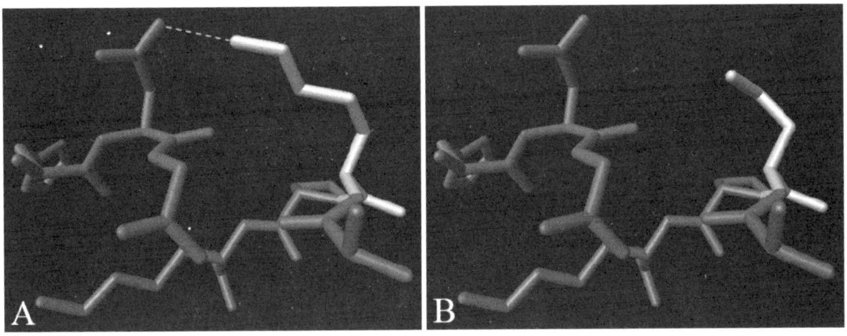

Fig. 6.12. Interaction of aspartic acid 319 with lysine (A) and glutamine (B) at position 472 of barley β-amylase. Aspartic acid is shown in magenta and residue 472 is in grey. The interchain hydrogen bond formed between NZ of Lys 472 and OD2 of Asp319 is shown as a green dashed line

(Evans *et al.*, 2005, 2007). This suggests that modification of α-amylase activity may have significant potential for optimising and controlling the rate or efficiency of starch degradation during mashing.

6.4.2.1 Biochemistry of α-amylase

α-amylase consists of two discreet families, AMY1 and AMY2. The two groups are distinguished by differences in isoelectric points, with pI 4.7~5.0 for AMY1 (low pI) and pI 5.9~6.4 for AMY2 (high pI), (Jacobsen and Higgins, 1982; Macgregor *et al.*, 1971; Muthukrishnan *et al.*, 1984; Svensson *et al.*, 1985). The combination of Amy2 and its inhibitor produces slightly higher pI's (6.4~6.6). Unfortunately, for the α-amylase genes have been given the reverse number to the proteins biochemical characterisation. That is the low pI (AMY1) group is expressed from $\hat{\alpha} - \hat{a}my2$ gene while the high pI (AMY2) is expressed from the $\hat{\alpha} - \hat{a}my1$ gene. The properties and nomenclature of the two α-amylase isoenzymes are summarised in Table 6.3. The AMY1 and AMY2 families share 80% amino acid identities (Rogers and Milliman, 1983; Rogers, 1985), and have the same substrate specificities and action patterns, however the Amy1 isoforms degrade raw starch granules faster than the high pI group (MacGregor and Balance, 1980; Sissons and MacGregor, 1994b).

The AMY2 isoforms are expressed earlier in germination and are also the most abundant, accounting for up to 80%~90% of the total α-amylase activity in malt (MacGregor *et al.*, 1988; MacGregor *et al.*, 2002a). Bertoft *et al.* (1984) observed that these isoforms differed in thermostability, with the AMY2 having the higher temperature tolerance but lower temperature optimum in that investigation. AMY2 is more sensitive to Ca^{2+} than AMY1 with its crystal structure being found to contain 3 Ca^{2+} atoms (Kadziola *et al.*, 1994). The activity of the AMY2 type is also improved by 2~4 fold

Table 6.4. Physical properties of the two α-amylase isoenzymes

Property	AMY1	AMY2
Apparent M.W.	45,000 Da[a]	45,000 Da[a]
Calculated M.W.	45,447 Da[b]	
pI	~4.5[a]	~5.8[a]
No. Isoforms	6[c]	8[d]
Gene Symbol	$\alpha - amy2$	$\alpha - amy1$
Chromosome	7H[e]	6H[e]
No. Genes	3[f]	6[g]

[a]Callis and Ho (1983), [b]Rogers and Milliman (1983), [c]Søgaard *et al.* (1991), [d]Simon and Jones (1988), [e]Brown and Jacobsen (1982), [f]Muthrukrishnan *et al.* (1984), [g]Khurseed and Rogers (1988).

by the inclusion of between 1-15 mM $CaCl_2$, while the activity of AMY1 is essentially constant (Søgaard *et al.*, 1993). It is thus not surprising that most brewing protocols call for the inclusion of ~40~100mg/L of calcium (as $CaCl_2$ or $CaSO_4$) in the mash (Briggs *et al.*, 2004).

The activity of the AMY2 group of isoforms is also modified by the bifunctional α-amylase/subtilisin inhibitor (BASI). BASI is a 19.8 kDa protein synthesised during grain filling (Munck *et al.*, 1985), and is a fast reacting, tight binding inhibitor of AMY2 (Sidenius, 1995). Henson and Stone (1988) have observed that the wide range in α-amylase activity in germinating barley is the result of variation in the levels of both the enzyme and BASI. However, the inhibitory activity of BASI is lost between 50 and 70°C, so it is not considered to have a substantial effect on starch hydrolysis during mashing (Munck *et al.*, 1985). As indicated earlier in this chapter, temperatures of around 65°C are required for starch gelatinisation to enable efficient starch hydrolysis during mashing.

6.4.2.2 Measurement of α-amylase activity

Similar to the measurement of β-amylase activity (section 6.4.1), the measurement of α-amylase activity by assay of the production of reducing sugars is problematic in malt due to the presence of other amylolytic enzymes. Fortunately again an assay has been developed, Ceralpha, that uses a specific substrate (blocked p-nitrophenyl maltoheptaoside - BPNPG7) that upon cleavage produces a product that can be measured spectrophotometrically (McCleary and Sheehan, 1987). The maltoheptaoside based substrate is large enough to enable efficient α-amylase binding but the chemical blocking of the non-reducing end precludes β-amylase or α-glucosidase cleavage. Recent investigations however have indicated some caution for the application of the Ceralpha assay as high concentrations of small carbohydrates or substrate lower the activity potentially measured by this assay (Baks *et al.*, 2006a, 2006b). Such conditions can potentially occur when studying the level of α-amylase activity during the mashing process, particularly when there are low levels of activity. However, the effect of these conditions on measuring α-amylase in extracts of malt used for quality evaluation are most likely minimal.

6.4.2.3 Genetics of α-amylase

The genetic variation for α-amylase has been assessed in barley but to date no specific alleles or targets for breeders have been identified. For instance, three alleles for $\alpha - amy2$ have been reported (Takano and Takeda, 1985), but there was no significant relationship between the alleles and total α-amylase activity (Takano *et al.*, 1988). The $\hat{\alpha} - \hat{a}my1$ gene loci (AMY2) are located on long arm of chromosome 7H, while the $\hat{\alpha} - \hat{a}my2$ gene loci (AMY1) are located on the long arm of chromosome 6H (Brown and Jacobsen, 1982). Recently it has been observed that there was substantial variation in the thermostability of α-amylase between varieties (Fig. 6.13) although the question still remains to be resolved how this variation impacts on wort fermentability. The basis and genetics of this variation in thermostability also remain to be resolved. There has also been found to be substantial genetic heterogeneity in isoelectric focusing (IEF) band pattern for both the AMY1 and AMY2 isoforms, although the potential influence of these enzyme characteristics during mashing has not yet been characterised (Brown and Jacobsen, 1982; Takano *et al.*, 1988; Weining and Henry, 1994). Since it has been found that there are at least three AMY1 (Muthrukrishnan *et al.*, 1984) and six AMY2 genes (Khurseed and Rogers, 1988), the scope for finding useful genetic variation should be good. Overall, the finding by Evans *et al.* (2005) that α-amylase has a key and modifiable role in starch degradation indicated that α-amylase biochemistry and genetics should be fruitful areas for characterisation and exploitation by barley breeders.

6.4.3 Limit Dextrinase Biochemistry and Genetics

Limit dextrinase, α-dextrin 6-glucanohydrolase (EC 3.2.1.41), cleaves the α-(1,6)-D-glucosidic linkages to produce small oligosaccharides and fermentable sugars (Enevoldsen and Schmidt, 1973) The enzyme does not hydrolyse α-1,6 bonds linking a single glucosyl residue to a maltosaccharide, as it requires at least one α-1,4 linkage an each side of the susceptible bond. Limit dextrinase is the only malt enzyme capable of hydrolysing α-1,6 bonds, therefore it complements the activity of the amylases in starch hydrolysis. Limit dextrinase has been purified to homogeneity from malt and is a monomeric protein of 104 kDa consisting of six isoforms with apparent PIS ranging from 4.2~4.6 and has optimal activity at pH 5.5 (Sissons *et al.*, 1992a). Barley limit dextrinase is the product of a single gene that encodes for a protein with a calculated molecular mass of 97 kDa (Burton *et al.*, 1999; Kristensen *et al.*, 1999). Limit dextrinase has been reported both in developing and germinating grain (Sissons *et al.*, 1992b). The gene is predominantly expressed in the aleurone layers

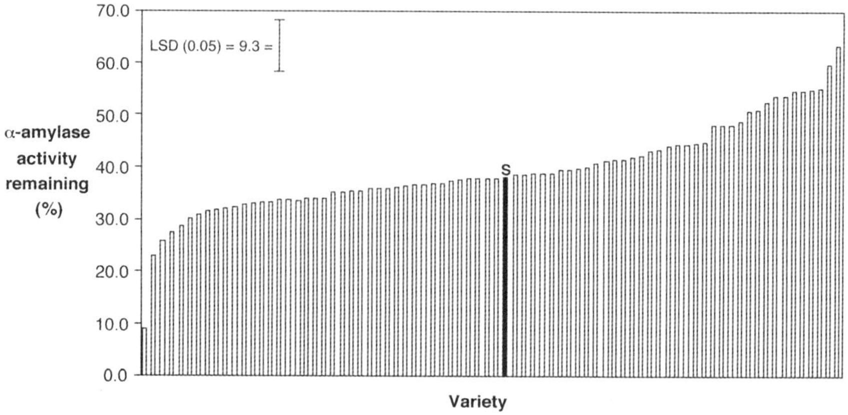

Fig. 6.13. α-amylase activity remaining in Ceralpha® extracts of malt from 78 Australian and international varieties and breeders lines incubated at 72.5°C for 10min. S = Schooner malt, the control variety. (From Evans *et al.*, 2003)

of germinated grain and is regulated by gibberellic acid (Hardie, 1975; Lee and Pyler, 1984; Sissons *et al.*, 1995; Schroeder and MacGregor, 1998; Burton *et al.*, 1999). The limit dextrinase gene has been mapped to the short arm of chromosome 7H (Li *et al.*, 1999).

6.4.3.1 Measurement of limit dextrinase activity

The absence of a simple, reliable and accurate assay for limit dextrinase, until relatively recently, has slowed the progress in the understanding of limit dextrinase biochemistry and genetics. The development of the Limit-DextriZyme assay, (Azurine-CL-Pullulan, McCleary, 1992) has resulted in a substantial increase in the progress in understanding this. Limit dextrinase action releases into solution the azurine portion that can be measured spectrophotometrically. The levels of total limit dextrinase are measured by a relatively long extraction at either 25°C (16 hrs) or 40°C (5~6 hrs) with the inclusion of dithiothreitol (25mM) in the extraction buffer or a short extraction at room temperature without reducing agent to extract the free form of the enzyme (McCleary, 1992; Evans *et al.*, 2005). Similar to the Ceralpha substrate, it has recently been shown that starch hydrolysis products can interfere with the Limit-DextriZyme assay resulting in over estimation of the level of the enzymes activity (MacGregor *et al.*, 2002b). Such conditions are most likely to occur when measuring limit dextrinase activity during mashing. However, as concluded for the Ceralpha assay, high levels of starch hydrolysis products are most likely minimal in the extractions used for malt quality evaluation.

6.4.3.2 Biochemistry of limit dextrinase

Limit dextrinase has free, latent and bound forms. Limit dextrinase has a 'free' form that is extractable in buffer alone, a 'bound' form that is extracted in the presence of proteases or reducing agents (MacGregor, 2004). The 'latent' form of limit dextrinase is more complex in that it is bound to a low molecular weight inhibitor (12.7~12.9 kDa), so that the enzyme-inhibitor complex is soluble in buffer (MacGregor, 2004). This complex is slowly disrupted by proteases, reducing agents or presumably heat during the mashing (Sissons *et al.*, 1995; Stenholm *et al.*, 1996; Walker *et al.*, 2001). Total limit dextrinase are the sum of the free, latent and bound fractions. In malt typically only 8%~20% of limit dextrinase is in the free form (Evans *et al.*, 2005). Synthesis of limit dextrinase begins around six days after anthesis and the enzyme is stored in the mature grain as a bound protein but is present in relatively low levels (Manners and Yellowlees, 1973; Lenoir *et al.*, 1984; Sissons *et al.*, 1993c). The first five days of germination see a rapid increase in the level of bound limit dextrinase due to additional enzyme synthesis (Longstaff and Bryce, 1993), that is promoted by gibberellin (Schroeder and MacGregor, 1998). As germination proceeds the bound enzyme is converted to an active free form through proteolytic modification by cysteine endoproteinases (Longstaff and Bryce, 1993).

Two inhibitors of limit dextrinase are present in both mature grain (MacGregor *et al.*, 1994) and at lower levels in germinated barley (Macri *et al.*, 1993). The two inhibitors are heat stable proteins of approximately 13 kDa and have isoelectric points of 6.7 and 7.2 (Macri *et al.*, 1993). Both inhibitors are active over a wide pH range, and they are most effective at the pH optima of their target enzyme (MacGregor *et al.*, 1994). Recently transgenic lines with down regulated levels of the inhibitor were found to influence starch granule size distribution and the proportion of α-(1,6)-D-glucosidic linkages in starch (Stahl *et al.*, 2004). This is consistent with the hypothesis that limit dextrinase and its inhibitor have important roles in regulating starch biosynthesis and starch granule formation (Burton *et al.*, 1999; Stahl *et al.*, 2007).

Wort and beer contain significant levels of branched dextrins (Enevoldsen and Bathgate, 1969; Enevoldsen and Schmidt, 1973), suggesting incomplete hydrolysis by limit dextrinase. Increasing the effectiveness of limit dextrinase during mashing therefore leads to higher levels of fermentable sugars in the wort (MacGregor *et al.*, 1999; Stenholm and Home, 1999). Limit dextrinase is only slightly less thermostable than α-amylase during kilning, with between 0%~25% of activity lost under standard kilning conditions (Sjöholm *et al.*, 1995). Under mashing conditions limit dextrinase exhibits a similar but superior thermostability profile to β-amylase (Fig. 6.3). Approximately 50% of activity remains after 1 hour at 62.5°C, 50% of activity remains after only 30 minutes at 65°C, and there is rapid inactivation as temperatures approach 70°C (Lee and Pyler, 1984; Sjöholm *et al.*, 1995).

During mashing, most limit dextrinase is inactive as 78.8%∼91.9% of total activity (Walker *et al.*, 2001) is in the bound or latent state complexed to its inhibitor. The free activity is relatively heat labile at mashing temperatures of 65°C while the bound and latent forms are relatively stable (Sissons *et al.*, 1995; Stenholm *et al.*, 1996; Walker *et al.*, 2001) suggesting that the binding to the inhibitor protects the enzyme, allowing it to be steadily released during mashing (MacGregor, 2004). It has also been concluded that complexing with the inhibitor largely limits the effectiveness of limit dextrinase during mashing (MacGregor *et al.*, 2002b). At least part of this observation is most likely due to over estimation due to the influence of starch hydrolysis products (MacGregor *et al.*, 2002b). Similarly, the thermostability of limit dextrinase during mashing may also be over-estimated. The significance of the limit dextrinase inhibitors during mashing requires further investigation, as lower levels of these proteins may also be a viable approach to increasing the effectiveness of limit dextrinase during brewing.

6.4.3.3 Genetics of limit dextrinase

Wide genetic variation of the enzyme activity and thermostability has been reported in the barley germplasm (Evans *et al.*, 2003; Yang *et al.*, 2008). Evans *et al.* (2003) observed that there was substantial variation in the thermostability of limit dextrinase (Fig. 6.14), although the impact of this variation on wort fermentability remains to be illucidated (Evans *et al.*, 2005). An early study was carried out by detecting five limit dextrinase alleles in a sample of 31 barley varieties based on sequence variation in intron 3 (Li *et al.*, 1999). Yang *et al.* (personal communication) studied 60 barley varieties showing a wide range of variation for enzyme activity and thermostability. Seven amino acid substitutions were identified from different barley cultivars for limit dextrinase with Thr233→Ala, Ala885→Ser and Gly888→Cys substitutions associated with enzyme thermostability based on single strand conformation polymorphism (SSCP) analysis. For barley varieties with known LD thermostability, Maud has Thr223 and Ala885 with the lowest enzyme thermostability (26.5%); AC Oxbow has Thr233 and Ser885 with medium enzyme thermostability (53.7%); Chinook and Pompadour have Ala233 and Ser885 with the higher enzyme thermostability (63% and 76.6%). At this stage is has not been resolved if these enhancements to thermostability improve limit dextrinase activity at true mashing temperatures and conditions. In addition to Ala233 and Ser885, Galleon also has a Gly888→Cys substitution and the highest enzyme thermostability (82.3%) recorded in that study. Potentially the Cys888 in Galleon could provide an additional -SH to form disulphide linkage with the enzyme inhibitor as has been shown for the Arg115VCys substitution in β-amylase (Li *et al.*, 2002). Thus the Cys888 substitution could result in lower levels of free limit dextrinase and could possibly explain the observation that there is variation in the level of free limit dextrinase (Ross *et al.*, 2003).

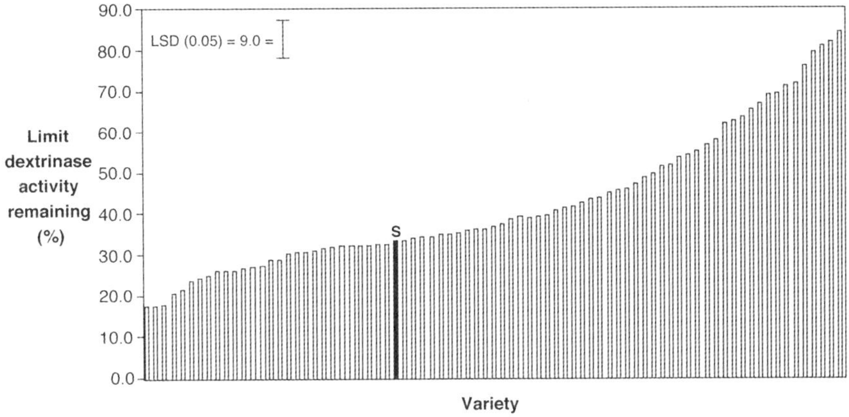

Fig. 6.14. Limit dextrinase activity remaining in Limit DextiZyme® extracts of malt from 79 Australian and international varieties and breeders lines incubated at 57.5°C for 15min. S = Schooner malt, the control variety. (From Evans *et al.*, 2003).

The value of increased limit dextrinase activity, lower levels of its inhibitor and increased thermostability would be determined by the brewing application. For instance, the use of rice adjunct results in increased levels of amylopectin (increased α-(1,6)-D-glucosidic linkages, Zhou *et al.*, 2002). Consequently, brewers who use rice adjunct (i.e., China, Japan) may desire higher levels of limit dextinase that would be expected to result in higher wort fermentability. The higher levels of limit dextrinase activity result in higher fermentability presumably because more potential substrate is being made available to β-amylase, in addition to the ability of limit dextrinase to slowly produce fermentable sugars (Enevoldsen and Schmidt, 1973). In whiskey production, the mash is not lautered, nor the wort boiled so that persistent (thermostable) limit dextrinase activity that survives into the fermenter would continue to produce fermentable sugars from remaining limit dexrins. This results in increase in fermentability and subsequent spirit yield (Walker *et al.*, 2001; McCafferty *et al.*, 2004). However, brewers producing beer may need to trade-off this extra fermentability for a loss of mouth-feel as limit dextrins potentially contribute to mouth-feel (Ragot *et al.*, 1989). Alternatively, the reduction in carbohydrates in beer would be desirable in the production of low carbohydrate or "lite" beers.

6.4.4 α-glucosidase Biochemistry and Genetics

α-glucosidase (EC 3.2.1.20) is the least well characterised of the starch degrading enzymes in germinating barley. α-glucosidase primarily catalyses the release of glucose from maltose and other small dextrins, although it has been shown to be able to attack intact starch granules (Sun and Henson, 1991).

It is important during germination for the production of glucose which can be assimilated and metabolised by the growing embryo. Glucose is utilised preferentially during the early stages of fermentation, however yeasts rapidly metabolise maltose, and therefore do not require hydrolysis to glucose for fermentation.

α-glucosidase is found in the mature grain, and after the initiation of germination there is further synthesis of the enzyme (Jorgensen, 1965; MacGregor and Lenoir 1984). Two α-glucosidase isoenzymes have been purified from malt and designated as G1 (high pI) and G2 (low pI) (Sissons and MacGregor, 1994b). The α-glucosidase gene encodes a polypeptide of 877-9 amino acids with a calculated molecular weight of approximately 97 kDa (Tibbot and Skadsen, 1996; Frandsen *et al.*, 2000). The gene was expressed from the aleurone and scutellum that respond to gibberellin, with peak transcription being reached 48 hrs after germination. α-glucosidase activity is easily measured by the procedure developed by Sissons and MacGregor (1994b) using a p-nitrophenyl-α-D-glucopyranoside substrate, a similar strategy to the Betamyl, Ceralpha and Limit DextriZyme assays.

It has been suggested that α-glucosidase may increase the effectiveness of α-amylase and β-amylase during mashing by removing maltose, which is a possible competitive inhibitor (Sun and Henson, 1991; MacGregor, 1996). There is also a synergistic effect between G1 and α-amylases in the hydrolysis of intact starch granules, while G2 and α-amylase exhibit additive effects (Sissons and MacGregor, 1994b).

In the β-amylase-less mutants from Tibet barley, the α-glucosidase activity level was high in the initial period of germination, and the glucose ratio in the sugar conformation was remarkably high in the germinated seed (Kaneko *et al.*, 2000). A putative clone of α-glucosidase was isolated from a cDNA library constructed from mRNA of the barley aleurone treated with GA (Tibbot *et al.*, 1996). The distribution and expression of the mRNA for this gene showed a similar pattern as the α-glucosidase proteins (MacGregor and Lenoir, 1987). However, Southern blot analysis indicated that only a single copy of this gene exists in the barley genome, and the deduced polypeptide had 877 amino acids with a molecular mass of 110 kD. Therefore, additional studies are needed to characterise the gene's possible product with respect to the 65 and 32 kD isoforms previously characterised (MacGregor *et al*, 1987; Stark and Yin, 1987; Sun *et al*, 1992). One of the α-glucosidase genes was mapped to barley chromosome 2H (Li, 1998).

To date the consensus of opinion is that α-glucosidase has still to be demonstrated to be of significance in malting and mashing (Bamforth, 2003; Evans *et al.*, 2005). Muslin *et al.* (2002, 2003) suggested that the inclusion of thermostable, recombinant α-glucosidase could increase the yield of fermentable sugars and fermentability. However, as previously noted, such a change in the fermentable sugar profile, increased proportion of glucose, would presumably influence yeast metabolism and the subsequent ester content and flavour of the beer (Verstrepen *et al.*, 2003) which would be undesirable for brewing. However, for the production of fuel alcohol, where flavour is not an issue, increased α-glucosidase activity/thermostability could result in higher fuel yields (Muslin *et al.*, 2003). Whether advances in malt quality can be made through altering the levels or properties of α-glucosidase will be determined with a more detailed understanding of the role of this enzyme during germination, and in malting and brewing.

6.5 Summary

The foundation of the production of alcohol for making beer is the hydrolysis of starch to simple sugars such as glucose, maltose and maltotriose that yeast can ferment. The primary enzymes that accomplish this task in malt are the diastatic power enzymes β-amylase, limit dextrinase, α-amylase and α-glucosidase. The interaction between the levels of the DP enzymes, their biochemical properties (i.e., thermostability) and the mashing program selected by the brewer determines the fermentability of the resultant wort (Evans *et al.*, 2005, 2007). Thus the real degree of brewhouse fermentability is clearly subject to the complex interaction of substrates and enzymes, in which recent biochemical studies have demonstrated both qualitative and quantitative variation is possible within the specific aspects of many of the individual components. This is a significant consideration for barley breeding as the genetic background of the recurrent germplasm may significantly influence the response to selection for altered fermentability.

Currently maltsters and brewers primarily rely on DP determination, and more recently in some cases α-amylase activity, to predict the potential fermentability of malt. The DP measure has been shown to be essentially a measure of β-amylase activity with limited contributions from α-amylase, limit dextrinase and other yet to be defined factors. However, selection of DP by barley breeders to improve fermentability is inefficient as it is a quantitative genetic character (controlled by multiple genes) and as already stated, it is not the best predictor of malt fermentability. Furthermore, the dilemma facing barley breeders selecting for DP can be appreciated when consulting the consensus genetic map of malt quality traits which locates multiple quantitative trait loci (QTL) for DP on each of the seven barley chromosomes (Fox *et al.*, 2003). In terms of DP enzyme selection targets, the structural genes, haplotype and their location are important for barley breeders devising selection strategies to improve or optimise malt quality.

Of the DP enzymes, β-amylase is by far the best characterised in both biochemical and genetic terms. As such, selection for specific β-amylase alleles has become an important component of routine selection within the Australian barley breeding programs. The importance of this selection tool is exemplified by the differences in preference for low and high fermentability between sugar and starch based adjunct brewers. The current paradigm suggests varieties carrying the *Bmy1*-Sd2L allele will suit brewers requiring low levels of brewhouse fermentability, while the *Bmy1*-Sd2H and Sd3 alleles will be most suited to high adjunct brewing styles.

Recently an improved, combined extraction protocol in an 8x12 micro well format has been devised that satisfactorily estimates the levels of β-amylase, α-amylase and limit dextrinase in malt (Evans, 2008). The increased labour and cost efficiencies described increase the potential for DP enzyme analysis to be undertaken as a routine assay in the malting quality evaluation laboratories of barley breeding programs, maltsters and grain traders.

The new DP enzyme analysis method was applied to survey more than 1000 commercial malting samples from Australia and internationally, for the levels of the DP enzymes, α-amylase, β-amylase and limit dextrinase (Evans *et al.*, 2008). The survey showed that there was more variation within a variety for DP and DP enzymes than between varieties. This and other studies (Evans *et al.*, 2007, 2009), demonstrated that varietal genetics selects the "ballpark" for DP enzyme level while the barley growing environment and malting conditions fine tune the resultant level of DP enzymes. A new malting paradigm was thus outlined where maltster selection of appropriate variety, growing location and malting conditions provide the maltster sufficient latitude to produce malt within an appropriate range of fermentibility characteristics to satisfy brewers varying requirements. The basis for this new paradigm of breeding, malting and brewing practice is a shift towards routine evaluation of malt quality based on the level of the DP enzymes, total β-amylase, α-amylase and total limit dextrinase, which are related to the degree of modification (KI) and β-amylase thermostability. The challenge remains to lift the characterisation of the genetic variation of α-amylase, limit dextrinase and other key components of fermentability to that available for β-amylase and harness these differences to guide malting barley improvement.

Acknowledgements

The authors acknowledge funding from the Australian Grains Research and Development Corporation and critical assistance from the Australian brewing and malting industry. The authors are indebted to Ray Anderson (UK) for a precis of the history of diastatic power.

References

Allison M J (1973) Genetic Studies on the β-amylase Isozymes of Barley Malt. Genetica, 44: 1-15

Allison M J (1978) Amylase Activity and Endosperm Hardness of High Lysine Barleys. J Inst Brew, 84: 231-232

Allison M J, Swanston J S (1974) Relationship between β-amylase Polymorphisms in Developing, Mature and Germinating Grains of Barley. J Inst Brew, 80: 285-291

American Society of Brewing Chemists (1992) ASBC Methods of Analysis. American Society of Brewing Chemists: Pilot Knob, Mn.

Anon (1906) Diastatic Activity (Lintner Value). J Inst Brew, 12: 1-10

Anon (1992) Recommended Rules for Nomenclature and Gene Symbolization in Barley. Barley Genet News, 21: 11-13

Ao Z, Jane J L (2007) Characterisation and Modelling of the A- and B-granule Starches of Wheat, Triticale and Barley. Carbohydr Polym, 67: 46-55

Arakawa T, Timasheff S N (1982) Stabilization of Protein Structure by Sugars. Biochem, 21: 6536-6544

Arends A M, Fox G P, Henry R J, Marschke R J, Symons M H (1995) Genetic and Environmental Variation in the Diastatic Power of Australian Barley. J Cereal Sci, 21: 63-70

Back J F, Oakenfull D, Smith M B (1979) Increased Themal Stability of Proteins in the Presence of Sugars and Polyols. Biochem, 18: 5191-5196

Baks T, Janssen A E M, Boom R M (2006a) A Kinetic Model to Explain the Maximum in A-amylase Activity Measurements in the Presence of Small Carbohydrates. Biotechnol Bioeng, 94: 431-440

Baks T, Janssen A E M, Boom R M (2006b) The Effect of Carbohydrates on A-amylase Activity Measurements. Enzyme Microb Technol, 39: 114-119

Bamforth C W (1986) Enzymes from the Grist. Brew, 72: 427-434

Bamforth C W (2001) Beer Flavour: Mouthfeel. Brew Guard, 130: 18-19

Bamforth C W (2003) Barley and Malt Starch in Brewing: a General Review. Master Brew Ass Am Tech Quart, 40: 89-97

Bekkers A, Vissenaekens J, Baks T (2007) Attenuation Limit of Malt as a Function of Gelatinisation Characteristics of Starch. In Eur Brew Conv Cong Proc Venice Vol.31, in Press

Bendelow V M (1964) Inheritance of Free β-amylase in Barley. Can J Plant Sci, 44: 550-554

Bertoft B, Andtfolk C, Kulp S E (1984) Effect of pH, Temperature, and Calcium Ions on Barley Malt α-amylase Isoenzymes. J Inst Brew, 90: 298-302

Bilderback D E (1974) Amylases from Aleurone Layers and Starchy Endosperm of Barley Seeds. Plant Physiol, 53: 480-484

Brandee L, Duensing W, Halstad S, Klimovitz R, Laidlaw A (1999) Adjuncts, In The Practical Brewer. In: McCabe J T (edt.) Master Brewers Association of the Americas: Wautosa: WI. 75-98

Briggs D E, Boulton C A, Brookes P A, Stevens R (2004) Brewing: Science and Practice. CRC Press, New York

Brown A H D, Jacobsen J V (1982) Genetic Basis and Natural Variation of α-amylase Isozymes in Barley. Genet Res Cambridge, 40: 315-324

Burton R A, Zhang X Q, Hrmova M, Fincher G B (1999) A Single Limit Dextrinase Gene Is Expressed Both in the Developing Endosperm and in Germinated Grains of Barley. Plant Physiol, 119: 859-871

Callis J, Ho T D (1983) Multiple Molecular Forms of the Gibberellin-induced Alpha-amylase from the Aleurone Layers of Barley Seeds. Arch Biochem Biophys, 224: 224-234

Chalmers K J, Waugh R, Watters J, Foster B P, Nero E, Abbott R J, Powell W (1992) Grain Isozyme and Ribosomal DNA Varibility in Hordeum Spontaneum Populations from Israel. Theor Appl Genet, 84: 313-322

Chiapparinio E, Donini P, Reeves J, Tuberosa T, O'Sullivan (2006) Distribution of β-amylase I Haplotypes Amoung European Cultivated Barleys. Mol Breeding, 18: 341-354

Chou P Y, Fasman G D (1978) Prediction of Secondary Structure of Proteins from Their Amino Seqence. Adv Enzymol Relat Areas Mol Biol, 47: 45-147

Clark S E, Hayes P M, Henson C A (2003) Effects of Single Nucleotide Polymorphisms in β-amylase 1 Alleles from Barley on Functional Properties of the Enzymes. Plant Physiol Biochem, 41: 798-804

Clark S E, Hayes P M, Henson C A (2005) Characterisation of Barley Tissue-ubiquitous β-amylase 2 and Effects of the Single Nucleotide Polymorphisms on the Enzymes Thermostability. Crop Sci, 45: 1868-1876

Coventry S, Collins H M, Barr A R, Jefferies S P, Chalmers K J, Logue S J, Langridge P (2003) Use of Putative QTLs and Structural Genes in Marker Assisted Selection for Diastatic Power in Malting Barley (*Hordeum Vulgare* L.). Aust J Agric Res, 54: 1241-1250

Daussant J S, Sadowski J, Ziegler P (1994) Cereal β-amylases: Diversity of the β-amylase Isozyme Status Within Cereals. J Plant Physiol, 143: 585-590

Delcour J A, Verschaeve S G (1987) Malt Diastatic Activity. Part 2. A Modified EBC Diastatic Power Assay for the Selective Estimation of β-amylase Activity. Time and Temperature Dependence of the Release of Reducing Sugars. J Inst Brew, 93: 296-301

Dowhanick T M (1999) Yeast - Strains and Handling Techniques, In The Practical Brewer. In: McCabe, J T (ed.) Master Brewers Association of the Americas: Wauwatosa: WI. 263-298

Eckstein P E, Li C D, Hay D, Rossnagel B G, Lance R C M, Scholes G J (2004) Allele-specific Markers and Molecular Diversity at the *b-amy1* Locus Determining Enzyme Thermostability. In Proc Intl Barley Genet Symp Bruno, 9: 263-269

Edney M J, Eglinton J K, Collins H M, Barr A R (2004) Biochemical and Genetic Basis of Wort Fermentability. In: Proc Intl Barley Genet Symp Brno: 418-424

Edney M J, Eglinton J K, Collins H M, Barr A R, Legg W G, Rossnagel B G (2007a) Effects of the High Heat Stable Beta-amylase Allele on Fermentability of Malts with Low Vs High Diastatic Power. In: Eur Brew Conv Cong Proc Venice, 31: 50

Edney M J, Eglinton J K, Collins H M, Barr A R., Legge W G , Rossnagel B G (2007b) Importance of Endosperm Modification for Malt Wort Fermentability. J Inst Brew, 113: 228-238

Eglinton J K (2003) Novel Alleles from Wild Barley for Breeding Malting Barley (*Hordeum Vulgare* L.), in School of Agriculture and Wine. University of Adelaide: Adelaide.

Eglinton J K, Evans D E (1997) Identification of the Beta-amylase Isoelectric Focusing Band Pattern in Barley (*Hordeum Vulgare* L.). Barley Genet News, 27: 8-12

Eglinton J K, Langridge P, Evans D E (1998) Thermostability Variation in Alleles of Barley Beta-amylase. J Cereal Sci, 28: 301-309

Enevoldsen B S, Bathgate G N (1969) Structural Analysis of Wort Dextrins by Means of β-amylases and the Debranching Enzymes Pullanase. J Inst Brew, 75: 433-443

Enevoldsen B S, Schmidt F (1973) Dextrins in Brewing. II. Distribution of Oligo- and Megalo- Saccharides during Mashing, in Wort, and in Beer. In Eur Brew Conv Cong Proc Salzberg, 14: 135-148

Engh R, Huber R (1991) Accurate Bond and Angle Parameters for X-ray Protein Structure Refinement. Acta Crystallogr, Sect. A: Found Crystallogr, 47: 392-400

Erdal K, Jensen M O, Kristensen M, Krough J J, Riis P, Vaag P (1993) Total β-amylase in Barley Used as a Screening Criteria for Combined Amylolytic Activity in Malt. In Eur Brew Conv Cong Proc Oslo, 24: 147-154

Erkkila M J, Leah R, Ahokas H, Cameron-Mills V (1998) Allele Dependant Barley Grain β-amylase Activity. Plant Physiol, 117: 679-685

Etokakpan O U, Palmer G H (1990) A Simple Di-amylase Procedure for The Estimation of Alpha-amylase and Diastatic Activity. J Inst Brew, 96: 89-91

European Brewery Convention (1998) Analytica - EBC, Nurnberg, Germany: Verlag Hans Carl

Evans D E (2008) A More Cost and Labour Efficient Assay for the Combined Measurement of the Diastatic Power Enzymes, β-amylase, α-amylase and Limit Dextrinase. J Am Soc Brew Chem, 66: 215-222

Evans D E, Hamet M A G (2005) The Selection of a Dried Yeast Strain for Use in the Apparent Attenuation Limit Malt Analysis (AAL) Procedure. J Inst Brew, 111: 209-214

Evans D E, Collins H M, Eglinton J K, Wilhelmson A (2005) Assessing the Impact of the Level of Diastatic Power Enzymes and Their Thermostability on the Hydrolysis of Starch during Wort Production to Predict Malt Fermentability. J Am Soc Brew Chem, 63: 185-198

Evans D E, Li C, Eglinton J K (2007) A Superior Prediction of Malt Attenuation. In: Eur Brew Conv Cong Proc Venice, 31: 54-66

Evans D E, Li C, Eglinton J K (2008a) Improved Prediction of Malt Fermentability by the Measurement of the Diastatic Power Enzymes, β-amylase, α-amylase and Limit Dextrinase. I. Survey of the Level so of Diastatic Power Enzymes in Commercial Malts. J Am Soc Brew Chem, 66: 223-232

Evans D E, Li C, Harasymow S, Roumeliotis S, Eglinton J K (2009) Improved Prediction of Malt Fermentability by the Measurement of the Diastatic Power Enzymes, β-amylase, β-amylase and Limit Dextrinase. II. Impact of Barley Genetics and Gibberellin on Levels of α-amylase and Limit Dextinase in Malt. J Am Soc Brew Chem, 67: in Press

Evans D E, MacLeod L C, Eglinton J K, Gibson C E, Zhang X, Wallace W, Skerritt J S, Lance R C M (1997a) Measurement of β-amylase in Malting Barley (*Hordeum Vulgare* L.). Part 1: Development of a Quantitative ELISA for Beta-amylase. J Cereal Sci, 26: 229-239

Evans D E, Van Wegen B, Ma Y, Eglinton J K (2003) The Impact of the Thermostability of α-amylase, β-amylase and Limit Dextrinase on Potential Wort Fermentability. J Am Soc Brew Chem, 61: 210-218

Evans D E, Wallace W, Lance R C M, MacLeod L C (1997b) Measurement of Beta-amylase in Malting Barley (*Hordeum Vulgare* L.). Part 2: The Effect of Germination and Kilning on β-amylase. J Cereal Sci, 26: 241-250

Filner F, Varner J E (1967) A Test for the *de Novo* Synthesis of Enzyme: Density Labeling with H_2O^{18} of Barley α-amylase Induced by Gibberellic Acid. Proc Natl. Acad Sci USA, 58: 1520-1526.

Forster B P, Thompson D M, Walters J, Powell W (1991) Water-soluble Proteins of Mature Barley Endosperm: Genetic Control, Polymorphism and Linkage with -amylase and Spring/winter Habit. Theor Appl Genet, 81: 787-792

Fox G P, Panozzo J F, Li C D, Lance R C M (2003) Molecular Basis of Barley Quality. Aust J Agric Res, 54: 1081-1101

Frandsen J P, Lok F, Mirgorodskaya E, Roepstorff P, Svensson B, (2000) Purification, Enzymatic Characterisation and Nucleotide Sequence of a High Isoelectric Point Alpha Glucosidase from Barley Malt. Plant Physiol, 123: 275-286

Gibson T S, Solah V, Glennie Holmes M R, Taylor H R (1995) Diastatic Power in Malted Barley: Contributions of Malt Parameters to Its Development and the Potential of Barley Grain β-amylase to Predict Malt Diastatic Power. J Inst Brew, 101: 277-280

Giese H, Hejgaard J (1984) Synthesis of Salt-soluble Proteins in Barley. Pulse Labelling Study of Grain Filling in Liquid-cultured Detached Spikes. Planta, 161: 172-177

Giese H, Hopp H E (1984) Influence of Nitrogen on the Amount of Hordein, Protein Z and β-amylase Messanger RNA in Developing Endosperms of Barley. Carlsberg Res Commun, 49: 365-383

Goldsmith M R (2007) Malt Attenuation - An Australian Brewers Perspective. In Aust Barley Technical Symposium. Fremantle, WA, 13: 107-115

Guerin J R, Lance R C M, Wallace W (1992) Release and Activation of Barley β-amylase by Malt Endopeptidases. J Cereal Sci, 15: 5-14

Gunkel J, Voetz M, Rath, F (2002) Effect of the Malting Barley Variety (*Hordeum Vulgare* L.) on Fermentability. J Inst Brew, 108: 355-361

Hamilton D M, Lewis M J (1974) Factors Affecting Wort Extract and Attenuation. Master Brew Ass Am Tech Quart, 11: 31-34

Hara-Nishimura M, Daussant J (1986) Conversion of Free β-amylase to Bound β-amylase on Starch Granules in Barley Endosperm during Desiccation Phase of Seed Development. Protoplasma, 134: 149-153

Hardie D G (1975) Control of Carbohydrate Formation by Gibberellic Acid in Barley Endosperm. Phytochemistry, 14: 1719-1722

Hayden M J, Nguyen T M, Waterman A, McMichael G L, Chalmers K J (2007) Application of Multiplex-ready PCR for Fluorescence-based SSR Genotyping in Barley and Wheat. Mol Breeding, 21, in Press

Hayes P M, Jones B L (2000) Malting Quality from a QTL Perspective. In Intl Barley Genet Symp VIII. Adelaide, Vol.I: 99-106

Hayes P, Castro A, Marquez-Cedillo L, Corey A, Henson C, Jones B L, Kling J, Mather D, Matus I, Rossi C, Sato K (2003) Genetic Diversity Inherited Agronomic and Malting Quality Traits. In: Von Bothmer R, Von Hintum T, Knupffer H, Sato K (eds.) Diversity in Barley (*Hordeum Vulgare*). Elsevier: Amsterdam, 201-226

Haywood R (1996) Brewing Adjuncts. Brew, 82(981):309-311

Hejgaard J (1976) Free and Protein-bound β-amylases of Barley Grain. Characterization by Two Dimentional Immunoelectrophoresis. Plant Physiol, 38: 293-299

Hejgaard J, Boisen S (1980) High-lysine Proteins in Hiproly Breeding: Identification, Nutritional Significance and New Screening Methods. Hereditas, 93: 311-320

Hejgaard J, Carlsen S, (1977) Immunoelectric Identification of a Heterodimer -amylase in Extracts of Barley Grain. J Sci Food Agric, 28: 900-904

Henson C A, Duke S H (2008) A Comparison of Standard and Nonstandard Measures of Malt Quality. J Am Soc Brew Chem, 66: 11-19
Henson C A, Stone J M (1988) Variation in Alpha-amylase and Alpha-amylase Inhibitor Activities in Barley Malts. J Cereal Sci, 8: 39-46
Holtekjolen A K, Uhlen A K, Brathen E, Knutsen S H (2006) Contents of Starch and Non-starch Polysaccharides in Barley Varieties of Different Origin. Food Chem, 94: 348-358
Ingverson J, Englyst A, Jorgensen K G (1989) Evaluation of Malting Quality in a Barley Breeding Program. Use of α-amylase and β-glucan Levels in Mlat as Preselection Tools. J Inst Brew, 95: 99-103
Institute Of Brewing (1998) Methods of Analysis. London: Institute of Brewing.
Jacobsen J V, Higgins T J V (1982) Characterization of the Alpha-amylase Synthesized by Aleurone Layers of Himalaya Barley in Response to Gibberellic Acid. Plant Physiol, 70: 1647-1653
Jorgensen O (1965) Barley Malt α-glucosidase Vi. Localization and Development During Barley Germination. Acta Chem Scand, 19: 15-16
Jung W, Skadsen R W, Peterson D M (2001) Characterisation of a Novel Barley β-amylase Gene Expressed Only during Early Grain Development. Seed Sci Res, 11: 325-333
Kadziola A, Abe J I, Svensson B, Haser R (1994) Crystal and Molecular Structure of Barley α-amylase. J Mol Biol, 239: 104-121
Kaneko T, Asakura T, Ito K, Takeda K (2000) Studies on Breeding of β-amyase Activity in Barley. II. Brewing Property of β-amylase-less Barley. Breeding Sci(Japan), 2, 156
Kaneko T, Kihara M, Ho K (2000) Genetic Analysis of β-amylase, Thermostability to Develop a DNA Marker for Malt Fermentability Improvement in Barley *Hordeum Vulgare*. Plant Breeding, 119: 197-201
Kaneko T, Zhang W S, Ito K, Takeda K (2001) World Wide Distribution of Beta-amylase Thermostability in Barley. Euphytica, 121: 223-228
Kessler M, Zarnkow M, Kreiss S, Bach W (2005) Gelatinization Properties of Different Cereals and Pseudocereals. Monat Brau, 58: 82-88
Khursheed B, Rogers J L (1988) Barley α-amylase Genes: Quantitative Comparison of Steady-state MRNA Levels from Individual Members of the Two Different Familiies Expressed in Aleurone Cells. J Biol Chem, 263: 18953-18960
Kihara M, Kaneko T, Ito K (1998) Genetic Variation of β-amylase Thermostability among Varieties of Barley,*Hordeum Vulgare* L., and Relation to Malting Quality. Plant Breeding, 117: 425-428
Kihara M, Kaneko T, Ito K, Aida Y, Takeda K (1999) Geographical Variation of β-amylase Thermostability among Varieties of Barley (*Hordeum Vulgare*) and -amylase Deficiency. Plant Breeding, 118: 453-455
Kihara M, Okada Y, Kuroda H, Saeki K, Yoshigi N, Ito K (1997) Generation of Fertile Transgenic Barley Synthesising Thermostable β-amylase. In Proc Eur Brew Conv, Cong Maastrict Vol.26
Kjeldahl J (1879) Diastase. Comptes Rendus Laboratory Carlsberg, 1, 109
Kreis M, Williamson M, Buxton B, Pywell J, Hejgaard J, Svenssen I (1987) Primary Structure and Differential Expression of β-amylase in Normal and Mutant Barleys. Eur J Biochem, 169: 517-525

Kreis M, Williamson P, Shrewry P R, Sharp P, Gale M (1988) Identification of Sccond Locus Encoding β-amylasc on Chromosomc 2 of Barley. Genet Res Cambridge, 51: 13-16

Kristensen M, Lok F, Planchot V, Svendsen I, Leah R, Svensson B (1999) Isolation and Characterisation of the Gene Encoding the Starch Debranching Enzyme Limit Dextrinase from Germinating Barley. Biochim Biophys Acta, 1431: 538-546

La B D E, Marchylo B A (1983) Heterogeneity of The Beta-amylase Enzymes of Barley. J Am Soc Brew Chem, 41: 120-124

Langstaff S A, Lewis M J (1993) The Mouthfeel of Beer - a Review. J Inst Brew, 99: 31-37

Lee W J, Pyler R E (1984) Barley Malt Limit Dextrinase: Varietal, Environmental and Malting Effects. J Am Soc Brew Chem, 42: 11-17

Lenoir P, MacGregor A W, Moll M, Daussant J (1984) Indentification of De La R-enzyme De'orge Et Du Malt Par Focalisation Isoelectrique. Comptes Rendus De L'Academie Des Sciences, Paris, 298: 243-248

Lewis M J, Young T W (2001) Brewing. Gaithersberg, Mayland: Aspen Publishers Inc

Li C (1998) Genetic Control of Hydrolytic Activity in Germinated Barley (*Hordeum Vulgare* L.), in *Ph.D. Thesis*. University of Adelaide: Adelaide, Australia

Li C D, Langridge P, Zhang XQ, Eckstein P E, Rosnagel B G, Lance R C M, Lefol E B, Lu M Y, Harvey B L, Scoles G J (2002) Mapping of Barley (*Hordeum Vulgare* L.) β-amylase Alleles in which an Amino Acid Substitution Determines β-amylase Isozymes Type and Level of Free β-amylase. J Cereal Sci, 35: 39-50

Li C D, Zhong X Q, Eckstein P, Rossragel B G, Scoles G J (1999) A Polymorphic Micro Satellite in the Limit Dextrinase Gene of Barley(Hordeum Vulgare L.). Mol Breeding, 5: 569-577

Lintner C J (1886) The Nature and Properties of Diastase. Wochenschrift Fur Brauerei: 733-736; 753-755.

Longstaff M A, Bryce J H (1993) Development of Limit Dextrinase in Germinated Barley (*Hordeum Vulgare* L.). Plant Physiol, 101: 881-889

Lund D (1984) Influence of Time, Temperature, Moisture, Ingredients, and Processing Conditions on Starch Gelatinization. Crit Rev Food Sci Nut, 20: 249-273

Lundgard R, Svensson B., (1986) Limited Proteolysis in The Carboxy-terminal Region of Barley β-amylase. Carlsberg Res Commun, 51: 487-491

Lundgard R, Svensson B (1987) The Four Major Forms of Barley β-amylase. Purification, Characterization and Structural Relationship. Carlsberg Res. Commun, 52: 313-326

Ma YF, Eglinton J K, Evans D E, Logue S J, Langridge P (2000) Removal of the Four C-terminal Glycine Rich Repeats Enhances the Thermostability of Barley β-amylase. Biochem, 39: 13350-13355

Ma Y F, Langridge P, Logue S J, Evans D E (2001) The Amino Acid Substitutions of Barley β-amylase Allellic Forms that Improve Thermostability and Substrate-binding Affinity. Mol Gen Genet, 266: 345-352

Ma Y F, Langridge P, Logue S J, Evans D E (2002) A Single Amino Acid Substitution that Determines the IEF Band Pattern of Barley β-amylase. J Cereal Sci, 35: 79-85

MacGregor A W (1996) Malting and Brewing Science: Challenges and Opportunities. J Inst Brew, 102: 97-102

MacGregor A W, Ballance D C (1980) Hydrolysis of Large and Small Starch Granules from Normal and Waxy Barley Cultivars by Alpha-amylases from Barley Malt. Cereal Chem, 57: 397-402

MacGregor A W, Lenoir C (1987) Studies on α-glucosidase in Barley and Malt. J Inst Brew, 93: 334-337

MacGregor A W, Bazin S L, Izydorczyk M S (2002a) Gelatinization Characteristics and Enzyme Susceptibility of Different Types of Barley Starch in the Temperature Range 48-75 °C. J Inst Brew, 108: 43-47

MacGregor A W, Bazin S L, Schroeder S W (2002b) Effect of Starch Hydrolysis Products on the Determination of Limit Dextrinase and Limit Dextrinase Inhibitors in Barley and Malt. J Cereal Sci, 35: 17-28

MacGregor A W, Bazin S L, Macri L J, Babb J C (1999) Modeling the Contribution of α-amylase, β-amylase and Limit Dextrinase to Starch Degradation during Mashing. J Cereal Sci, 29: 161-169

MacGregor A W, LaBerge D E, Meredith W O S (1971) Separation of α- and β-amylase Enzymes from Barley Malt by Ion-exchange Chromatography. Cereal Chem, 48: 490-498

MacGregor A W, Macri L J, Bazin S L, Sadler G W (1995) Limit Dextrinase Inhibitor in Barley and Malt and Its Possible Role in Malting and Brewing. In Eur Brew Conv Cong Proc Brussels, 25: 185-192

MacGregor A W, Macri L J, Schroeder S W, Bazin S L (1994) Purification and Characterisation of Limit Dextrinase Inhibitors from Barley. J Cereal Sci, 20: 33-41

MacGregor A W, Marchylo B A, Kruger J E (1988) Multiple α-amylase Components in Germinated Cereal Grains Determined by Isoelectric Focusing and Chromatofocusing. Cereal Chem, 65: 326-333

MacGregor E A (2004) The Proteinaceous Inhibitor of Limit Dextrinase in Barley and Malt. Biochim Biophys Acta, 1696: 165-170

Macri L J, MacGregor A W, Schroeder S W, Bazin S L (1993) Detection of a Limit Dextrinase Inhibitor in Barley. J Cereal Sci, 18: 103-106

Malysheva L, Ganal M W, Roder M S (2004) Evaluation of Cultivated Barley (*Hordeum Vulgare*) Germplasm for the Presence of Thermostable Alleles of β-amylase. Plant Breeding, 123: 128-131

Manners D J, Yellowlees D (1973) Studies on Debranching Enzymes. Part 1. The Limit Dextrinase Activity of Extracts of Certain Higher Plants and Commercial Malts. J Inst Brew, 79: 377-385

McCafferty C A, Jenkinson H R, Brosnan J M, Bryce J H (2004) Limit Dextrinase - Does Its Malt Activity Relate to Its Activity during Brewing. J Inst Brew, 110: 284-296

McCleary B V (1992) Measurement of the Content of Limit-dextrinase in Cereal Flours. Carbohydr Res, 227: 257-268

McCleary B V, Codd R (1989) Measurement of β-amylase in Cereal Flours and Commercial Enzyme Preparations. J Cereal Sci, 9: 17-33

McCleary B V, Sheehan H (1987) Measurement of Cereal β-amylase: a New Assay Procedure. J Cereal Sci, 6: 237-251

Meredith W O S (1970) I Effects of Amylases and Metals on the Pasting Properties of Wheat Flour, Determined by the Amylograph and by Hagberg's Falling-Number Method. Cereal Chem, 47: 483-491

Mikami B, Degano M, Hehre E, Sacchettini J C (1994) Crystal Structures of Soybean β-amylase Reacted with β-maltose and Maltol : Active Site Components and Their Apparent Roles in Catalysis. Biochem, 33: 7779-7787

Moll M, Flayeux R, Lipus G, Marc A (1981) Biochemistry of Mashing. Master Brew Ass Am Tech Quart, 18: 166-173

Mousia Z, Balkin R C, Pandiella S S, Webb C (2004) The Effect of Milling Parameters on Starch Hydrolysis of Milled Malt in the Brewing Process. Proc Biochem, 39: 2213-2219

Munck L, Mundy J, Vaag P (1985) Characterization of Enzyme Inhibitors in Barley and Their Tentitive Role in Malting and Brewing. J Am Soc Brew Chem, 43: 35-38

Muslin E H, Clark S E, Henson C A (2002) The Effect of Proline Insertions on the Thermostability of a Barley α-glucosidase. Protein Eng, 15: 29-33

Muslin E H, Karpelenia C B, Henson C A (2003) The Impact of Thermostable α-glucosidase on the Production of Fermentable Sugars during Mashing. J Am Soc Brew Chem, 61: 142-145

Muthukrishnan S, Gill B S, Swegle M, Chandra C R (1984) Structural Genes for α-amylases Are Located on Barley Chromosomes 1 and 6. J Biol Chem, 259: 13637-13639

Myllarinen P, Schulman A A, Salovaara H, Poutanen K (1998) The Affect of Growth Temparature on Gelatinization Properties of Barley Starch. Acta Agricultural Scandinavia, Section B, Soil Plant Sci, 48: 85-90

Nielsen G, Johansen H B (1986) Proposal for the Identification of Barley Varieties Based on the Genotypes for 2 Hordein and 39 Isoenzyme Loci of 47 Reference Varieties. Euphytica, 35: 717-728

Nielsen G, Johansen H, Jemsen J, Hejgaard J (1983) Localization on Barley Chromosome 4 of Genes Coding for β-amylase (Bmy 1) and Protein Z (Paz 1). Barley Genet News: 13: 55-57

Niku-Paavola M L, Sakoun A, Nummi M, Daussant J (1973) The Polymorphism of Barley β-amylase. Biochim Biophys Acta, 322: 281-184

Norris K, Lewis L (1985) Application of Commercial Barley β-amylase in Brewing. J Am Soc Brew Chem, 43: 96-101

Okada Y, Yoshigi N, Sahara H, Koshino S (1995) Increase in Thermostability of Recombinant Barley β-amylase by Random Mutagenesis. Bios Biotechnol Biochem, 59: 1152-1153

Oliveira A B, Rasmusson D C, Fulcher R G (1994) Genetic Aspects of Starch Granule Traits in Barley. Crop Sci, 34: 1176-1180

Ovesna J, Machova-Polakova K, Kucera L, Vaculova K, Miltova J (2006) Evaluation of Czech Spring Malting Barleys with Respect to B-amylase Allele Incidence. Plant Breeding, 125: 236-242

Palmer J J (2006) How the Mash Works, In How to Brew: Everything You Need to Know to Brew Beer Right the First Time. Brewers Association: Boulder: Co. 141-152 .

Paris M, Eglinton J K (2002) Genotyping Single Nucleotide Polymorphisms for Selection of Barley β-amylase Alleles. Plant Mol Biol Rept, 20: 149-159

Payen A, Persoz J F (1833) Memoire Sur La Diastase, Les Principaux Produits De Ses Reactions, Et Leurs Applications Aux Arts Industriels. Ann Chim, 53, 73-92

Piendl A (1973) Malt Modification and Mashing Conditions as Influencing Factors on the Carbohydrates of Wort. Brew Dig, 48: 58-84

Polakova K, Laurie D, Vaculova K, Ovesna J (2003) Charactrization of B-amylase Alleles in 79 Varieties with Pyrosequencing. Plant Mol Biol Rept, 21: 439-447

Powling A, Islam A K M R, Shepard K W (1981) Isozymes in Wheat-barley Hybrid Derivitive Lines. Biochem. Genet, 19: 225-232

Ragot F, Guinard J X, Shoemaker C F, Lewis M J (1989) The Contribution of Dextrins to Beer Sensory Properies. Part 1 Mouthfeel. J Inst Brew, 95: 427-430

Ramachandran G N, Ramakrishnan V, Sasisekharen V (1963) Stereochemistry of Polypeptide Chain Conformations. J Mol Biol, 7: 95-99

Richardson J S, Richardson D C (1988) Amino-acid Preferences for Specific Locations at the Ends of Alpha-helices. Science, 240: 1648-1652

Rogers J C (1985) Two Barley Alpha-amylase Gene Families Are Regulated Differently in Aleurone Cells. J Biol Chem, 260: 3731-3738

Rogers J C, Millinan C (1983) Isolation and Sequence Analysis of A Barley Alpha-amylase CDNA Clone. J Biol Chem, 258: 8164-8174

Ross H A, Sungurtas L, Ducreux J S, Swanston H V, Davies G J, McDougall (2003) Limit Dextrinase in Barley Cultivars of Differing Malting Quality: Activity, Inhibitors, and Limit Dextrin Profiles. J Cereal Sci, 38: 325-334

Runkel U D (1975) Malt Kilning and Its Influence on Malt and Beer Quality. Eur Brewers Convention Monograph, II(Zeist): 222-233

Sallans H R, Anderson J A (1940) Varietal Differences in Barley and Malts. X. Correlations of Carbohydrates with Nitrogen Fractions and with Malt Extract, Steeping Time and Malting Loss Can J Res, 18: 219-229

Sandegren E, Klang N (1950) On Barley Amylase and Proteinase. J Inst Brew, 56: 313-318

Santos M M M, Riis, P (1996) Optimized McCleary Method for Measurement of Total β-amylase in Barley and Its Applicability. J Inst Brew, 102: 271-275

Schroeder S W, MacGregor A W (1998) Synthesis of Limit Detrinase in Germinated Barley Kernals and Aleurone Tissues. J Am Soc Brew Chem, 56: 32-37

Schwartz P B, Li Y, Barr J, Horsley R D (2007) Effect of Operational Paramenters on the Determination of Laboratory Extract and Associated Wort Quality Factors. J Am Soc Brew Chem, 65: 219-228

Shen-Miller J, Kreis M, Shrewry P R, Springall D R (1991) Spatial Distribution of β-amylase in Germinating Barley Seeds Based on the Avidin-biotin-peroxidase Complex Method. Protoplasma, 163: 162-173

Shewry P R, Parmar S, Buxton B, Gale M D, Liu C J, Hejgaard J, Kreis M (1988) Multiple Molecular Forms of β-amylase in Seeds and Vegitative Tissues of Barley. Planta, 176: 127-134

Sidenius U, OlsenK, Svensson B, Christensen V (1995) Basi-Amyz Complex Occurs in a Two Step Reaction. FEBS Lett, 361: 250-254

Simon P, Jones R L (1988) Synthesis and Secretion of Catalytically Active Barley α-amylase Isoforms by *Xenopus* Oocytes Injected with Barley MRNAs. Eur J Cell Biol, 47: 213-221

Sissons M J, MacGregor A W (1994b) Hydrolysis of Starch Granules by α-glucosidases from Malt. J Cereal Sci, 19: 161-169

Sissons M J, Lance R C M, Sparrow D H B (1992a) Studies on Limit Dextrinase in Barley 1. Purification of Malt Limit Dextrinase and Production of Monospecific Antibodies. J Cereal Sci, 16: 107-116

Sissons M J, Lance R C M, Sparrow D H B (1992b) Studies on Limit Dextrinase in Barley 2. Application of an ELISA and Immunoblotting to Studies of Genetic Variability and Malting Effects. J Cereal Sci, 16: 117-128

Sissons M J, Lance R C M, Sparrow D H B (1993c) Studies on Limit Dextinase in Barley. 3. Limit Dextrinase in Developing Kernals. J Cereal Sci, 17: 19-24

Sissons M J, Taylor M, Proudlove M (1995) Barley Malt Limit Dextrinase: Its Extraction, Heat Stability and Activity during Malting and Mashing. J Am Soc Brew Chem, 53: 105-110

Sjöholm K, Macri L J, MacGregor A W (1995) Is There a Role for Limit Dextrinase in Mashing? In Eur Brew Conv Cong Proc Bruss Vol. 25: 277-284

Sogaard M, Adensen J S, Roepstoff P, Svensson B (1993) Electrospray Mass Spectrometry Characterization of Post-translational Modifications of Barley Alpha-amylase-1 Produced in Yeast. Biotechnol, 11: 1162-1164

Sogaard M, Olsen F L, Svensson B (1991) C-terminal Processing of Barley Alpha-amylase 1 in Malt, Aleurone Protoplasts and Yeast. Proc Natl Acad Sci USA, 88: 8140-8144

Sopanen T, Lauriere C (1989) Release and Activity of Bound β-amylase in a Germinating Barley Grain. Plant Physiol, 89: 244-249

Sorensen S B, Muller M, Skerritt J, Simpson D (1996) Hordein Promotor Methylation and Transcriptional Activity in Wild Type and Mutant Barley Endosperm. Mol Gen Genet, 250: 750-760

Stahl Y, Alexander R D, Coates S, Bryce, J H, Jenkinson, H R, Morris P L (2007) The Barley Limit Dextrinase Inhibitor: Gene Expression, Protein Location, and Interaction with 14-3-3 Protein. Plant Sci, 172: 452-461

Stahl Y, Coates S, BryceJ H, Morris P L (2004) Antisense Downregulation of the Barley Limit Dextrinase Inhibitor Modulates Starch Granule Size Distribution, Starch Composition and Amylopectin Structure. Plant J, 39: 599-611

Stark J R, Yin X S (1987) Evidence for the Presence of Maltase and α-glucosidase Isoenzymes in Barley. J Inst Brew, 93: 108-112

Stenholm K, Home S (1999) A New Approach to Limit Dextrinase and Its Role in Mashing. J Inst Brew, 105: 205-210

Stenholm K, Home S, Pietila K, Jaakkola N, Leino E (1996) Are the Days of Congress Mashing Over? In Proc Barley Malt Wort Symp, Institute of Brewing, (Central and South African Section). Zimbabwe:149-163

Stenholm K, Home S, Pietila K, Macri L J, MacGregor A W (1996) Starch Hydrolysis in Mashing. In Proc Inst Brew Conv (Asia/Pacific). Singapore, 24: 142-145

Stenholm K, Wilhelmson A, Home S (1998) Starch Gelatinization Temperature as a Malt Quality Characteristic. In Proc Avimore Conference on Malting, Brewing and Distilling. Avimore: Institute of Brewing, London, Vol. 5: 242-245

Sun Z, Henson C A (1991) A Quantitative Assessment of the Importance of Barley Seed α-amyalse, β-amyalse, Debranching Enzyme and α-glucosidase in Starch Degradation. Arch. Biochem Biophys, 284: 298-305

Sun Z, Henson C A (1992) Extraction of Alpha-glucosidase from Germinated Barley Kernels. J Inst Brew, 98: 289-292

Svensson B, Mundy J, Gibson R M, Svendson I (1985) Partial Amino Acid Sequences of α-amylase Isozymes from Barley Malt. Carlsberg Res Commun, 50: 15-22

Takano T, Takeda G (1985) Polymorphism in A-amylase in Germinating Seed and Malt of Barley Varieties Detected by Iso-electric Focusing Gel Electrophoresis. Jpn J Breeding, 35: 9-16

Takano T, Kiribuchi C, Takeda G (1988) Genetic Analysis on Banding Patterns and Activity of α-amylase Isozymes in Barley (*Hordeum Vulgare* L.). Jpn J Breeding, 38: 65-71

Tester R F, Karkalas J, Qi X (2004) Starch - Composition, Fine Structure and Architecture. J Cereal Sci, 39: 151-165

Tester R F, South J B, Morrison W R, Ellis R P (1991) The Effects of Ambient Temperature during the Grain Filling Period on Composition and Properties of Starch from Four Barley Genotypes. J Cereal Sci, 13: 113-127

Tibbot B K, Skadsen R W (1996) Molecular Cloning and Characterization of Gibberellin-induced Putative α-glucosidase Gene from Barley. Plant Mol Biol, 30: 229-241

Verstrepen K J, Derdelinckx G, Winderickx J, Thevelein J M, Pretorius I S, Delvaux F R (2003) Flavor-active Esters: Adding Fruitiness to Beer. J Biosci Bioeng, 96: 110-118

Visuri K, Nummi M (1972) Purification and Characterization of Crystalline β-amylase from Barley. Eur J Biochem, 28: 555-565

Walker J W, Bringhurst T A, Broadhead A L, Brosnan J M, Pearson S Y (2001) The Survival of Limit Dextrinase during Fermentation in the Production of Scotch Whiskey. J Inst Brew, 107: 99-106

Watson S A (1964) Determination of Starch Gelatinization Temperature. In: Whistler R J S R L, Wolfrom M L (eds.) Methods in Carbohydrates Chemistry, Academic Press, New York, 240-242

Weining S, Henry R J (1994) Polymorphisms in the Alpha-amy 1 Gene of Wild and Cultivated Barley Revealed by Polymerase Chain Reaction. Theor Appl Genet, 89: 509-513

Yoshigi N, Okada Y, Maeba H, Sahara H, Tamaki T (1995) Construction of A Plasmid Used for The Expression of A Sevenfold-mutant Barley β-amylase with Increased Thermostability in Escherichia Coli and Properties of the Sevenfold-mutant β-amylase. J Biochem, 118: 562-567

Yoshigi N, Okada Y, Sahara H, Koshino S (1994) PCR Cloning and Sequencing of the β-amyalse CDNA from Barley. J Biochem, 115: 47-51

Zhang W, Kaneko T, Takeda K (2004a) B-Amylase Variation in Wild Barley Accessions. Breeding Sci, 54: 41-49

Zhang W, Kaneko T, Ishii M, Takeda K (2004b) Differentiation of β-amylase Phenotypes in Cultivated Barley. Crop Sci, 44: 1608-1614

Zhang W S, Li X, Lin J B (2007) Genetic Variation of Bmy1 Alleles in Barley (*Hordeum Vulgare* L.) Investigated by CAPS Analysis. Theor Appl Genet, 114: 1039-1050

Zhou Z, Robards K, Helliwell S, Blanchard C (2002) Composition and Functional Properties of Rice. Intl J Food Sci Technol, 37: 849-868

7

The Role of Endosperm Cell Walls in Barley Malting Quality

R. A. Burton, H. M. Collins and G. B. Fincher
Australian Centre for Plant Functional Genomics, School of Agriculture, Food and Wine, University of Adelaide, Waite Campus, Glen Osmond, SA 5064, Australia
corresponding author: Geoffrey B. Fincher, geoffrey.fincher@adelaide.edu.au

7.1 Introduction

The cell walls of the starchy endosperm of barley are relatively minor components of the grain in comparison to starch and protein, but exert considerable influence over many malting quality characteristics. Wall components such as (1,3;1,4)-β-D-glucans, which are commonly but less precisely known as mixed linkage β-glucans, can have direct and often detrimental effects on various processes in the brewery and on the quality and shelf life of the final beer. The same wall components can have indirect effects on other quality characteristics, such as malt extract and fermentability. The undesirable nature of many of these direct and indirect effects has led breeders, maltsters and brewers to develop precise assays to monitor levels of (1,3;1,4)-β-D-glucans in barley grain and malt, and in many cases to select against high levels of this polysaccharide. While commercial interest in endosperm cell wall components has been focused on (1,3;1,4)-β-D-glucans, the walls contain other components that have similar chemical and physicochemical properties and which merit further attention in the context of malting quality of barley. Here, we will examine the properties and roles of all components of barley endosperm walls in malting quality and relate these to the biological functions of cell walls in grain development and germination. There is a good deal of information available on the physiology and enzymology of wall degradation in germinated barley grain, but only recently have we been able to identify enzymes involved in the biosynthesis of key wall constituents. The latter processes will also be reviewed, in particular with respect to potential enhancement of barley malting quality in the future.

Cell walls are central to the growth and development of the endosperm of cereal grains and to the remobilization of the stored reserves during germination. The endosperm consists of two distinct tissues, the aleurone and the starchy endosperm, that have different functions during grain development, dormancy and after germination (Fincher and Stone, 2004). Cells of the starchy endosperm of barley and other cereals are non-living. They are packed with starch and storage protein reserves that are used to support the growth of the developing seedling following grain germination. Walls of the starchy endosperm are usually relatively thin compared with those of other organs and tissues, presumably because of the need for rapid removal of the walls during reserve mobilization after germination, and because the walls of the starchy endosperm do not have a major structural role in the grain, which is supported largely by the surrounding pericarp and seedcoat tissues.

In contrast to cells of the starchy endosperm, the thick-walled, nucleated, aleurone cells are living at grain maturity, and are packed with protein bodies and lipid droplets (Fulcher *et al.*, 1972). The reserve proteins, carbohydrates, lipids and other storage components of aleurone cells, together with hydrolytic enzymes and mitochondria, enable the rapid proliferation of the intracellular membrane system and the protein synthesising machinery, and provide amino acids for the synthesis of the hydrolytic enzymes that mediate starchy endosperm mobilization following germination (Fincher, 1989). The walls of the aleurone layer consist of an inner, relatively thin layer and a thicker layer, but again have a relatively low cellulose content (Bacic and Stone, 1981).

In the following sections we will describe the chemical composition of walls from the starchy endosperm of barley and the structures of individual constituents.

7.2 Composition of Walls of Barley Endosperm Cells

Cell walls have been isolated in non-aqueous media from the starchy endosperm (Fincher, 1975; Ballance and Manners, 1978; Fig. 7.1) and aleurone (Bacic and Stone, 1981) and have been subjected to detailed analyses. The walls from barley endosperm are composed mainly of non-starchy polysaccharides, together with smaller amounts of protein and phenolic acids. The non-starchy polysaccharides of barley grains include cellulose, (1,3;1,4)-β-D-glucans, heteroxylans, which are mainly in the form of arabinoxylans, glucomannans, xyloglucans, pectic polysaccharides, callose, fructans, and arabinogalactan-proteins. With the exception of the fructans and the arabinogalactan-proteins, all are key components of the walls of cells in the various tissues that comprise the grain. Together, the non-starchy polysaccharides usually constitute less than 10% by weight of the grain but, as mentioned above, can be key determinants of grain quality. The precise physical relationships between individual non-starchy polysaccharides and other wall components have not been described, although it is generally considered that microfibrils of cellulose are embedded in a matrix phase of non-cellulosic polysaccharides and protein (Carpita and Gibeaut, 1993; McCann and Roberts, 1991). Wall integrity is maintained predominantly through extensive non-covalent interactions, especially hydrogen bonding, between the matrix phase and microfibrillar constituents. Recently, Hrmova *et al.* (2007) showed that enzymes purified from barley seedlings, known as xyloglucan endotransglycosylases (XETs), could covalently link polysaccharides such as xyloglucan, cellulosic substrates and (1,3;1,4)-β-D-glucans with a range of acceptor substrate molecules, including oligo-xyloglucosides and cellodextrins, at least *in vitro*. If the barley XETs catalyse similar covalent cross-linking reactions in the walls of endosperm cells *in vivo*, this will carry implications for the mobilization of the starchy endosperm in germinated barley grain.

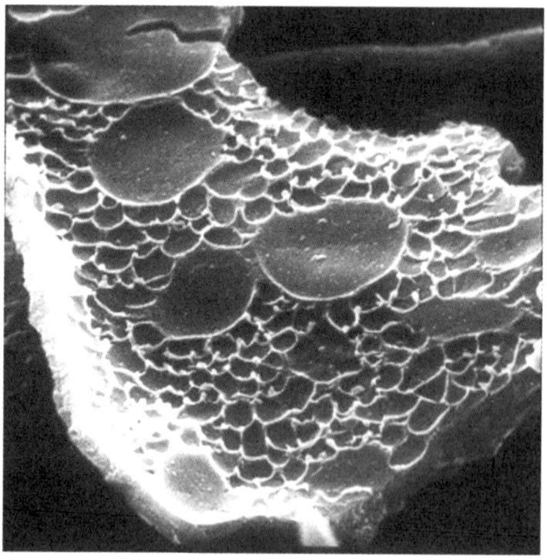

Fig. 7.1. Isolated cell walls from the starchy endosperm of barley grain showing a scanning electron micrograph of a cell wall fragment isolated from the starchy endosperm of mature barley grain (Fincher, 1975)

Non-cellulosic polysaccharides, especially heteroxylans and (1,3;1,4)-β-D-glucans, constitute a relatively high proportion of the walls of the aleurone and starchy endosperm of barley grain (Table 7.1). Again, the generally low cellulose content of these walls, together with the fact that they contain no lignin, might be related to a limited requirement for structural rigidity of walls in central regions of the grain, and to a requirement to rapidly depolymerise wall components following germination of the grain.

In the next section, the physicochemical properties of these major components of barley endosperm walls will be compared.

Table 7.1. Major polysaccharide constituents of cell walls from barley

Source of Walls	*Major Polysaccharides*	*References*
aleurone	71% arabinoxylan	Bacic and Stone (1981)
	26% (1,3;1,4)-β-glucan	
	2% cellulose	
	2% glucomannan	
starchy endosperm	75% (1,3;1,4)-β-glucan	Fincher (1975)
	20% arabinoxylan	
	2% cellulose	
	2% glucomannan	
coleoptiles (4 days)	10% (1,3;1,4)-β-glucan	Gibeaut *et al.* (2005)
	28% arabinoxylan	
	35% cellulose	
	10% pectin	
	10% xyloglucan	

7.3 Properties of Major Wall Components

The properties of cell walls of barley grains are directly related to germination and malting procedures, and they are mainly dependent on chemical composition.

7.3.1 (1,3;1,4)-β-D-Glucans

(1,3;1,4)-β-D-Glucans (thereafter refers to β-Glucan) are distributed almost exclusively in members of the monocotyledon family Poaceae, to which the cereals and grasses belong, and in related families of the order Poales (Trethewey *et al.*, 2005). The concentration of β-glucans in barley grains depends on the genotype, the position of the grain on the spike and environmental factors such as planting location, climatic conditions during development, and soil nitrogen.

Barley β-glucans are invariably polydisperse with respect to molecular mass and this is illustrated by a weight average to number average molecular mass ratio (Mw/Mn) of 1.18 for a water-soluble (1,3;1,4-β-D-glucan extracted from barley flour (Woodward *et al.*, 1983). The extractability of β-glucans from walls of barley endosperm will be a function of their degree of self-association and their association with other wall polysaccharides and proteins (Fincher and Stone, 2004). In particular, extractability depends on the molecular mass and linkage distribution in the (1,3;1,4)-β-D-glucan chains (Woodward *et al.*, 1983). Extensive association with other polymers and very high molecular masses will render the (1,3;1,4)-β-D-glucans more difficult to extract from the grain. Approximately half of the β-glucan from barley flour is extracted by water at pH 7.0 and 40 °C; further fractions can be solubilized at higher temperatures (Woodward *et al.*, 1983). Waxy, or high amylose, barleys have a higher proportion of water-soluble β-glucan than normal barleys. Complete extraction of β-glucan from barley grain requires the use of alkaline extractants such as 4M NaOH or aqueous $Ba(OH)_2$ (Fincher, 1975). Once extracted with hot water or alkali, the β-glucans are usually soluble at neutral pH and room temperature, although they will slowly aggregate and precipitate upon cooling (Fincher and Stone, 2004).

In aqueous media barley β-glucan adopt a partially extended, somewhat rigid, reptate conformation with a cross-sectional diameter of 0.45 nm and an axial ratio (length : width) of about 100 (Woodward *et al.*, 1983). The asymmetrical conformations of barley β-glucan have clearly evolved to allow the polysaccharide to perform its key role as the gel-like matrix phase component of the wall that ensures that the wall, particularly the primary wall, remains strong yet flexible, pliable and sufficiently porous to permit the transfer of water, nutrients, phytohormones and other small molecules across the wall and therefore between cells. Precisely the same properties are responsible for their high viscosities in aqueous media and hence for the undesirable characteristics attributed to β-glucan in malting and brewing processes, and to the beneficial effects of barley and other β-glucan on human health and nutrition (Brennan and Cleary, 2005). Thus, the viscosity-generating properties of soluble β-glucan are critical determinants in many aspects of barley processing.

The intrinsic viscosities of β-glucan solutions are dependent on the cereal species from which the polysaccharide is extracted and on the solvent used. Intrinsic viscosity values for barley β-glucan in water range from 4.6-6.9 dl g^{-1}. The extractability, solubility, and viscosity characteristics of barley β-glucan can be understood by reference to their fine structures (Fincher and Stone, 2004). Long stretches of contiguous (1,4)-β-D-glucosyl units in the (1,3;1,4)-β-D-glucan chain will form extended ribbon-like 'cellulosic' regions, while the insertion of a (1,3)-linked glucosyl unit between these cellulosic regions introduces a molecular "kink" in the polysaccharide chain (Buliga *et al.*, 1986). Chain flexibility arises principally from the isolated (1,3)-β-linkages and because the (1,3)-linked β-glucosyl units do not occur regularly along the chain, the polysaccharides are not able to align over extended regions. Hence, they remain "dispersed" and soluble in aqueous media (Woodward *et al.*, 1983). The overall extended polysaccharide conformation is due to the predominance of (1,4)-linkages and results in the occupancy of a high volume of solvent (high hydrodynamic volume). This in turn results in aqueous solutions of high viscosity (Fincher and Stone, 2004).

7.3.2 Arabinoxylans

The heteroxylans, or pentosans, of barley are of two main types, namely the arabinoxylans and the glucuronoarabinoxylans. Heteroxylans are characteristically abundant in walls of vegetative tissues of cereals and grasses, compared with other monocots and dicots. In walls of the starchy endosperm cells and of the aleurone layer of barley and other cereal grains, heteroxylans usually have arabinosyl residues as their predominant substituent and are therefore referred to here as arabinoxylans. In contrast, the xylan backbones of glucuronoarabinoxylans are substituted with both arabinosyl and glucuronosyl residues and are located mainly in walls of the pericarp and seedcoat tissues (Fincher and Stone, 2004).

Although much of the chemical and physicochemical information available for cereal grain arabinoxylans has been accumulated through studies on the water-soluble arabinoxylans from wheat flour, it is likely that the characteristics of wheat and barley endosperm arabinoxylans will often be similar. Average molecular masses of cereal heteroxylans range from 65,000 to 5,000,000 (DP 500 to 38,000) and, as is the case with the β-glucan, the value depends on the cereal species, cell-wall type, extraction procedure, and the method used for molecular mass determination. The cereal heteroxylans, including those from barley, are highly polydisperse with respect to molecular mass as indicated by Mw/Mn values (Fincher and Stone, 2004). As with the β-glucan, a portion of the arabinoxylans from walls of cereal grains is soluble in water at 40 °C and higher temperatures, but other extractants such as NaOH are required to bring all the arabinoxylans into solution.

Unsubstituted (1,4)-β-D-xylans adopt a twofold, helical conformation with one hydrogen bond between adjacent Xylp units. This conformation is more flexible than the twofold helix of the (1,4)-β-D-glucan, cellulose, which has two hydrogen bonds between adjacent glucosyl units (Kroon-Batenburg and Kroon, 1997). The absence of cooperative inter-residue associations in xylans allows the molecules to be conformationally more flexible, such that in the solid state an unsubstituted (1,4)-β-D-xylan chain preferentially exists as a fully-extended, threefold (i.e., three Xylp units per helical turn), left-handed helix or a "slowly twisted ribbon" (Atkins, 1992; Fincher and Stone, 2004).

The length of the backbone chains of arabinoxylans is approximately 3~5 nm (Andrewartha *et al.*, 1979). In solution, the arabinoxylans behave as partially stiff, worm-like, cylindrical molecules whose flexibility is largely unaffected by Araf:Xylp ratios, which are generally in the range 0.39-0.82 : 1. The arabinosyl substituents sterically hinder the aggregation of the linear (1,4)-β-D-xylan backbones into fibrillar structures and hence enable the polysaccharide to remain soluble in aqueous media. As with the β-glucans, the asymmetrical nature of the arabinoxylans and their high DP result in molecules that form solutions of high viscosity.

In the context of cell wall biology in the barley grain, the arabinoxylans, which are also believed to form a gel-like matrix between cellulosic microfibrils in the wall, have functional properties that are similar to those of (1,3;1,4)-β-D-glucans, insofar as they impart a degree of structural rigidity, but they also remain strong yet flexible, pliable and sufficiently porous to permit the transfer of small molecules across the wall. Thus, similar functional properties are achieved by the cell, using different chemical structures. Again the asymmetrical conformations and high DPs are responsible for the high viscosities of arabinoxylans in aqueous media. Furthermore, many of the undesirable characteristics attributed to (1,3;1,4)-β-D-glucans in malting and brewing processes, and to the beneficial effects of barley and other (1,3;1,4)-β-D-glucans on human health and nutrition (Brennan and Cleary, 2005) might partly result from the contributions of arabinoxylans in grain and malt extracts. Thus, incompletely degraded arabinoxylans from malted barley and cereal adjuncts could also contribute to wort and beer viscosity and to problems in wort separation and beer filtration. Arabinoxylans are also components of some beer hazes (Coote and Kirsop, 1976).

The intrinsic viscosities of wheat arabinoxylans range from 0.8 to 5.5 dL g^{-1}, depending on the degree of substitution. These values may be compared with 6.9 dL g^{-1} for barley (1,3;1,4)-β-D-glucans (Woodward *et al.*, 1983). The behaviour of arabinoxylans in solution will be influenced not only by the overall asymmetry of the molecules and their DP, but also by the specific arrangement of Araf units along the xylan backbone. Thus, stretches of unsubstituted Xylp units may permit intermolecular alignment over these sequences and lead to the formation of chain associations that are stabilized by hydrogen bonds.

7.3.3 Glucomannans

Glucomannans are linear copolymers of β-D-glucopyranosyl residues (approximately 30%) and its 2-epimer, β-D-mannopyranose (about 70%), joined by (1,4)-linkages to form linear chains with DPs ranging from less than 100 to several thousand (Fincher and Stone, 2004). In some examples, the backbones of glucomannans have (1,6)-linked β-D-galactopyranosyl substituents and are often esterified with acetyl groups. The conformation of glucomannans is similar to that of cellulose and it has therefore been suggested that the chains may associate strongly with surfaces of cellulose microfibrils. Consistent with this possibility, glucomannans are difficult to extract from walls (Fincher, 1975; Fincher and Stone, 2004). Glucomannans are minor constituents of walls of aleurone and starchy endosperm cells of barley, where they usually account for 2% or less by weight of the wall (Table 7.1).

7.3.4 Xyloglucans

Xyloglucans are a family of substituted and side-chain-branched heteroglycans that have a cellulose-like (1,4)-β-D-glucan backbone substituted at regular intervals by β-D-xylopyranosyl units at C(O)6 of the glucosyl residues. Some side-branch xylosyl units carry β-D-galactopyranosyl and α-L-fucopyranosyl substituents. Compared with xyloglucans from dicotyledonous plants, those from grasses are less substituted with xylose, contain little galactose, and have stretches of unsubstituted backbone, as shown by Gibeaut *et al.* (2005) in analyses of cell walls from barley coleoptiles. Xyloglucans have DPs of 600$\sim$700 and readily associate with surfaces of cellulose microfibrils. Xyloglucans have been reported in walls of rice endosperm (Shibuya and Misaki, 1978) but have not been detected in barley endosperm walls. However, methylation analyses used (Fincher, 1975; Bacic and Stone, 1981) may have missed their existence in barley endosperm wall preparations. In any case, xyloglucans are generally considered to be of minor significance in walls of barley grain tissues.

7.3.5 Pectic Polysaccharides

The pectic polysaccharides represent a diverse group of branched heteroglycans in which the backbone is a copolymer of alternating (1,4)-α-galacturonosyl and (1,2)-linked α-L-rhamnopyranosyl residues. This rhamnogalacturonan backbone carries a variety of side-branches on the rhamnosyl units, including arabinan, arabinogalactan and more complex branched oligosaccharide chains. In addition, variable amounts of (1,4)- α-linked homogalacturonan may be present in the same backbone chain and these can form Ca^{2+} complexes that promote gelation. Pectic polysaccharides are minor constituents of the walls of vegetative tissues of barley (Gibeaut *et al.*, 2005) and other cereals, although measurable amounts have been reported in walls of rice endosperm cells (Shibuya and Nikane, 1984).

7.3.6 Callose

Callose is a (1,3)-β-D-glucan that is a linear homopolymer and is insoluble in water. Callose can be seen in tissue sections by an intense yellow, UV-induced fluorescence when the tissues are stained with the aniline blue fluorochrome, which is a sulphonated 4,4'-dianilinobenzophenone. In mature barley endosperm, callose occurs as small bead-like deposits on the inner surfaces of walls throughout the starchy endosperm but especially at the aleurone/subaleurone interface (Fulcher *et al.*, 1977). The deposition of callose is a well-known consequence of stress and wounding. Thus, the endosperm deposits may arise as a consequence of plasmolysis during the drying of the grain (Fincher and Stone, 2004). The callose content of barley endosperm wall preparations is about 1% (Fincher, 1975; Bacic and Stone, 1981). Callose is found transiently as a major component of the developing walls of barley endosperm during the early stages of cellularization, but disappears soon afterwards and is not a major component of mature walls (Wilson *et al.*, 2006).

7.3.7 Cellulose

Cellulose is a linear polysaccharide consisting exclusively of (1,4)-linked β-glucopyranosyl residues. Individual molecules of cellulose are long, with up to 6,000 residues in primary walls and up to 14,000 residues in secondary walls (Fincher and Stone, 2004). The cellulose molecules have extended ribbon-like conformations and pack into three-dimensional, fibrillar aggregates that are stabilized by extensive intermolecular hydrogen bonding and van der Waal's interactions. These aggregates are termed microfibrils and may reach diameters of 3 nm in primary walls and 5–10 nm in secondary walls. Within microfibrils, individual cellulose molecules are believed to be packed into parallel, highly ordered, crystalline arrays, although there are usually regions where molecular alignment is less ordered. Within the wall, cellulose microfibrils may associate with water and with matrix polysaccharides such as (1,3;1,4)-β-D-glucans, arabinoxylans and glucomannans. As mentioned above, cellulose is a relatively minor constituent of walls in the endosperm of barley grain (Table 7.1).

7.4 Biosynthesis of Walls during Barley Grain Development

The importance of wall polysaccharides in malting barley quality has been recognised for some time. In particular, wall (1,3;1,4)-β -D-glucans have been implicated as major contributors to wort and beer viscosity, and to processing difficulties associated with high beer and wort viscosity. It is clear that the final concentration of (1,3;1,4)-β-D-glucan in malt extracts or beer will depend not only on the initial level of wall (1,3;1,4)-β-D-glucan in the ungerminated grain, but also on the ability of the germinated grain to rapidly produce large amounts of enzymes that depolymerize the (1,3;1,4)-β-D-glucans during malting and mashing. As a result, attention in the area has been focused both on the selection of breeding lines with low (1,3;1,4)-β-D-glucan contents and/or on the selection of barley varieties that quickly produce large amounts of (1,3;1,4)-β-D-glucanases after the grain is germinated (Bamforth, 1993). As mentioned above, wall arabinoxylans have physicochemical properties that are similar to the (1,3;1,4)-β-D-glucans and hence might also contribute to wort and beer viscosity. There is a good deal of information on the hydrolytic enzymes that are produced in the germinated barley grain for the depolymerisation of arabinoxylans (Simpson *et al.*, 2003). Thus, for both classes of polysaccharide, work on the hydrolases has been more fruitful than work on the enzymes involved in arabinoxylan and (1,3;1,4)-β-D-glucan biosynthesis, because the hydrolases are soluble and relatively stable, and hence amenable to purification and biochemical characterization.

In contrast, the properties of the synthases required for arabinoxylan and (1,3;1,4)-β-D-glucan synthesis in walls of the starchy endosperm cells of barley and other cereals have not been defined, because of extreme difficulties associated with biochemical studies of membrane-bound enzymes that lose activity very quickly following lysis of cells. Indeed, despite the attempts of many groups around the world there have been no published reports of purified and characterized (1,3;1,4)-β-D-glucan or xylan synthases from barley or other cereals or grasses. As a corollary of this, the identity of the genes that mediate in (1,3;1,4)-β-D-glucan or arabinoxylans biosynthesis have remained unknown until very recently. However, the emerging techniques of functional and comparative genomics have now allowed a family of genes involved in (1,3;1,4)-β-D-glucan synthesis in barley and rice to be identified (Burton *et al.*, 2006, 2008) and it is likely that the genes responsible for xylan biosynthesis will soon be identified (Mitchell *et al.*, 2007; Pena *et al.*, 2007; Brown *et al.*, 2007). As a result, one can be confident that information on the enzymic and cellular systems that are required for the biosynthesis, deposition and remodelling of the major cell wall polysaccharides in barley grain will soon emerge, and that this information will prove useful for understanding the genetic and environmental factors that affect wall composition and for manipulating levels of these polysaccharides in barley grain.

In the sections that follow, we will outline recent advances in our knowledge on the development of walls in barley grain at the microscopic level and on the gene families that might participate in arabinoxylan and (1,3;1,4)-β-D-glucan biosynthesis.

7.4.1 Morphology of Barley Grain Development

The endosperm of barley follows a nuclear pattern of development that has been well described for other cereals and angiosperms (Brown *et al.*, 1996; Brown *et al.*, 1997; Olsen, 2004). Following fusion in the embryo sac of the nucleus of a single egg cell or oospore, with a single male nucleus, a diploid zygote is formed. The first division of this diploid zygote produces the apical cell, which ultimately forms the embryo, and a basal cell (Duffus and Cochrane, 1993). The fusion of the second male generative nucleus and the two polar nuclei gives rise to the triploid endosperm (Bewley and Black, 1983). The triploid nucleus of the endosperm divides repeatedly to create a syncytium in which numerous free nuclei are embedded in a peripheral cytoplasmic layer that lines the perimeter of a large vacuolated central cell. About three days after pollination (DAP), when approximately 2,000 endosperm nuclei have been formed (Duffus and Cochrane, 1993), the initial extension of anticlinal walls between adjacent nuclei occurs to form a layer of open-ended alveoli. Following mitosis, these alveoli are subsequently compartmentalized into individual cells by the deposition of periclinal cell walls between daughter nuclei. The pattern of anticlinal wall extension, followed by mitosis and the deposition of the periclinal wall across the next tier is repeated centripetally until, at approximately 6 DAP, the alveoli converge and the endosperm is fully cellularized (Brown *et al.*, 1994; Wilson *et al.*, 2006). At this point starch and protein begin to accumulate in the central endosperm cells but division continues until about half way through grain filling; thereafter cell enlargement and eventually programmed cell death are the major events (Cochrane and Duffus, 1981; Berger, 1999). Around 9–10 DAP in early grain development division of the peripheral endosperm cells results in the formation of a proaleurone layer and an inner, sub-aleurone layer (Hoshikawa, 1984). These are resolved into a clearly discernible aleurone layer by 15 DAP (Cochrane and Duffus, 1981), in which small, dense, isodiametric cells are surrounded by a relatively thicker wall that is compositionally distinct from that of the starchy endosperm cells (Table 7.1; Bacic and Stone, 1981; Fincher 1975).

The temporal and spatial appearance of key polysaccharides in the cell walls of early developing endosperm of barley up to 8 DAP have been defined at the light and electron microscope levels (Wilson *et al.*, 2006). Monoclonal antibodies were used for several of the polysaccharide constituents, while the enzyme cellobiohydrolase II was used to detect cellulose (Wilson *et al.*, 2006). Both (1→3)-β-D-glucan (callose) and cellulose were detected in the first cell walls formed, between 3 and 4 DAP. However, the presence of callose in these walls was transient and by 7 DAP it was only detected in collars surrounding the plasmodesmata linking the endosperm cells (Wilson *et al.*, 2006). Given that cellulose comprises 2% of the final wall (Fincher, 1975), it is assumed that it persists from the early stages of endosperm development until grain maturity, although it cannot be specifically traced past 5 DAP because the cellobiohydrolase also binds to (1→3,1→4)-β-D-glucans.

Deposition of (1→3,1→4)-β-D-glucans was not detected in the developing cell walls until approximately 5 DAP and hetero-(1→4)-β-D-mannans followed at 6 DAP (Wilson *et al.*, 2006). The deposition of arabinoxylans was first detected after cellularization was complete and once starch had begun to accumulate. At 8 DAP arabinoxylan was found only in walls closest to the vascular strand, although it is possible that it is deposited earlier in a highly substituted form that is not recognised by the monoclonal antibody. By 9 DAP arabinoxylans were found uniformly throughout the starchy endosperm tissue (Wilson *et al.*, 2006).

There is little published information concerning the deposition of the major cell wall polysaccharides in the latter stages of endosperm development in barley. It is not known whether deposition proceeds steadily throughout development, although data from Coles (1979) on the formation of non-starchy polysaccharides suggested that these polysaccharides accumulate in a biphasic fashion in the cultivar Bomi and in a linear manner from 13 to 40 DAP in other cultivars. These conclusions were supported by Prentice and Faber (1981) from work on North American barleys. Levels of barley grain (1→3,1→4)-β-D-glucans vary significantly as a result of both genetic and environmental factors; moisture availability and temperature fluctuations are the most notable environmental factors (Coles *et al.*, 1991). Wallwork (1997) examined the effect of high temperature on grain filling and on subsequent malting quality. She showed that temperatures of 32°C during the mid-grain filling stage of grain development stimulated an increase in (1,3;1;4)-β-D-glucan synthesis, while temperatures over 35°C resulted in a decrease in the rate of (1,3;1;4)-β-D-glucan synthesis to half that observed in control plants.

7.4.2 Enzymes Involved in Wall Polysaccharide Biosynthesis

The enzymes responsible for polysaccharide biosynthesis in plants can be classified into two groups, the polysaccharide synthases and the glycosyltransferases, which are in turn classified within the overall "glycosyl transferase" (GT) class of carbohydrate-modifying enzymes (Coutinho and Henrissat, 1999). The GTs are among the largest groups of enzymes known and have been divided into more than 70 families based on sequence similarities, the existence of certain motifs, hydrophobic cluster analysis (HCA) and their catalytic specificity (Campbell *et al.*, 1997; Imberty *et al.*, 1999; Ross *et al.*, 2001; Rosen *et al.*, 2004). Although there are a large number of glycosyltransferases in any plant, the biochemical activity of relatively few plant GTs has been demonstrated so far (Keegstra and Raikhel, 2001). It has been estimated that about 1.6% of the Arabidopsis genome is made up of genes that encode glycosyltransferases (Egelund *et al.*, 2004). This value accounts for more than 400 genes from 41 GT families, compared to approximately 200 genes in the human genome (Breton *et al.*, 2001; Scheible and Pauly, 2004). Many of the plant GTs are believed to be involved in the synthesis of complex cell wall polysaccharides (Coutinho *et al.*, 2003).

Within the overall GT group, two quite distinctive classes of glycosyltransferases can be identified. The type I polysaccharide synthases catalyse the iterative incorporation of glycosyl residues from sugar nucleotide donors into molecular backbones that constitute the main chains of both homo-polysaccharides, such as cellulose and (1,3;1;4)-β-D-glucans, and hetero-polysaccharides such as arabinoxylans (Perrin *et al.*, 2001; Vergara and Carpita, 2001; Bonetta *et al.*, 2002). These are integral membrane proteins with multiple transmembrane helices, and have been difficult, if not impossible, to purify by traditional biochemical methods (Delmer, 1999).

The second, type II class of glycosyltransferases within the overall GT group contains enzymes that are also referred to as glycosyltransferases, but in a more restricted sense insofar as they transfer glycosyl residues from the donor to a polysaccharide backbone, in a single event. This second class of GT enzymes is often designated GlyT. They typically have a single transmembrane helix that spans the membrane and functions as an anchor, together with a short cytosolic NH_2-terminus, an extended hydrophilic stem region, and a globular catalytic domain within the lumen of the Golgi (Kleene and Berger, 1993; Breton and Imberty, 1999; Gibeaut, 2000; Keegstra and Raikhel, 2001; Perrin *et al.*, 2001; Sterling *et al.*, 2001; Pagny *et al.*, 2003). In the current context, a key example of the GlyT class of enzyme would be the α-L-arabinosyltransferases, which are assumed to add single α-L-arabinofuranosyl substituents to the (1,4)-β-D-xylan backbone of arabinoxylans, while the xylan synthase required for the synthesis of the (1,4)-β-D-xylan backbone itself is assumed to be a type I polysaccharide synthase (Farrokhi *et al.*, 2006).

The type I polysaccharide synthases involved in wall biosynthesis are believed to be encoded by multi-gene families of the cellulose synthase (CesA) and cellulose synthase-like (Csl) groups, the glucan synthase-like (GSL) genes and possibly by other genes (Richmond and Somerville, 2000). The CesA and Csl enzymes are large proteins, with about 850 to 1,000 amino acid residues, and contain three to six transmembrane helices towards the COOH-terminus and one to two towards the NH_2-terminus. They share a common D_D_D_QxxRW motif, which is believed to be involved in sugar binding and catalytic activity of the enzyme (Richmond and Somerville, 2001; Bonetta *et al.*, 2002).

We are now able to assign biosynthetic enzymes to a significant proportion of the cell wall polysaccharides found in the barley endosperm, although our knowledge of the mechanisms and control of the biosynthetic steps, coupled with a limited understanding of the physical and chemical interactions between wall components suggests that attempts to manipulate such processes will not be straightforward.

7.4.2.1 (1,3;1;4)-β-D-glucan synthesis

There is emerging evidence that cellulose synthase-like *CslF* genes mediate the synthesis of (1,3;1,4)-β-D-glucans in the cell walls of the commelinoid monocotyledon group of land plants, which include barley (Burton *et al.*, 2006). Seven members of the barley *HvCslF* gene family have now been characterized (Burton *et al.*, 2008). The initial identification of the *CslF* genes as potential participants in (1,3;1,4)-β-D-glucan synthesis was based upon comparative genomics of rice and barley. Molecular markers flanking a major QTL on chromosome 2H for (1,3;1,4)-β-D-glucan content in un-germinated barley grain (Han *et al.*, 1995) were used to locate the syntenous region of the rice genome, where a cluster of six rice OsCslF genes represented prime candidates for a role in (1,3;1,4)-β-D-glucan synthesis. Expression of these genes in Arabidopsis, which does not have (1,3;1,4)-β-D-glucans in its walls and which do not have *CslF* genes, resulted in the deposition of (1,3;1,4)-β-D-glucans in the walls of the transgenic Arabidopsis lines (Burton *et al.*, 2006). This was taken as proof that the rice *CslF* genes were involved in the synthesis of (1,3;1,4)-β-D-glucans, but the possibility that other genes might also be involved could not be precluded (Burton *et al.*, 2006). The *HvCslF* genes have now been mapped.

In terms of the synthesis of (1,3;1,4)-β-D-glucans specifically in barley grain, transcript analyses of all the HvCslF genes showed that transcriptional levels of the *HvCslF6* and *HvCslF9* genes predominated in the developing endosperm (Burton *et al.*, 2008). Transcription of the *HvCslF9* gene is maximal at 8 DAP, at a stage when cellularization of the endosperm is complete and starch deposition has commenced. (1,3;1,4-β-D-Glucans are detectable in endosperm walls from 5 DAP onwards (Wilson *et al.*, 2006; Burton *et al.*, 2008; Fig. 7.2). In contrast, *HvCslF6* transcripts are detected at much higher levels throughout endosperm development, even before (1,3;1,4)-β-D-glucans appear in the walls (Wilson *et al.*, 2006). From about 12 DAP to 20 DAP and beyond there are further increases in the abundance of *HvCslF6* transcripts (Burton *et al.*, 2008), consistent with earlier work on the deposition of (1,3;1,4)-β-D-glucans in barley grain, where a late increase in (1,3;1,4)-β-D-glucan content was reported (Coles, 1979). Variations in transcript levels of the *HvCslF6*, *HvCslF9* and *HvCslF8* genes were also observed between different barley varieties (Burton *et al.*, 2008). Transcript levels of the HvCslF9 gene are much higher in the malting variety Sloop than in the hulless variety Himalaya during the cellularization stage of initial wall development, while levels of the most abundant mRNA, from the *HvCslF6* gene, are higher in Himalaya than in Sloop during most of endosperm development, and particularly at 20 days post pollination (Burton *et al.*, 2008). The *HvCslF6* transcript levels are consistent with higher levels of (1,3;1,4)-β-D-glucan in Himalaya grain (typically 6% by weight) than in Sloop, where 3%~4% by weight (1,3;1,4)-β-D-glucan is normally found (Box, Burton, Collins, Jobling, unpublished data).

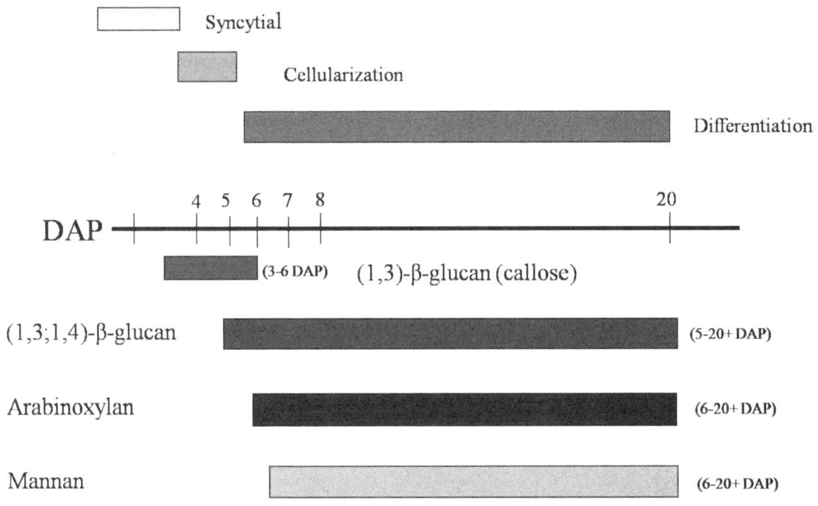

Fig. 7.2. The temporal appearance of key polysaccharides during endosperm development (3-20 DAP) in barley. *Between 5 and 6 DAP callose is progressively lost from the cell wall and only found surrounding plasmodesmata

It is likely that the biosynthesis of (1,3;1,4)-β-D-glucan requires a multi-enzyme complex and that an individual *CslF* isoenzyme represents just one component of such a complex (Burton *et al.*, 2008). Furthermore, there are a number of fine structural features of barley (1,3;1,4)-β-D-glucans, in particular the arrangements of (1,3)-β-D-glucosyl and (1,4)-β-D-glucosyl residues along the chain, that have not yet been characterized at the biosynthetic level. The complexity of (1,3;1,4)-β-D-glucan (section 3.1.1) is suggestive of a similar complexity in the enzymology of (1,3;1,4)-β-D-glucan biosynthesis

(Buckeridge *et al.*, 2001), where one might again infer that multiple enzymes are required (Burton *et al.*, 2008).

Burton *et al.* (2008) also raised the possibility that (1,3;1,4)-β-D-glucan endohydrolases might participate in (1,3;1,4)-β-D-glucan synthesis. If this were true, transcripts for both genes would be expected in the developing endosperm. Burton *et al.* (2008) showed that both genes are indeed transcribed in developing barley endosperm, although the (1,3;1,4)-β-D-glucan endohydrolase gene that encodes isoenzyme EI (*HvGlb1*) was transcribed transiently. (1,3;1,4)-β-D-Glucanase activity was not tested in experiments conducted by Burton *et al.* (2008), but Finnie *et al.* (2006) did detect (1,3;1,4)-β-D-glucan endohydrolase isoenzyme EI during proteome profiling of developing barley grain. Similarly, genes encoding glycoside hydrolase family GH1 β-glucosidases, which are more correctly known as (1,4)-β-D-glucan glucohydrolases (Hrmova and Fincher, 2002), are transcribed in the maturing endosperm, and activity does not increase after germination (Simos *et al.*, 1994; Leah *et al.*, 1995). The enzymes are active in developing grain when (1,3;1,4)-β-D-glucans are being deposited in walls of the starchy endosperm (Leah *et al.*, 1995), and could participate in the trimming or turnover of wall (1,3;1,4)-β-D-glucans during synthesis. There is evidence from other sources that hydrolytic enzymes participate in polysaccharide biosynthesis during wall assembly. For example, mutations in the Arabidopsis *KOR* gene, which encodes a cellulase, lead to large reductions in wall cellulose. The hydrolases could be involved in trimming or "editing" nascent cellulose chains (Szyjanowicz *et al.*, 2004), or possibly in releasing newly synthesised chains from the biosynthetic enzymes (Farrokhi *et al.*, 2006).

Whether or not an acceptor molecule is required for initiation of (1,3;1,4)-β-D-glucan chain synthesis is unknown, but the presence of some covalently linked protein on a water-extracted (1,3;1,4)-β-D-glucan from barley endosperm (Forrest and Wainright, 1977) suggests that this might be possible.

7.4.2.2 Arabinoxylan biosynthesis

At this stage the genes and enzymes that mediate the biosynthesis of arabinoxylans have not been identified unequivocally. It might be expected, based on other wall polysaccharide biosynthetic systems, that a type I polysaccharide synthase would be involved in the synthesis of the (1,4)-β-D-xylan backbone, and that a type II arabinosyl transferase would be involved in the addition of the α-L-arabinofuranosyl substituents to the (1,4)-β-D-xylan backbone (Farrokhi *et al.*, 2006). Similarly, a type II glucuronyl transferase could be involved in the addition of the α-D-glucuronopyranosyl substituents to the (1,4)-β-D-xylan backbone.

However, there is as yet no firm evidence that the xylan backbone is synthesised by a type I polysaccharide synthase. Given the chemical similarities between a (1,4)-β-D-xylan and a (1,4)-β-D-glucan, one might predict that an enzyme from the *Csl* families might be involved. Despite determined analyses of the various *Csl* gene sub-families from several species, *Csl* genes have not been linked with (1,4)-β-D-xylan biosynthesis (Farrokhi *et al.*, 2005). Indeed, recent analyses of mutant lines and bioinformatical information on transcript profiles during periods of arabinoxylan synthesis suggest an involvement of various type I glycosyl transferases, notably from families GT8, GT43 and GT47, in the synthesis of (1,4)-β-D-xylan and in some cases it has been suggested that these type I enzymes are responsible for the synthesis of the (1,4)-β-D-xylan backbone itself (Brown *et al.*, 2007, Pena *et al.*, 2007, Persson *et al.*, 2007). These suggestions await confirmation through proof-of-function data.

The demonstration that GAXs from various sources have an oligosaccharide consisting of 4-β-D-Xylp-(1,4)-β-D-Xylp-(1,3)-α-L-Rhap-(1,2)-α-D-GalpA-(1,4)-D-Xylp at their reducing termini (Pena *et al.*, 2007) suggest that additional glycosyl transferases would be required for the synthesis of this portion of the polysaccharide, if it were shown to exist on arabinoxylans from barley.

It is also possible that hydrolytic enzymes could participate in the biosynthetic process, because there are indications that changes in the fine structure of arabinoxylans might occur during grain development. In arabinoxylans of developing barley coleoptiles, the ratio of substituted to unsubstituted 4-linked xylosyl units changed from about 4:1 to 1:1 over about three days (Gibeaut *et al.*, 2005). About 80% of backbone xylopyranosyl residues in the newly synthesised arabinoxylan were substituted with arabinosyl residues, but these were progressively removed during growth (Gibeaut *et al.*, 2005). The removal of the arabinosyl residues from newly synthesised coleoptile arabinoxylan is believed to be mediated by the action of arabinoxylan arabinofuranohydrolases (Lee *et al.*, 2001), although it has not yet been demonstrated that similar alterations to arabinoxylan fine structure occur in the developing walls of the endosperm.

7.4.2.3 Cellulose biosynthesis

In the developing barley endosperm it is likely that one or more of the cellulose synthase (*CesA*) genes will be responsible for the synthesis of the cellulose in the early anticlinal and periclinal walls, and that found also in the walls of mature endosperm cells.

The *CesA* gene family from barley has at least nine members, which are distributed across the genome (Burton *et al.*, 2004). Q-PCR has been used to determine the relative abundance of mRNA transcripts for individual *HvCesA* genes in vegetative and floral tissues, at different stages of development. In the developing endosperm, *HvCesA1*, *HvCesA2*, and *HvCesA6* predominate, and their relative abundance is very similar in many tissues. As a result, Burton *et al.* (2004) concluded that they are co-ordinately transcribed and that the three encoded enzymes represent candidate components of the so-called terminal rosettes that are involved in cellulose biosynthesis at the plasma membrane.

7.4.2.4 Callose biosynthesis

The callose synthases of higher plants are probably encoded by the glucan synthase-like (*GSL*) gene family (Saxena and Brown, 2000; Cui*et al.*, 2001). Plant *GSL* genes are usually members of multi-gene families, with at least eight in barley (Schober, Jacobs, Burton and Fincher, unpublished data). Each of the barley genes is probably responsible for callose synthesis in a different location within the plant and/or at different stages of growth and development (Hong *et al.*, 2001a; Cui *et al.*, 2001; Doblin *et al.*, 2001).

Biochemical approaches have been used to show that GSL proteins are involved in callose synthesis. A direct link between the barley *HvGSL1* gene and callose synthase activity was established by Li *et al.* (2003), who used peptide mass fingerprinting to identify HvGSL1 as the major polypeptide in a callose-synthesizing region of a native gel. However, the GSL proteins do not contain a recognised UDP-Glc binding motif and there has been doubt expressed as to whether these proteins constitute the catalytic subunit, or whether they are a pore-forming unit in a larger callose synthase complex (Douglas *et al.*, 1994; Cui *et al.*, 2001; Hong *et al.*, 2001a). Nevertheless, in the developing barley endosperm it is likely that one or more of the glucan synthase-like (GSL) gene family will be involved in the synthesis of the callose in the early anticlinal and periclinal walls, and also during the synthesis of the callosic deposits that are scattered through the starchy endosperm. Specific primers designed for quantitative real-time PCR (Q-PCR) show that transcripts of several *HvGSL* genes are expressed at significant levels in developing barley endosperm between 2 and 11 DAP (Schober, Burton, Jacobs, Fincher, unpublished data).

7.4.2.5 Glucomannan biosynthesis

Although the genes and enzymes that participate in glucomannans biosynthesis in barley have not been identified, it is likely that the *CslA* genes will be involved. It has been shown that a *CslA* gene from guar (Cyamopsis tetragonoloba) encodes a (1→4)-β-mannan synthase (Dhugga *et al.*, 2004). This result was confirmed by Liepman *et al.* (2005), who expressed Arabidopsis *CslA* genes heterologously in Drosophila S2 cells and showed that several members of the *CslA* gene families encode β-mannan synthases. Recombinant *CslA* proteins produced β-linked mannan polymers when supplied with GDP-mannose and also produced β-linked glucomannan heteropolymers when supplied with both GDP-mannose and GDP-glucose (Liepman *et al.*, 2005).

7.5 Germination

Following germination of the grain, it is important that the storage components of the starchy endosperm are rapidly mobilised, to support vigorous growth of the young seedling so that it will become photosynthetic and self-sufficient as quickly as possible. The major storage components of the grain are starch and various proteins, which are enclosed mostly within the starchy endosperm cells. Many of the degradative enzymes that hydrolyse these storage polymers are synthesised de novo in the aleurone and scutellar epithelial cells, and subsequently secreted into the starchy endosperm. Despite earlier assertions that the aleurone is the sole source of the key hydrolytic enzymes in germinated barley (Palmer and Sattler, 1996), the overwhelming weight of evidence indicates that the scutellar epithelium is also responsible for the secretion of several hydrolytic enzymes, probably during the early stages of endosperm mobilisation, and that the aleurone subsequently becomes the major source of hydrolytic enzymes that are secreted into the starchy endosperm to effect storage polymer mobilization (Gibbons, 1981; Briggs and MacDonald, 1983; Mundy *et al.*, 1985; Stuart *et al.*, 1986; McFadden *et al.*, 1988; MacGregor *et al.*, 1994; Sugimoto *et al.*, 1998).

Once α-amylases and endo-peptidases are secreted from the aleurone and scutellar epithelial cells, the walls of the starchy endosperm cells represent a physical barrier between the enzymes and their starch and protein substrates within the starchy endosperm cells. For this reason, rapid degradation of wall polysaccharides and "perforation" of the walls is advantageous for providing enzymic access to the storage polymers and for the rapid establishment of the young seedling.

There are numerous reviews in which the physiology, cell biology, biochemistry and molecular biology of barley grain germination are described (Morrall and Briggs, 1978; Fincher, 1989; Gubler *et al.*, 1987; Kuntz and Bamforth, 2007) and Evans (2008) describes more generally the enzymes involved in carbohydrate-degradation in germinated barley grain in Chapter 6 of this volume. Thus, we will focus here on the enzymic degradation of the major wall polysaccharides of the starchy endosperm following germination of the grain. It is worth noting that sugars released during wall degradation can make a significant contribution to the total energy available for growth of the young seedling. Morrall and Briggs (1978) calculated that sugars released from endosperm wall polysaccharides account for up to 18.5% of the total carbohydrate available to the young seedling, including that provided by the hydrolysis of the major storage polysaccharide, starch.

7.5.1 Hydrolysis of (1,3;1,4)-β-D-glucan

A battery of glycoside endo- and exohydrolases is required for the complete depolymerization of (1,3;1,4)-β-D-glucans to glucose in germinated barley grain. The enzymes include (1,3;1,4)-β-D-glucan endohydrolases (EC 3.2.1.73), which are classified as family GH17 glycoside hydrolases and β-D-glucan glucohydrolases from glycoside hydrolase families GH1 and GH3 (Hrmova and Fincher, 2002). The enzymes have been purified from germinated grain or young barley seedlings, and their kinetics, substrate specificities, and three-dimensional (3D) structures characterised in detail.

A putative carboxypeptidase, referred to as β-glucan solubilise, has been implicated in the initiation of (1,3;1,4)-β-D-glucan hydrolysis and its release into solution (Kuntz and Bamforth, 2007). It has been shown that barley (1,3;1,4)-β-D-glucan is covalently linked to a small amount of protein (Forrest and Wainright, 1977), but it is difficult to envisage how a carboxypeptidase could 'release' the polysaccharide through the sequential removal of amino acid residues from the carboxy terminus of such a protein. The enzyme has not been purified and therefore has not been characterized, so we will not discuss it further here. Indeed, MacGregor *et al.* (1996) suggested that the so-called β-D-glucan solubilase activity is a result of microbial contamination, while Hrmova and Fincher (2002) raise the possibility that it might be a cellulase.

7.5.1.1 (1,3;1,4)-β-D-glucan endohydrolases

The (1,3;1,4)-β-D-glucan endohydrolases are particularly important in the complete degradation of walls in germinated barley grain. They exclusively hydrolyse (1,4)-β-glucosidic linkages, but only where these linkages are adjacent to (1,3)-β-D-glucosyl residues, as follows:

$$4\,G\,4\,G\,3\,G\,\overset{\downarrow}{4}\,G\,4\,G\,3\,G\,\overset{\downarrow}{4}\,G\,4\,G\,4\,G\,4\,G\,3\,G\,\overset{\downarrow}{4}\,G\,4\,\cdots red,$$

where G represents a β-D-glucosyl residue, 3 and 4 are (1,3)- and (1,4)-linkages, respectively, and red indicates the reducing terminus (Anderson and Stone, 1975; Woodward and Fincher, 1982). Thus, the EC 3.2.1.73 enzymes require adjacent (1,3)- and (1,4)-β-D-glucosyl residues, they release (1,3;1,4)-β-D-tri- and tetrasaccharides (G4G3Gred and G4G4G3Gred) as major hydrolysis products, but also release higher oligosaccharides of up to 10 or more (1,4)- β-D-glucosyl residues with a single reducing terminal (1$\rightarrow$3)-β-D-glucosyl residue (e.g. G4G4G4G4G4G4G3Gred) from the longer regions of adjacent (1$\rightarrow$4)-linkages described in section 3.1.1 (Woodward *et al.*, 1983; Wood *et al.*, 1994).

In barley, two (1,3;1,4)-β-D-glucan endohydrolase isoenzymes are detected in germinated grain. Expression patterns of the genes encoding the two isoenzymes are subject to independent, tissue-specific regulation. It appears that (1,3;1,4)-β-D-glucan endohydrolase isoenzyme EI is expressed in the scutellar epithelium soon after germination and later in both the aleurone and scutellum (McFadden *et al.*, 1988; Slakeski and Fincher, 1992). (1,3;1,4)-β-D-Glucan endohydrolase isoenzyme EII appears to be expressed only in the aleurone of germinated grain, where transcript levels of the gene are high (Slakeski and Fincher, 1992). The genes encoding both of the barley (1,3;1,4)-β-D-glucan endohydrolase isoenzymes have been isolated and mapped (Figure 2), and nucleotide sequence motifs linked with gibberellic acid induction have been identified in their promoters (Litts *et al.*, 1990; Slakeski *et al.*, 1990). Gibberellic acid induces the transcription of many genes involved in endosperm mobilization, including α-amylases and endo-peptidases (Fincher, 1989).

The 3D structure of barley (1→3,1→4)- β-D-glucan endohydrolase isoenzyme EII has been defined by X-ray crystallography to 2.2-2.3Å resolution (Varghese *et al.*, 1994). The enzyme is a family GH17 glycoside hydrolase (Henrissat, 1998) that folds into a $(\beta/\alpha)_8$ barrel. The substrate-binding region consists of a deep cleft that extends across the surface of the enzyme and is long enough to accommodate 6-8 glucosyl-binding subsites (Varghese *et al.*, 1994). The cleft across the enzyme's surface is consistent with its endo-action pattern, because the enzyme can bind at most positions along the polysaccharide substrate and hydrolyse internal glycoside linkages. The availability of the 3D structure of this enzyme has enabled the rational re-design of the enzyme, through site-directed mutagenesis of the corresponding cDNA, to increase its heat stability under conditions similar to those encountered in malting and brewing (Stewart *et al.*, 2001).

7.5.1.2 (1→4)-β-D-glucan glucohydrolases

The (1,4)-β-D-glucan glucohydrolases found in extracts of young barley seedlings could hydrolyse the β-D-oligoglucosides released from wall (1,3;1,4)-β-D-glucans by the endohydrolases described above. The preferred substrates for (1,4)-β-D-glucan glucohydrolases are (1,4)-β-D-oligoglucosides, and (1,3)-β-D-oligoglucosides are hydrolysed very slowly (Hrmova and Fincher, 1996). Oligosaccharides released by (1,3;1,4)-β-D-glucan endohydrolases have (1,4)-β-D-glucosyl residues at their non-reducing termini and a single (1,3)-β-D-glucosyl residue at their reducing termini. Thus, an oligosaccharide G4G4G-4G3Gred can be hydrolysed by barley (1,4)-β-D-glucan glucohydrolases (Hrmova *et al.*, 1996) and their ability to hydrolyse laminaribiose, albeit slowly, suggests that these (1,3;1,4)-β-D-oligoglucosides could be completely depolymerized to glucose by this group of enzymes (Hrmova and Fincher, 2002). It must be noted however that the precise location of the (1,4)-β-D-glucan glucohydrolases in young seedlings has not been defined, so it remains possible that they are actually expressed in vegetative tissues and are not involved in (1,3;1,4)-β-D-glucan hydrolysis in the germinated barley grain.

7.5.1.3 β-D-glucan exohydrolases

The broad specificity β-D-glucan exohydrolases from barley can hydrolyse glycosidic linkages in several β-D-glucans, in β-D-oligoglucosides containing (1→2)-, (1→3)-, (1→4)- or (1→6)-linkages, in aryl β-D-glucosides such as 4-nitrophenyl β-D-glucoside (4NPGlc), and in some β-D-oligoxyloglucosides (Hrmova and Fincher, 1998; Kim *et al.*, 2000). Single glucose molecules are released from the non-reducing termini of these substrates, with retention of anomeric configuration (Hrmova *et al.*, 1996). They belong to the GH3 family of glycoside hydrolases (Henrissat, 1998) but their broad substrate specificity precludes their assignment to existing EC classes (Hrmova and Fincher, 1998). The β-D-glucan exohydrolases rapidly hydrolyze polysaccharide substrates such as laminarin and (1→3,1→4)-β-D-glucans, although their preferred substrates are (1→3)-β-D-glucans (Hrmova and Fincher, 1998).

The 3D structure of barley β-D-glucan exohydrolase isoenzyme ExoI has been determined by X-ray crystallography to 2.2Å resolution (Varghese *et al.*, 1999). The enzyme consists of two distinct domains that are connected by a 16-amino acid helix-like linker. The first domain is a $(\beta/\alpha)_8$ barrel of 357 amino acid residues, while the second domain consists of a six-stranded β-sheet flanked on either side by three α-helices (Varghese *et al.*, 1999). A long antiparallel loop of 42 amino acid residues is found at the COOH-terminus of the enzyme. The broad specificity of the barley enzymes has been rationalized from the X-ray crystallographic data, which show that the glucosyl residue occupying binding subsite -1 is tightly locked into a fixed position through extensive hydrogen bonding with six amino acid residues near the bottom of an active site pocket (Hrmova *et al.*, 2002). In contrast, the glucosyl residue at subsite +1 is located between two tryptophan residues at the entrance of the pocket, where it is less tightly constrained. The relative flexibility of binding at subsite +1, coupled with the projection of the remainder of bound substrate away from the enzyme's surface, means that the overall active site can accommodate a range of substrates with variable spatial dispositions of adjacent β-D-glucosyl residues (Hrmova *et al.*, 2002).

Two family GH3 β-D-glucan exohydrolases, designated isoenzymes ExoI and ExoII, have been purified from extracts of young barley seedlings and characterized (Hrmova and Fincher, 1998), although the total number of genes in the family has not been determined. Again, it is not known if the enzymes are actually located in the starchy endosperm of germinated grain but, as indicated above, they have the potential to participate in (1,3;1,4)-β-D-glucan hydrolysis and to hydrolyse (1,3;1,4)- β-D-oligoglucosides released from (1,3;1,4)-β-D-glucans by the action of endohydrolases (Hrmova *et al.*, 2002).

7.5.2 Hydrolysis of arabinoxylans

Our understanding of the enzymology and physiology of arabinoxylan degradation after germination is not yet complete, although it is likely that the depolymerization of arabinoxylans to their constituent sugars requires the concerted action of (1,4)-β-D-xylan endohydrolases (Slade *et al.*, 1989; Caspers *et al.*, 2001), arabinoxylan arabinofuranohydrolases (Ferre *et al.*, 2000; Lee *et al.*, 2001), α-L-arabinofuranosidases and β-D-xylosidases (Taiz and Honigman, 1976; Lee *et al.*, 2003). The roles of each of these enzymes are shown in Figure 6 and their properties are described in more detail below. Feruloyl esterases may be required for the hydrolysis of ester linkages between the α-L-arabinofuranosyl residues of the arabinoxylan and ferulic acid (Moore *et al.*, 1996) and α-glucuronosidases might be required for the removal of α-glucuronosyl residues from the xylan backbone, but these enzymes have not been purified and will not be discussed further here.

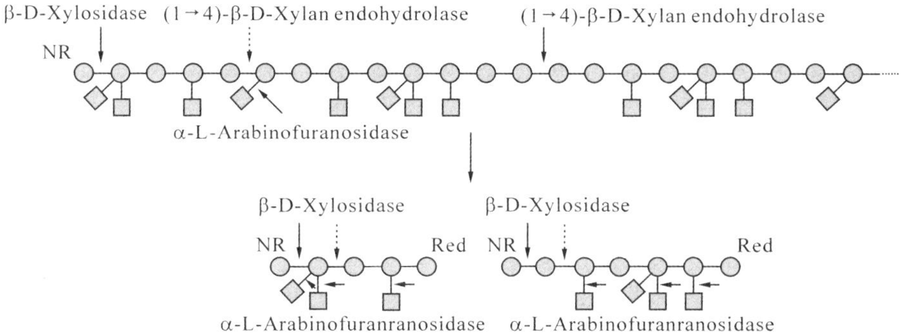

Fig. 7.3. Structure of arabinoxylan, showing proposed sites of action of major hydrolytic enzymes and sequential stages of depolymerisation. The (1,4)-linked xylan backbone (circles) is shown with arabinofuranosyl residues linked via C(O)2 (perpendicular squares), C(O)3 (angled squares) and C(O)2,3. The non-reducing end (NR) of arabinoxylan and both reducing (Red) and non-reducing ends of oligosaccharides, generated by partial enzymic degradation, are indicated. Solid arrows indicate bonds susceptible to enzymic hydrolysis and broken arrows indicate bonds that may be hydrolysed after prior removal of other arabinofuranosyl and/or xylosyl residues. As discussed in the text, the removal of other arabinofuranosyl residues might also be mediated by arabinoxylan arabinofuranohydrolases (AXAHs). Figure kindly provided by Rob Lee.

7.5.2.1 (1,4)-β-D-Xylan endohydrolases

(1,4)-β-D-Xylan endohydrolases (EC 3.2.1.8) hydrolyse the internal (1,4)-β-D-xylosidic linkages of the xylan backbone of arabinoxylans (Fig. 7.3). In extracts of germinating barley, xylanase activity appears several days later than (1,3;1,4)-β-D-glucanases (Slade *et al.*, 1989), in fact the endo-xylanase genes are transcribed approximately 24 h after those of the (1,3;1,4)-β-D-glucanases (Banik *et al.*, 1997). Similarly, (1,4)-β-D-xylan endohydrolase activity is detected in the medium surrounding isolated aleurone layers somewhat later than other hydrolytic enzymes (Taiz and Honigman, 1976; Ashford and Gubler, 1984; Caspers *et al.*, 2001).

In an examination of the xylanolytic activities of four different barley cultivars, Sungurtas *et al.* (2004) found significant differences in both the absolute levels and in the profiles of (1,4)-β-D-xylan endohydrolases present in the germinating barley grain. In early work, (1,4)-β-D-xylan endohydrolase activity was detected in isolated barley aleurone layers and could be induced by treatment with the phytohormone gibberellic acid; the activity was believed to be attributable to a 29 kDa protein (Dashek and Chrispeels, 1977). Using the same aleurone layer experimental system, Benjavongkulchai and Spencer (1986) partially purified a 34 kDa (1,4)-β -D-xylan endohydrolase. In contrast, Slade *et al.* (1989) partially purified a 41 kDa (1,4)-β-D-xylan endohydrolase from extracts of germinated barley grain and obtained NH_2-terminal amino acid sequence for the enzyme. The complete primary structure of the barley (1,4)-β-D-xylan endohydrolase was subsequently deduced from corresponding cDNAs and genes, although there was some uncertainty about the position of the NH_2-terminal residue of the protein (Banik *et al.*, 1996; 1997). Caspers *et al.* (2001) suggested that the primary translation product of the barley (1,4)-β-D-xylan endohydrolase was a protein with a molecular weight of about 61 kDa.

Thus, a major concern with this early work on the barley (1,4)-β-D-xylan endohydrolase related to the apparent discrepancies in the molecular sizes reported for the various enzyme preparations, which ranged from 29 kDa to over 60 kDa. Computer analyses of their first 70 amino acid residues predicted cytosolic locations for the barley (1,4)- β-D-xylan endohydrolases (Simpson *et al.*, 2003) and this might explain why the enzymes are detected in soluble extracts of grain relatively late after the initiation of germination (Slade *et al.*, 1989; Fincher, 1989; Simpson *et al.*, 2003; Kuntz and Bamforth, 2007). The thin, inner layer of the aleurone cell wall remains intact following germination (Taiz and Jones, 1973; Gubler *et al.*, 1987) and this is probably important in maintaining the structural integrity of the aleurone cells as they synthesize and secrete hydrolytic enzymes into the starchy endosperm (Fincher, 1989; Simpson *et al.*, 2003). If the inner wall layer of the aleurone were composed principally of arabinoxylans, as suggested by Fincher and Stone (2004), restricting the (1,4)-β-D-xylan endohydrolases to the cytosol of aleurone cells would protect the inner aleurone wall layer from degradation. Secreted (1,3;1,4)-β-D-glucan endohydrolases and other enzymes would degrade the outer layer of the aleurone walls and most of the walls of the starchy endosperm, so that diffusion of starch- and protein-degrading enzymes through the starchy endosperm could proceed (Simpson *et al.*, 2003). When the aleurone reserves were depleted and enzyme secretion completed, the cells were thought to undergo programmed cell death (Jones and Jones, 1991; Kuo *et al.*, 1996), which would lead to their rupture and to the release of cellular contents (Caspers *et al.*, 2001; Simpson *et al.*, 2003). The 61 kDa cytosolic (1,4)-β-D-xylan endohydrolases could be released during this process and exposed to endo-peptidases that might cleave away NH_2- and COOH-terminal regions to generate the 41 kDa and approximately 30 kDa forms that are detected in aleurone layers and germinated grain. While it must be acknowledged that elements of this proposal remain speculative, such a strategy would ensure that (1,4)-β-D-xylan endohydrolases were synthesized, but not released until the final stages of endosperm mobilization. As mentioned above, the late appearance of active (1,4)-β-D-xylan endohydrolases in extracts of germinated grain is well-documented (Caspers *et al.*, 2001; Simpson *et al.*, 2003), and there is some evidence for a similar mechanism for delaying the release of limit dextrinase from barley aleurone cells (Burton *et al.*, 1999). In the light of these possibilities, Van Campenhout *et al.* (2007) cloned and expressed 34, 41 and 61.5 kDa forms of a barley (1,4)- β-D-xylan endohydrolase in *Escherichia coli*. They could only recover the unprocessed (61.5 kDa) form in the cytoplasmic soluble fraction, but found it to be active on both large arabinoxylans and on xylo-oligosaccharides (Van Campenhout *et al.*, 2007).

The deduced amino acid sequences confirmed that the barley (1,4)-β-D-xylan endohydrolases belong to the GH10 family of glycoside hydrolases (Banik *et al.*, 1996; Courtinho and Henrissat, 1998; Simpson *et al.*, 2003). There are at least three (1,4)-β-D-xylan endohydrolase genes in barley and these are very closely linked on the long arm of chromosome 5H (Banik *et al.*, 1997). No recombination was detected between the genes in 234 doubled haploid lines in two mapping populations, which suggested that they have arisen by tandem duplication of a single ancestral gene (Banik *et al.*, 1997).

At this stage it is not known whether endo-xylanase inhibitors in the germinated barley grain might partially explain some of the observations on the development of (1,4)-β-D-xylan endohydrolase activity (Goesaert *et al.*, 2001). Cereal xylanase inhibitors have been studied in detail in wheat grain, where a number of distinct classes of inhibitors are found, including the Triticum aestivum xylanase inhibitor (TAXI), the xylanase inhibitor protein (XIP), and the thaumatin-like xylanase inhibitor (TLXI) (Goesaert *et al.*, 2001; Fierens *et al.*, 2007, 2008). However, it would appear that these inhibitors predominantly target (1,4)-β-D-xylan endohydrolases of microbial origin, many of which belong to the family GH11 group of enzymes (Juge *et al.*, 2004; Fierens *et al.*, 2007, 2008).

7.5.2.2 Arabinoxylan arabinofuranohydrolases

A family GH51 arabinoxylan arabinofuranohydrolase, designated AXAH-I, has been purified from extracts of 7-day old barley seedlings (Ferre *et al.*, 2000; Lee *et al.*, 2001). The enzyme releases L-arabinose from cereal cell wall arabinoxylans and, while hydrolysis of α-L-arabinofuranosyl residues linked to C(O)3 of backbone (1,4)-β-D-xylosyl residues proceeds at the fastest rate, α-L-arabinofuranosyl residues on doubly substituted xylosyl residues are also hydrolyzed, at lower rates (Figure 6). The complete amino acid sequence of the enzyme was deduced from near full-length cDNA and a second cDNA, which was 81% identical to the cDNA encoding AXAH-I, encodes another barley arabinoxylan arabinofuranohydrolase, designated AXAH-II (Lee *et al.*, 2001).

The most important finding of this work was that the barley AXAH-I could be distinguished from several α-L-arabinofuranosidases in the seedling extracts because of its ability to hydrolyze arabinoxylan. The α-L-arabinofuranosidases are unable to do this, although both types of enzyme can hydrolyze 4-nitrophenyl α-L-arabinofuranoside (Lee *et al.*, 2001).

The barley arabinoxylan arabinofuranohydrolases are likely to play key roles in wall metabolism in barley. Thus, the enzymes could participate in the modification of arabinoxylan fine structure during wall deposition, maturation or expansion (Gibeaut *et al.*, 2005), or in wall turnover and the hydrolysis of arabinoxylans in germinated grain (Ferre *et al.*, 2000; Lee *et al.*, 2001). As mentioned in sections 3.2.2 and 3.2.3, enzymes that catalyze the hydrolytic removal of α-L-arabinofuranosyl residues from arabinoxylans can generate changes in arabinoxylan fine structure that will have dramatic effects on physicochemical properties such as solubility and interactions with other wall polysaccharides. However, more information will be necessary before we can confidently conclude that the AXAH enzymes do indeed modify newly synthesized arabinoxylans or play a significant role in the depolymerization of arabinoxylans in germinated barley grain.

7.5.2.3 α-L-Arabinofuranosidases and β-D-Xylosidases

An α-L-arabinofuranosidase and a β-D-xylosidase, designated ARA-I and XYL, respectively, and both members of the GH3 family of glycoside hydrolases, have been purified from extracts of 5-day old barley seedlings and characterized (Lee *et al.*, 2003). The catalytic efficiency of ARA-I during the hydrolysis of 4-nitrophenyl α-L-arabinofuranoside is only slightly higher than during the hydrolysis of 4-nitrophenyl β-D-xyloside (Lee *et al.*, 2003). Thus, the enzyme is more correctly referred to as a bifunctional α-L-arabinofuranosidase/ β-D-xylosidase. In contrast, the XYL enzyme preferentially hydrolyses 4-nitrophenyl β-D-xyloside and has only low activity on 4-nitrophenyl α-L-arabinofuranoside.

The enzymes hydrolyse wheat flour arabinoxylan very slowly, but rapidly hydrolyse oligosaccharide products released from this polysaccharide by (1,4)-β-D-xylan endohydrolases (Fig. 7.3). Thus, both enzymes hydrolyse (1,4)-β-D-xylopentaose and ARA-I can also degrade (1,5)-α-L-arabinohexaose. However, neither will hydrolyse substituted polysaccharides and the observation that no arabinose is removed from oligoarabinoxylosides released by (1,4)-β-D-xylan endohydrolase action suggests that the enzymes are unable to completely hydrolyse substituted oligomeric substrates.

The complete amino acid sequences of ARA-I and XYL have been deduced from cDNAs (Lee *et al.*, 2003). A discrepancy between the apparent Mr values obtained for the purified enzymes and those predicted from the cDNAs are attributable to COOH-terminal processing, through which about 130 amino acid residues are removed from the primary translation product. The genes encoding the ARA-I and XYL have been mapped to chromosomes 2H and 6H, respectively (Lee *et al.*, 2003).

The α-L-arabinofuranosidases and β-D-xylosidases are secreted from isolated aleurone layers much earlier than the (1,4)- β-D-xylan endohydrolases

(Banik *et al.*, 1997). Thus, the secretion from aleurone layers of endohydrolases, α-L-arabinofuranosidases and β-D-xylosidases involved in arabinoxylan degradation is clearly not coincident. Lee *et al.* (2003) detected little or no mRNA encoding ARA-I in the aleurone layer of germinated grain but nevertheless concluded that the two family GH3 enzymes play important roles in cell wall degradation in germinated barley grain. (1,4)-β-D-Xylan endohydrolases are synthesized in the aleurone of germinated barley grain and are eventually secreted into the starchy endosperm, where they are likely to participate in cell wall degradation (Banik *et al.*, 1996; Caspers *et al.*, 2001). The ARA-I and XYL enzymes could participate in further hydrolysis of the oligosaccharides released by the endohydrolases (Lee *et al.*, 2003).

7.5.3 Hydrolysis of Glucomannans

A family GH5 (1,4)-β-D-mannan endohydrolase, designated HvMAN1, has also been purified from extracts of young barley seedlings (Hrmova *et al.*, 2006). The purified HvMAN1 releases mannose-containing oligosaccharides of degree of polymerization 2-6 from mannans, galactomannans and glucomannans. The complete sequence of 374 amino acid residues of the mature barley enzyme was deduced from the nucleotide sequence of a near full-length cDNA, and allowed a three-dimensional model of the *HvMAN1* to be generated (Hrmova *et al.*, 2006). The barley *HvMAN1* gene is a member of a family of at least four genes, and is transcribed at relatively low levels in developing endosperm in particular, but also in the basal region of young roots and in leaf tips (Hrmova *et al.*, 2006). A barley family GH1 β-D-mannosidase, designated Hv β MANNOS1, is also believed to participate in mannan or glucomannan depolymerization through its ability to hydrolyze manno-oligosaccharides to mannose with an action pattern typical of a (1,4)-β-D-mannan exohydrolase (Hrmova *et al.*, 2006).

While it is likely that the (1,4)-β-D-mannan endo- and exo-hydrolases play a role in the hydrolysis of mannans and glucomannans in the walls of the starchy endosperm following germination of barley grain, it must be remembered that the enzymes were purified from extracts of young seedlings, and that they therefore have not yet been shown unequivocally to be located in the starchy endosperm of germinated grain. The same is true of the AXAH enzymes described in section 5.2.2.

7.5.4 Hydrolysis of Callose

Despite the fact that (1,3)-β-D-glucan in ungerminated barley grain is limited to small deposits that are scattered throughout the starchy endosperm, levels of (1,3)-β-D-glucan endohydrolases increase markedly in the germinated grain (Ballance *et al.*, 1976) and are also secreted from isolated aleurone layers following treatment with gibberellic acid (Taiz and Jones, 1970). The enzymes are encoded by a gene family of at least six members and isoenzyme GII appears to be the most abundant in germinated barley grain (Hoj *et al.*, 1988; Xu *et al.*, 1992). While one of the roles of the enzymes could be to depolymerize the (1,3)- β-D-glucan deposits, the (1,3)-β-D-glucan endohydrolases might also be a component of a more extensive strategy to protect the germinated grain against pathogen attack (Fincher, 1989). Indeed, many pathogenesis-related proteins, including (1,3)- β-D-glucanases and chitinases, are found in germinated barley grain, together with toxic thionins and other peptides, inhibitors of microbial hydrolases, and inhibitors of protein synthesis (Fincher, 1989). The (1,3)-β-D-glucan endohydrolases are capable of hydrolyzing the (1,3)-β-D-glucans and (1,3;1,6)-β-D-glucans of fungal cell walls, and might thereby protect the grain from fungal invasion, although at this stage the evidence for such a role is circumstantial rather than proven.

7.6 Effects of Wall Components on Malting Quality

Cell walls contribute approximately 7% of the dry weight of the endosperm of barley grain, but this can vary considerably depending on the variety and the environmental growing conditions (Molino-Cano and Conde, 1982; Edney *et al.*, 1998; Table 7.2). Under the controlled conditions of germination used for the generation of malt, the breakdown and modification of cell walls is an important process, because partial or complete degradation of the wall removes the physical barrier that prevents hydrolytic enzymes secreted from the aleurone or scutellum from reaching carbohydrate and protein reserves in the cells of the starchy endosperm (Taiz and Jones, 1970; Ashford and Jacobsen, 1974; Gubler *et al.*, 1987; Fincher 1989). Depending on the temperatures used in the brewery during extraction of the malt (mashing), the remaining wall material in the malt may or may not be further degraded in the wort. Residual wall material in the wort will pass through the fermentation process and will be found essentially unchanged in the final beer. (1,3;1,4)-β-D-Glucans and arabinoxylans are known to form viscous solutions (sections 3.1.4 and 3.2.3), which slow down wort separation (lautering) and beer filtration (Forrest and Wainwright, 1977) and, in some cases, contribute to haze formation when the beer is cooled (Coote and Kirsop,1976). The addition of endogenous (1,3;1,4)-β-D-glucanases and (1,4)-β-D-xylanases to the mash or fermenting beer has been shown to reduce the viscosity and hence to reduce filtration problems (Ducroo and Felon, 1989).

7.6.1 Wall Components in Malting

7.6.1.1 (1,3;1,4)-β-D-glucans

Because up to 75% of starchy endosperm walls in the barley grain are composed of (1,3;1,4)-β-D-glucan (Fincher, 1975), the processes involved in the breakdown and removal of these polysaccharides in the malthouse have been studied extensively. During malting up to 90% of the grain (1,3;1,4)-β-D-glucan is degraded, but this is highly dependent on the variety, the growing environment, particularly late in grain development, and on the malting regime itself (Edney *et al.*, 1998, Molina-Cano *et al.*, 2002). At the completion of the malting process, as little as 0.2% but up to 1.5% of the dry weight of the malted barley grain consists of (1,3;1,4)-β-D-glucan, depending on the level of modification achieved (Table 7.2).

Table 7.2 Levels of component affecting wall metabolism in barley, malt, wort and beer

	Arabino-xylan*	(1→3)(1→4)-β-glucan*	Viscosity (cP)	β-glucanase (U/kg)	comment	reference
Barley	4.4-7.8	3.4-5.7				Henry, 1986
	6.6 (1.4% in endosperm)	4.2 (4.1% in endosperm)				Henry, 1987
		3.1-3.8				Morgan *et al.*, 1983
		2.5-5.2				Edney *et al.*, 1998
		4.1. 4.5				Molina-Cano *et al.*, 2002
	7.5	3.70-3.99				Han and Swarz, 1996
		2.6-3.9				Stuart *et al.*, 1988
Malt				approx 120-300	using different steeping program	Nischwitz *et al.*, 1999
		<0.2-0.7		225-350	laboratory and Industrial scale trials	Home *et al.*, 1993
		0.05-1.41			Pilot Malting	Edney *et al.*, 1998

(to be continued)

	Arabino-xylan*	(1→3)(1→4)-β-glucan*	Viscosity (cP)	β-glucanase (U/kg)	comment	reference
		1.5, 1.1				Molina-Cano *et al.*, 2002
		0.8-2.0		23.7-49.0 (U/g)		Stuart *et al.*, 1988
	3.6	0.50-0.57				Han and Swarz, 1996
		0.33-1.44		226-579		Wang *et al.*, 2004
	3.64-5.40					Li *et al.*, 2005
Wort		approx 40 to over 500			using different steeping program	Nischwitz *et al.*, 1999
		190-385	1.46-1.66		laboratory and Industrial Scale Trials	Home *et al.*, 1993
		14-438				Edney *et al.*, 1998
			1.51, 1.60			Molina-Cano *et al.*, 2002
	946-1791	98-189	1.49-1.52			Li *et al.*, 2005
Beer		12-296			Pilot-scale brewing	Edney *et al.*, 1998
	2540-3022mg/L	330-580				Han and Swarz, 1996
	621-932					Li *et al.*, 2005

*Units barley and malt % dry of flour, wort and beer mg/L

(1,3;1,4)-β-D-Glucan endohydrolases (EC 3.2.1.73) are key enzymes in the hydrolysis of (1,3;1,4)-β-D-glucans during the malting process and are key determinants of malt quality characteristics, because as endohydrolases they quickly reduce the viscosity of long (1,3;1,4)-β-D-glucan molecules (Brunswick *et al.*, 1987). Two (1,3;1,4)-β-D-glucanase isoenzymes, which are designated isoenzymes EI and EII, are synthesised in germinated barley (Woodward and Fincher, 1982). However, both isoforms, and in particular isoenzyme EI, are unstable at elevated temperatures. The activities of both are therefore reduced during kilning, when the malted barley grain is dried at progressively higher temperatures, up to 80 °C or more. After kilning, essentially no isoenzyme EI activity remains, and there is little or no activity of isoenzyme EII (Loi *et al.*, 1987; Barber *et al.*, 1994). Inactivation rates depend on both the temperature and moisture of the malt, beginning early in the kilning process when temperatures are relatively low and grain moisture content is high (Hamalainen and Reinikainen, 2007).

7.6.1.2 Arabinoxylans

Arabinoxylans make up about 20% of the wall in the starchy endosperm (Table 7.1) and have not been investigated as thoroughly as the (1,3;1,4)-β-D-glucans. It has been estimated that only about 50% of the arabinoxylans of the endosperm walls are degraded during malting (Han and Swarz, 1996; Han 2000) and large amounts of this polysaccharide therefore remain in malted barley (Table 7.2).

As noted in section 5.1, the depolymerisation of arabinoxylans in the germinated barley grain involves several enzymes. Again the (1,4)-β-D-xylan endohydrolases are likely to be the most important in reducing the undesirable effects of arabinoxylans that might be associated with the tendency of arabinoxylans to form highly viscous solutions following their extraction from the grain. However, there are several unanswered questions relating to the release of (1,4)-β -D-xylan endohydrolases from the aleurone layer in the germinated barley grain (section 5.1.1). Thus, it is not clear if and when the enzymes gain access to arabinoxylan substrates in the endosperm walls. In addition, there are contradictory reports on the processing of barley (1,4)-β-D-xylan endohydrolases, and final and/or active forms of the (1,4)-β-D-xylan endohydrolases have not been defined unequivocally (Kuntz and Bamforth, 2007; Van Campenhout *et al.*, 2007; section 5.1.1). Nevertheless, it has been shown that during malting the activity of (1,4)-β-D-xylan endohydrolases increases but that this activity is reduced by up to 50% during kilning (Li *et al.*, 2005; Table 7.2).

7.6.2 Wall Components in Brewing

Brewing of beer involves two main steps, namely mashing and fermentation. In the first, ground malt, called grist, is mixed with warm water to enable the enzymes produced in the controlled germination of the grain during malting to act upon the constituents of the malt, in particular starch and its partial degradation products generated during malting. A central objective of brewing is to release and ferment sugars from the starch reserves of the barley grain, but amylolytic enzymes can hydrolyse gelatinized starch much more rapidly than they can hydrolyse more highly structured and relatively impenetrable starch granules. Because barley starch gelatinizes at between 60°C and 65°C, mashing protocols usually involve raising the temperature of the malt extract to these temperatures. However, the important wall degrading enzymes of malted barley are relatively unstable at elevated temperatures.

In general there are three methods of mashing, namely infusion, decoction and temperature programmed mashing (Briggs *et al.*, 1981). Infusion mashing uses a single temperature system of around 65°C, for anywhere from 30 minutes to several hours. This temperature is highly destructive to many of the cell wall degrading enzymes, such that limited cell wall degradation occurs (Loi *et al.*, 1987; Barber *et al.*, 1994). For this reason brewers tend to use well modified malts with little cell wall material for infusion mashing. In decoction and temperature programmed mashing processes, the temperature of the mash is increased incrementally from about 35°C to about 65°C. The slower rates of temperature increases in these processes allow wall degrading enzymes a longer period of activity for the degradation of wall polysaccharides. Thus, decoction and temperature programmed mashing are more suitable for malts that are less well modified (Briggs *et al.*, 1981).

Finally, the wort is separated from the spent grain, using either a mash bed or a mash filter. The wort is cooled and moved to a fermentation vessel where yeast is added (Briggs *et al.*, 1981). Fermentation is the process by which the yeast converts simple sugars, produced during mashing, into ethanol and carbon dioxide. While glucose and maltose from starch degradation are the most abundant fermentable sugars, glucose released from wall (1,3;1,4)-β-D-glucan can make a significant contribution to total fermentable sugar levels (Morall and Briggs, 1978).

7.6.2.1 (1,3;1,4)-β-D-glucans

Residual (1,3;1,4)-β-D-glucan in the molecular weight range of 70 to 600 kDa has been detected in wort and beer, and is likely to be a major contributing factor to filtration problems experienced elsewhere in the brew house (Bamforth, 1982; 1985; 1994; MacGregor and Fincher 1994). While levels of (1,3;1,4)-β-D-glucan endohydrolases might be sufficiently high in germinated barley to effect rapid (1,3;1,4)- β -D-glucan depolymerisation, these enzymes are heat labile and are quickly inactivated at temperatures commonly used in both kilning (section 6.1.1) and mashing (Loi *et al.*, 1987; Barber *et al.*, 1994). At temperatures above 55°C very little (1,3;1,4)- β -D-glucanase activity remains after 5 minutes (Fig. 7.4; Stewart *et al.*, 2001; Home and Elamo, 1993).

At low mash temperatures (1,3;1,4)-β-D-glucan solubilised from the flour is rapidly degraded, but at higher temperatures where there is little or no (1,3;1,4)-β-D-glucan endohydrolase activity, (1,3;1,4)-β-D-glucans released into the wort are not degraded (Home and Elamo, 1993). In a study on the influence of different steeping times on (1,3;1,4)-β-D-glucan levels in wort it was found that longer steep times reduced wort (1,3;1,4)-β-D-glucan (Nischwitz

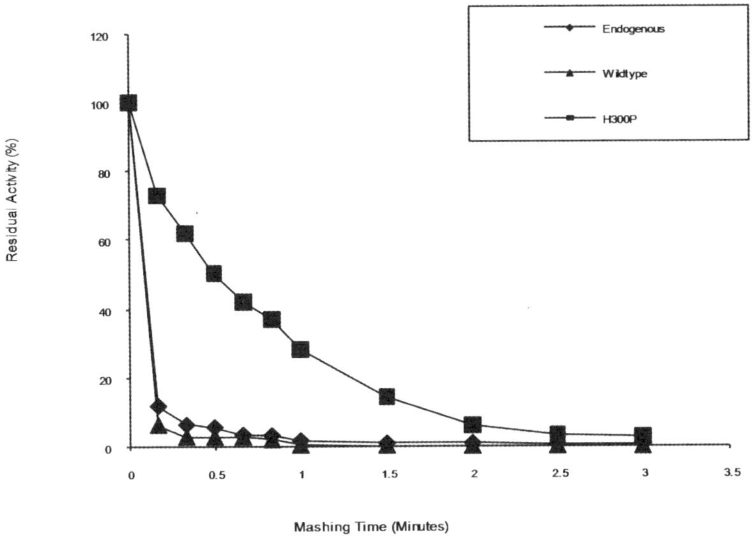

Fig. 7.4. Effects of temperature on (1,3;1,4)-β-D-glucanase activity during simulated mashing at 65°C. The diamonds and triangles show the loss of activity, within seconds, of endogenous (1,3;1,4)-β-D-glucanase activity in the green malt sample and of heterologously expressed (1,3;1,4)-β-D-glucanase added to the malt, respectively. The squares show the activity decay curve of the mutated (1,3;1,4)-β-D-glucanase (Stewart *et al.*, 1991), in which increased thermostability was engineered through the H300P mutation. Figure provided by Richard Stewart

et al., 1999). Large variations have been found in the levels of (1,3;1,4)-β-D-glucan remaining in both wort and beer (Table 7.2).

7.6.2.2 Arabinoxylans

As mentioned in section 6.1.2, only about 50% of the arabinoxylans are degraded during malting (Han and Swarz, 1996; Han 2000) and large amounts therefore remain in both wort and beer (Table 7.2). In one study as much as 3 g/L arabinoxylan was detected in beer, while (1,3;1,4)- β -D-glucan levels were relatively low at 0.5 g/L (Han and Swarz, 1996). It is well-documented that many enzymes are involved in the degradation of arabinoxylans during malting (Sungurtas *et al.*, 2004), but very little is known about the stability of these enzymes during mashing. Endo-(1,4)-β-D-xylanases are known to be heat labile and are degraded during kilning (Li *et al.*, 2005), so one might anticipate that (1,4)-β-D-xylanases would also be inactivated during mashing. However, the observation that the water extractable arabinoxylan content of wort is greater than the water extractable arabinoxylan content of malt suggests that (1,4)-β-D-xylanases or related enzymes are active for a least a short time during mashing (Li *et al.*, 2005).

7.6.3 Effects of Wall Components on Other Malting Quality Parameters

In consideration of the effects of cell wall polysaccharides on malting quality of barley, attention has usually been focused on the generally detrimental effects of high levels of (1,3;1,4)-β-D-glucans that form highly viscous solutions when extracted from malted barley grains. The high viscosity of such malt extracts reduces the rates of filtration rates in the brewery, either during lautering (Nischwitz *et al.*, 1999) or during filtration of the beer following fermentation (Narziss, 1993; Stewart *et al.*, 1998). Wall arabinoxylans have attracted less attention, but could be at least partly responsible for many of the filtration problems normally attributed to (1,3;1,4)-β-D-glucans (Vietor *et al.*, 1993; Li *et al.*, 2005).

However, wall polysaccharides might also be determinants of other malting barley quality characteristics. Malt extract is a measure of the amount of soluble material extractable from malted barley in the brewery, and the industry strives for high malt extract values. Extract values will be determined by the degree of mobilization and solubilisation of starch, storage proteins, nucleic acids and other molecules contained within the cells of the starchy endosperm, and it can therefore be argued that the rate of cell wall degradation will be one of the most important factors that influence extract. For example, walls that are particularly high in (1,3;1,4)-β-D-glucans, whether this results from genetic or environmental conditions during grain development, may take longer to degrade during malting. Given that the walls represent a general barrier to the free access of hydrolytic enzymes that degrade starch, proteins and other storage polymers in the grain, delayed degradation of the wall could slow the solubilisation of many other components of the endosperm and hence reduce extract values more generally. Similar effects would be observed if the levels of (1,3;1,4)-β-D-glucan were "normal" but if (1,3;1,4)-β-D-glucanase levels or activities were relatively low.

Similarly, high fermentability values are considered desirable and are often focused upon the degradation of starch in the grain. However, if carbon flux is re-directed into wall polysaccharides as a result of environmental conditions in the developing grain, the complete hydrolysis of the wall polysaccharides to glucose during malting and mashing will also be necessary to achieve optimal fermentability values. The (1,3;1,4)-β-D-oligoglucosides that represent partial degradation products of (1,3;1,4)-β-D-glucans are not fermentable. Thus, high (1,3;1,4)-β-D-glucans levels are often associated with lower levels of glucose in the wort, and hence reduced fermentability (Molino Cano, 2002; Edney *et al.*, 2007). Indeed, the level of endosperm modification of the grain is an important factor for high fermentability (Edney *et al.*, 2007), for the reasons outlined above.

Other key malting quality characteristics, such as diastatic power and Kolbach index, could also be indirectly affected by the effects of walls on the rate of starchy endosperm mobilization in the malting process. A study of (1,3;1,4)-β-D-glucan content and (1,3;1,4)-β-D-glucanase activity in both grain and malt of eight domesticated cultivars grown at seven different locations indicated that malt (1,3;1,4)-β-D-glucan content is more dependent on malt (1,3;1,4)-β-D-glucanase activity than on the original level of (1,3;1,4)-β-D-glucan in the grain. Although the malt quality parameters diastatic power, Kolbach index, viscosity and malt extract were correlated with high significance to malt (1,3;1,4)-β-D-glucan content, they were only weakly correlated to (1,3;1,4)-β-D-glucan content in the ungerminated grain (Wang *et al.*, 2004).

7.7 Enhancing Malting Quality through Manipulation of Cell Wall Metabolism

Emerging genetic manipulation and functional genomics technologies are providing new opportunities for enhancing the malting quality of barley. For example, until recently our ability for manipulating barley and malt (1,3;1,4)-β-D-glucan content was limited through lack of information on genes and the encoded enzymes that are responsible for the biosynthesis of these polysaccharides. Although a complete description of (1,3;1,4)-β-D-glucan biosynthesis has not yet been achieved, the implication of the *HvCslF* genes in the process (Burton *et al.*, 2006, 2008) presents new target genes and new possibilities for manipulating the levels of (1,3;1,4)-β-D-glucans in barley and malt.

In a more general way, once candidate genes have been identified, there are several strategies that can be invoked to test their impact on malting quality. One strategy is to use the gene sequence and derived molecular markers to better and more rapidly exploit natural variation in levels of the particular wall polysaccharide in barley grain (Burton *et al.*, 2008). Similarly, the gene sequences can be used to examine mutants with altered levels of those wall polysaccharides. Material identified in these ways potentially enables the most useful genetic variants to be selected for use in conventional, non-GM breeding programs. An alternative GM route would be to investigate the effects on cell wall composition of over- and under-expressing key genes in transformed barley.

7.7.1 Down-regulation of (1,3;1,4)-β-D-glucan Synthesis

In the case of the barley (1,3;1,4)-β-D-glucans, which are the predominant components of the starchy endosperm cell walls, it might be assumed that a reduction of levels in the mature, ungerminated barley grain would lead to a reduction of (1,3;1,4)-β-D-glucan in the final malt. If so, the recently identified *HvCslF* genes would be prime candidate genes for the manipulation of (1,3;1,4)-β-D-glucan levels in grain (Burton *et al.*, 2006). Transcript analyses show that the *HvCslF6* and HvCslF9 genes are transcribed at high levels in developing barley grain, and these two members of the gene family are therefore obvious targets for manipulation (Burton *et al.*, 2008). The definition of allelic diversity, or haplotypes, for the two *HvCslF* genes that are expressed in the developing grain, as correlated to (1,3;1,4)-β-D-glucan levels in the mature grain, may be a way of providing information for use with conventional breeding approaches to either reduce the final amount of (1,3;1,4)-β-D-glucan, or to create structural changes that might make it more susceptible to enzymic degradation.

However, the chances of finding a genetic variant that would make a substantial difference to (1,3;1,4)-β-D-glucan levels might be difficult if the germplasm pool were to cover a relatively narrow spectrum of variation. As expected, a study of the haplotype structure of seven barley genes from 20 domesticated barley lines (*Hordeum vulgare*) compared to the same genes in 25 wild forms of barley (*Hordeum spontaneum*) indicated that there are many more haplotypes with a higher nucleotide diversity in the wild lines (Kilian *et al*, 2006). This result is reflected in the equally reduced range of (1,3;1,4)-β-D-glucan contents in domesticated barley compared with wild barley. Henry (1986) found the (1,3;1,4)-β-D-glucan contents of 17 barley varieties grown at three different locations ranged from 3.4% to 5.7%. This may be compared with wild barley, *Hordeum spontaneum*, collected from 23 sites in Israel, which contained 4.5% to 13.2% (1,3;1,4)-β-D-glucan (Henry and Brown, 1987). While the level of natural variation was higher in the wild barleys, the domesticated barleys had a lower limit of (1,3;1,4)-β-D-glucan (3.4% compared with 4.5% in wild barleys), presumably as a result of barley breeders selecting against high (1,3;1,4)-β-D-glucan levels. Thus, the opportunities for lowering (1,3;1,4)-β-D-glucan contents of barley through exploitation of natural variation might be limited.

Whether or not down-regulation of the *HvCslF6* and HvCslF9 genes in transgenic barley generated through GM technologies would result in grain with lower (1,3;1,4)-β-D-glucan content remains to be demonstrated. Similarly, it is not known if the non-GM but targeted inactivation method called Targeted Local Lesions IN Genomes (TILLING) could be used successfully as an extension of conventional chemical mutagenesis and screening (McCallum *et al.*, 2000; Bentley *et al.*, 2000). TILLING enables the isolation of mutations, usually in the form of a single base-pair substitution, in a chosen target gene, such as the *HvCslF6* or *HvCslF9* genes. Multiple mutations can be isolated in the same gene and in this way an allelic series can be generated. This technology has already demonstrated its usefulness in wheat in producing novel waxy variants (Slade *et al.*, 2005) and barley TILLING populations are now also available. Another strategy to the down-regulation of *HvCslF* genes would be through double-stranded RNA interference (dsRNAi) (Waterhouse *et al.*,

2006), but while this method can be used to reduce expression levels of the target gene, it cannot be used to create true nulls, which might be better achieved by TILLING.

7.7.2 Increasing (1,3;1,4)-β-D-glucanase Activity

Another key target for manipulation of (1,3;1,4)- β-D-glucan levels in malt are the (1,3;1,4)-β-D-glucan endohydrolases genes (*HvGlbI* and *HvGlbII*), which appear to be involved in both biosynthesis and hydrolysis of (1,3;1,4)-β-D-glucan (Burton *et al.*, 2008). These genes are also attractive routes for malt (1,3;1,4)-β-D-glucan manipulation, based on the data of Wang *et al.* (2004), who concluded that malt (1,3;1,4)-β-D-glucan content depends more on malt (1,3;1,4)-β-D-glucanase activity than on the level of (1,3;1,4)-β-D-glucans in the ungerminated grains.

One approach to increase levels of (1,3;1,4)-β-D-glucanase activity in malt was to genetically engineer increased thermostability into barley (1,3;1,4)-β-D-glucanases, so that higher proportions of active enzyme would survive the heating steps in malting and mashing. The approach was based upon rational re-design of the enzymes and relied on the availability of the 3D structures of the enzymes (Chen *et al.*, 1994). Stewart *et al.* (2001) changed specific nucleotides in barley (1,3;1,4)-β-D-glucanase cDNAs and expressed the mutated genes in E. coli. One mutated enzyme, which contained a single histidine to proline amino acid substitution (H300P), displayed a significant increase in thermostability when compared with the wildtype recombinant (1,3;1,4)-β-D-glucanase isoenzyme EII under in vitro conditions, and during simulated mashing at 65 °C (Fig. 7.4). The histidine to proline substitution decreases the entropy of the unfolded state of the protein, which effectively increases the stability of the folded state (Stewart *et al.*, 2001). The manipulation of the thermostability of (1,3;1,4)-β-D-glucanase isoenzymes EI or EII might also be achieved through haplotype selection or TILLING.

A second approach to the reduction of (1,3;1,4)-β-D-glucan levels through the manipulation of (1,3;1,4)-β-D-glucanase activity would be to increase expression of the (1,3;1,4)- β-D-glucanase GlbI gene in the developing grain (Burton *et al.*, 2008). It is not known if this would lead to reduced (1,3;1,4)-β-D-glucan levels in walls of the starchy endosperm, but in an analogous situation it has been shown that mutations in the *Arabidopsis KOR* gene, which encodes a (1,4)-β-D-glucanase, lead to large reductions in wall cellulose (Szyjanowicz *et al.*, 2004).

7.8 Future Prospects

As indicated in preceding sections, the biosynthesis of the major polysaccharide components of barley endosperm cell walls is likely to involve many different types of enzymes and ancillary proteins, possibly organized into biosynthetic complexes. These complexes will be sensitive to many developmental cues, such as hormone levels, moisture availability and temperature, as well as the intrinsic controls exerted by the genetic make-up of each cultivar or line. The very early stages of wall development in barley may also be good targets for manipulation of malting quality characteristics of the mature grain, but the enzymes and processes active at this early stage have not been described. For example, transcription factors important in the initiation and control of wall biosynthesis have not been identified. Equally valid targets may lie within the site of enzymes involved in the hydrolysis of the various components of the cell walls during the germination or malting processes. In either case, recent progress in the identification of more of the genes directly implicated in the biosynthesis of integral endosperm cell wall polysaccharides and in their degradation, coupled with emerging technologies that facilitate the identification of partner proteins in multi-enzyme complexes, may provide more avenues to understanding wall biology in the developing grain and in the malting and brewing processes, and hence in malting quality.

General improvements in the analysis of transcript levels using methods such as real-time quantitative PCR, SAGE, MPSS and microarrays also offer opportunities to identify groups of genes involved in complex biosynthetic or degradative processes. For example, Potokina *et al.* (2004) used a "closed set" of 1400 ESTs on a cDNA microarray to examine ten barley genotypes, previously classified using six malting quality-associated traits, in an effort to identify key genes which might be contributing to those traits. White *et al* (2006) described the construction of seven Serial Analysis of Gene Expression (SAGE) libraries from grain during malting. Both methods identified specific genes, many of which are well-known to be malting related, but some previously unknown candidates were also identified. Potokina *et al.* (2004) mapped a selection of genes, over half of which displayed linkage to known QTLs for malt quality. Continuation of these functional association strategies might be expected to provide more candidates for use in barley malting quality improvement. The genome-wide analysis of expression quantitative trait loci (eQTL), which has recently been described for 16,000 barley genes using hybridization of Affymetrix arrays (Potokina *et al.*, 2008), might prove to be an even more powerful tool for the identification of candidate genes for malting quality in the future.

Other prospects for the immediate future include the definition of the role of wall arabinoxylans as determinants of malting quality. This polysaccharide has physicochemical properties that are similar to the (1,3;1,4)-β-D-glucans and are likely to undergo similar biosynthetic processes during grain development. Key differences in arabinoxylan depolymerisation have already been identified, but the mechanism of its depolymerisation in the germinated grain has not been clearly defined. One would expect that the identification of the genes involved in (1,4)-β-D-xylan biosynthesis and in arabinosyl transfer to the main (1,4)-β-D-xylan chain will be an important first step in understanding the biosynthesis of the arabinoxylans. One might also expect that the identities of these genes will become apparent in the immediate future.

Acknowledgements

The work described in this review has been supported by the Australian Research Council and the Grains Research and Development Corporation. We thank Maria Hrmova and Andrew Harvey for their assistance with various aspects of the work.

References

Anderson M A, Stone B A (1975) New Substrate for Investigating Specificity of Beta-glucan Hydolases. Febs Letters, 52: 202-207

Andrewartha K A, Phillips D R, Stone B A (1979) Solution Properties of Wheat-flour Arabinoxylans and Enzymatically Modified Arabinoxylans. Carbohydro-hydrate Res, 77: 191

Ashford A E, Jacobsen J V (1974) Cytochemical Localization of Phosphatase in Barley Aleurone Cells - Pathway of Gibberellic Acid Induced Enzyme Release. Planta, 120: 81-105

Ashford A E, Gubler F (1984) Mobilization of Polysaccharide Reserves from Endosperm. In: Murray D R (ed.) Seed Physiology. 2: Germination and Reserve Mobilization. Academic Press, Sydney, 117-162

Atkins E D T (1991) Three-dimensional Structure, Interactions and Properties of Xylans. Progress in Biotechnology International Symposium, Wageningen, Netherlands, Elsevier Science Publishers B.V, Amsterdam, Netherlands, New York, USA.

Bacic A, Stone B A (1981) Chemistry and Organization of Aleurone Cell Wall Components from Wheat and Barley. Aust J Plant Physiol, 8: 475-495

Ballance G M, Meredith W O S, Laberge D E (1976) Distribution and Developmen of Endo-beta-glucanase Activities in Barley Tissues during Germination. Can J Plant Sci, 56: 459-466

Ballance G M, Manners D J (1978) Structural Analysis and Enzymic Solubilization of Barley Endosperm Cell Walls. Carbohydr Res, 61: 107-118

Bamforth C W (1982) Barley β-glucans. Their Role in Malting and Brewing. Brewers Digest, 6: 22-27

Bamforth C W (1985) Cambridge Prize Lecture, Biochemical Approaches to Beer Quality. J Inst Brew, 91: 154-160

Bamforth C W (1993) β-Glucans and β-glucanases in Malting and Brewing: Practical Aspects. Brewers Digest, 69: 12-16

Banik M, Garrett T P J, Fincher G B (1996) Molecular Cloning of cDNA Encoding $(1\rightarrow4)$-β-xylan Endohydrolases from the Aleurone Layer of Germinated Barley (*Hordeum Vulgare*). Plant Mol Biol, 31: 1163-1172

Banik M, Li C D, Langridge P, Fincher G B (1997) Structure, Hormonal Regulation and Chromosomal Location of Genes Encoding Barley $(1\rightarrow4)$-β-xylan Endohydrolases. Mol Gen Genet, 253: 599-608

Barber M G, Jackson E A, Smith D B (1994) Total and Individual Barley (1-3), (1-4)-β-D-Glucanase Activities in Some Green and Kilned Malts. J Inst Brew, 100: 91-97

Benjavongkulchai E, Spencer M S (1986) Purification and Characterization of Barley Aleurone Xylanase. Planta, 169: 415-419

Bentley A, MacLennan B, Calvo J, Dearolf C R (2000) Targeted Recovery of Mutations in Drosophila. Genetics, 156: 1169-1173

Berger F (1999) Endosperm Development. Curr Opin Plant Biol, 2: 28-32

Bewley J D, Black M (1983) Physiology and Biochemistry of Seeds in Rclation to Germination. Volume I. Physiology and Biochemistry of Seeds in Relation to Germination., 306

Bonetta D T, Facette M, Raab T K, Somerville C R (2002) Genetic Dissection of Plant Cell Wall Biosynthesis. Biochem Soc Trans, 30: A25

Brennan C S, Cleary L J (2005) The Potential Use of (1→3,1→4)-β-D-glucans as Functional Food Ingredients. J Cereal Sci, 42: 1-13

Breton C, Imberty A (1999) Structure/function Studies of Glycosyltransferases. Curr Opin Struct Biol, 9: 563-571

Breton C, Mucha J, Jeanneau C (2001) Structural and Functional Features of Glycosyltransferases. Biochimie, 83: 713-718

Briggs D E, Macdonald J (1983) The Permeability of the Surface-layers of Cereal-grains, and Implications for Tests of Abrasion in Barley. J Inst Brew, 89: 324-332

Briggs D, Hough J, Stevens R, Young T (1981) Chapter 2, 9. Malting and Brewing Science. London, Chapman and Hall. Volume 1, Malt and Sweet Wort.

Brown R C, Lemmon B E, Olsen O A (1994) Endosperm Development in Barley - Microtubule Involvement in the Morphogenetic Pathway. Plant Cell, 6: 1241-1252

Brown R C, Lemmon B E, Olsen O A (1996) Development of the Endosperm in Rice (*Oryza Sativa L*): Cellularization. J Plant Res, 109: 301-313

Brown R C, Lemmon B E, Stone B A, Olsen O A (1997) Cell Wall (1→3)- and (1→3;1→4)-β-glucans during Early Grain Development in Rice Endosperm. Planta, 202: 414-426

Brown D M, Goubet F, Vicky W W A, Goodacre R, Stephens E, Dupree P, Turner S R (2007) Comparison of Five Xylan Synthesis Mutants Reveals New Insight into the Mechanisms of Xylan Synthesis. Plant J, 52: 1154-1168

Brunswick P, Manners D J, Stark J R (1987) The Development of Beta-D-Glucanases during the Germination of Barley and the Effect of Kilning on Individual Isoenzymes. J Inst Brew, 93: 181-186

Buckeridge M S, Vergara C E, Carpita N C (2001) Insight into Multi-site Mechanisms of Glycosyl Transfer in (1,4)-beta-D-glycans Provided by the Cereal Mixed-linkage (1,3),(1,4)-beta-D-glucan Synthase. Phytochemistry, 57: 1045-1053

Buliga G S, Brant D A, Fincher G B (1986) The Sequence Statistics and Solution Conformation of a Barley (1,3;1,4)-beta-D-glucan. Carbohydr Res, 157: 139-156

Burton R A, Zhang X Q, Hrmova M, Fincher G B (1999) A Single Limit Dextrinase Gene Is Expressed Both in the Developing Endosperm and in Germinated Grains of Barley. Plant Physiol, 119: 859-871

Burton R A, Shirley N A, King B J, Harvey A J, Fincher G B (2004) The *CesA* Gene Family of Barley. Quantitative Analysis of Transcripts Reveals Two Groups of Co-expressed Genes. Plant Physiol, 134: 224-236

Burton R A, Wilson S M, Hrmova M, Harvey A J, Shirley N J, Medhurst A, Stone B A, Newbigin E J, Bacic A, Fincher G B (2006) Cellulose Synthase-like *CslF* Genes Mediate the Synthesis of Cell Wall (1,3;1,4)-β-D-glucans. Science, 311: 1940-1942

Burton R A, Jobling S A, Harvey A J, Shirley N J, Mather D E, Bacic A, Fincher G B (2008) The Genetics and Transcriptional Profiles of the Cellulose Synthase-Like *HvCslF* Gene Family in Barley (*Hordeum Vulgare* L.). Plant Physiol, 146:1821-1833

Campbell J A, Davies G J, Bulone V, Henrissat B (1997) A Classification of Nucleotide-diphospho-sugar Glycosyltransferases Based on Amino Acid Sequence Similarities. Biochem J, 326: 929-939

Carpita N, Gibeaut D (1993) Structural Models of Primary-cell Walls in Flowering Plants - Consistency of Molecular-structure with the Physical-properties of the Walls during Growth. Plant J, 3: 1-30

Caspers M P M, Lok F, Sinjorgo K M C, Van Zeijl M J, Nielsen K A, Cameron-Mills V (2001). Synthesis, Processing and Export of Cytoplasmic Endo-beta-1,4-xylanase from Barley Aleurone during Germination. Plant J, 26: 191-204

Cochrane M P, Duffus C M (1981) Endosperm Cell Number in Barley. Nature, 289: 399-401

Coles G (1979) Relationship of Mixed Link Beta-glucan Accumulation to Accumulation of Free Sugars and Other Glucans in the Developing Barley Endosperm. Carlsberg Res Commun, 44: 439-453

Coles G D, Jamieson P D, Haslemore R M (1991) Effect of Moisture Stress on Malting Quality in Triumph Barley. J Cereal Sci, 14: 161-177

Coote N, Kirsop B H (1976) A Haze Consisting Largely of Pentosan. J Inst Brew, 82: 34

Coutinho P M: Henrissat B (1999) Carbohydrate-active Enzymes Server at URL: [http://afmb.cnrs-mrs.fr/ pedro/CAZY/db.html]

Coutinho P M, Deleury E, Davies G J, Henrissat B (2003) An Evolving Hierarchical Family Classification for Glycosyltransferases. J Mol Biol, 328: 307-317

Cui X, Shin H, Song C, Laosinchai W, Amano Y, Brown R M J (2001) A Putative Plant Homolog of the Yeast β-1,3-glucan Synthase Subunit FKS1 from Cotton (*Gossypium Hirsutum* L.) Fibers. Planta, 213: 223-230

Dashek W V, Chrispeels M J (1977) Gibberellic-acid-induced and Release of Cell-wall-degrading Endoxylanase by Isolated Aleurone Layers of Barley. Planta, 134: 251-256

Delmer D P (1999) Cellulose Biosynthesis: Exciting Times for a Difficult Field of Study. Annu Rev Plant Physiol Plant Mol Biol, 50: 245-276

Dhugga K S, Barreiro R, Whitten B, Stecca K, Hazebroek J, Randhawa G S, Dolan M, Kinney A J, Tomes D, Nichols S, Anderson P (2004) Guar Seed Beta-mannan Synthase Is a Member of the Cellulose Synthase Super Gene Family. Science, 303: 363-366

Doblin M S, De Melis L, Newbigin E, Bacic A, Read S M (2001) Pollen Tubes of Nicotiana Alata Express Two Genes from Different Beta-glucan Synthase Families. Plant Physiol, 125: 2040-2052

Douglas C M, Foor F, Marrinan J A, Morin N, Nielsen J B, Dahl A M, Mazur P, Baginsky W, Li W, El-Sherbeini M, Al E (1994) The Saccharomyces Cerevisiae FKS1 (ETG1) Gene Encodes an Integral Membrane Protein which Is a Subunit of 1,3-beta-D-glucan Synthase. Proc Natl Acad Sci USA, 91: 12907-12911

Ducroo P, Frelon P G (1989) Improvement of Beer Production by the Use of β-glucanase-pentosanase from Disporotrichum Dimorphosporum. Proceedings of the 22nd European Brewery Congress, Zurich.

Duffus C M, Cochrane M P (1993) Formation of the Barley Grain-morphology, Physiology, and Biochemistry. Barley: Chem Technol: 31-72

Edney M J, LaBerge D E, Langrell D E (1998) Relationships among the Beta-glucan Contents of Barley, Malt Congress Extract, and Beer. J Am Soci Brew Chem, 56: 164-168

Edney M J, Eglinton J K, Collins H M, Barr A R, Legge W G, Rossnagel B G (2007) Importance of Endosperm Modification for Malt Wort Fermentability. J Inst Brew, 113: 228-238

Egelund J, Skjot M, Geshi N, Ulvskov P, Petersen B L (2004) A Complementary Bioinformatics Approach to Identify Potential Plant Cell Wall Glycosyltransferase-encoding Genes. Plant Physiol, 136: 2609-2620

Evans D E (2008) Enzymes Involved in Carbohydrate-degradation in Germinated Barley Grain: Chapter 6 of this Volume.

Farrokhi N, Burton R A, Brownfield L, Hrmova M, Wilson S M, Bacic A, Fincher, G B (2006) Plant Cell Wall Biosynthesis: Genetic, Biochemical and Functional Genomics Approaches to the Identification of Key Genes. Plant Biotechnol J, 4: 145-167

Ferre H, Broberg A, Duus J O, Thomsen K K (2000) A Novel Type of Arabinoxylan Arabinofuranohydrolase Isolated from Germinated Barley - Analysis of Substrate Preference and Specificity by Nano-probe NMR. Eur J Biochem, 267: 6633-6641

Fierens E, Rombouts S, Gebruers K, Goesaert H, Brijs K, Beaugrand J, Volckaert G, Van Campenhout S, Proost P, Courtin C M, Delcour J A (2007) TLXI, a Novel Type of Xylanase Inhibitor from Wheat (*Triticum Aestivum*) Belonging to the Thaumatin Family. Biochem J, 403: 583-591

Fierens E, Gebruers K, Courtin C M, Delcour J A (2008) Xylanase Inhibitors Bind to Nonstarch Polysaccharides. J Agric Food Chem, 56: 564-570

Fincher G (1975) Morphology and Chemical Composition of Barley Endosperm Cell Walls. J Inst Brew, 81: 116-122

Fincher G B (1976) Ferulic Acid in Barley Cell Walls: A Fluorescence Study. J Inst Brew, 82: 347-349

Fincher G B, Stone B A (1986) Cell Walls and Their Components in Cereal Grain Technology. In: Pomeranz Y, Paul S T (eds.) Advances in Cereal Sience and Technology. American Association of Cereal Chemists VIII, Minn, USA, 364

Fincher G B (1989) Molecular and Cellular Biology Associated with Endosperm Mobilazation in Germinating Cereal-grains. Ann Rev Plant Physiol Plant Mol Biol, 40: 305-346

Fincher G B, Stone B A (2004) Chemistry of Nonstarch Polysaccharides. In: Wrigley C, Corke H, Walker C E (eds.) Encyclopedia of Grain Science, Oxford, Elsevier.

Finnie C, Bak-Jensen K S, Laugesen S, Roepstorff P, Svensson B (2006) Differential Appearance of Isoforms and Cultivar Variation in Protein Temporal Profiles Revealed in the Maturing Barley Grain Proteome. Plant Sci, 170: 808-821

Forrest I S, Wainwright T (1977) Differentation between Desirable and Troublesome β-glucans. Proceeding of the European Brewing Convention, Amsterdam.

Fulcher R, OBrien T, Simmonds D (1972) Localization of Arginine-rich Protein in Mature Seeds of Some Members of Gramineae. Aust J Biol Sci, 25, 487

Fulcher R G, Setterfield G, McCully M E, Wood P J (1977) Observations on Aleurone Layer. 2: Fluorescence Microscopy of Aleurone Junction with Emphasis on Possible Beta-1,3-glucan Deposits in Barley. Aust J Plant Physiol, 4: 917-928

Gibbons G C (1981) On the Relative Role of the Scutellum and Aleurone in the Production of Hydrolases during Germination of Barley. Carlsberg Res Commun, 46: 215-225

Gibeaut D M (2000) Nucleotide Sugars and Glycosyltransferases for Synthesis of Cell Wall Matrix Polysaccharides. Plant Physiol Biochem, 38: 69-80

Gibeaut D M, Pauly M, Bacic A, Fincher G B, (2005) Changes in Cell Wall Polysaccharides in Developing Barley (*Hordeum Vulgare*) Coleoptiles. Planta, 221: 729-738

Goesaert H, Debyser W, Gebruers K, Proost P, Van Damme J, Delcour J A (2001) Purification and Partial Characterization of an Endoxylanase Inhibitor from Barley. Cereal Chem, 78: 453-457

Gubler F, Ashford A E, Jacobsen J V (1987) The Release of Alpha-amylase through Gibberellin-treated Barley Aleurone Cell-walls - an Immunocytochemical Study. Planta, 172: 155-161

Hamalainen J J, Reinikainen P (2007) A Simulation Model for Malt Enzyme Activities in Kilning. J Inst Brew, 113: 159-167

Han F, Ullrich S E, Chirat S, Menteur S, Jestin L, Sarrafi A, Hayes P M, Jones B L, Blake T K, Wesenberg D M, Kleinhofs A, Kilian A (1995) Mapping of Beta-Glucan Content and Beta-Glucanase Activity Loci in Barley Grain and Malt. Theor Appl Genet, 91: 921-927

Han J, Schwarz P (1996) Arabinoxylan Composition in Barley, Malt, and Beer. Am Soc Brew Chem, 54: 216-220

Han J Y (2000) Structural Characteristics of Arabinoxylan in Barley, Malt, and Beer. Food Chem, 70: 131-138

Henrissat B (1998) Glycosidase Families. Biochem Soc Trans, 26: 153-156

Henry R J (1986) Genetic and Environmental Variation in the Pentosan and β-Glucan Contents of Barley, and Their Relation to Malting Quality. J Cereal Sci, 4: 269-277

Henry R J (1987) Pentosan and (1-3),(1-4)-β-Glucan Concentrations in Endosperm and Wholegrain of Wheat, Barley, Oats and Rye. J Cereal Sci, 6: 253-258

Henry R J, Brown A H D (1987) Variation in the Carbohydrate-composition of Wild Barley (Hordeum Spontaneum) Grain. Plant Breed, 98: 97-103

Hoj P B, Slade A M, Wettenhall R E H, Fincher G B (1988) Isolation and Characterization of (1-3)-beta-glucan Endohydrolase from Germinating Barley (hordeum Vulgare) - Amino Acid Sequence Similarity with Barley (1-3, 1-4)-beta-glucanases. FEBS Lett, 230: 67-71

Home S, Elamo E (1993) Evaluation of Malting Potential in Barley Breeding Programmes. Monatsschrift Fur Brauwissenschaft, 46: 216-220

Hong Z L, Zhang Z M, Olson J M, Verma D P S (2001) A Novel UDP-glucose Transferase Is Part of the Callose Synthase Complex and Interacts with Phragmoplastin at the Forming Cell Plate. Plant Cell, 13: 769-779

Hoshikawa K (1984) Development of the Endosperm Tissue with Special Reference to the Translocationof Reserve Substances in Cereals. 2: Modification of Cell-shape in the Developing Endosperm Parenchyma. Aleurone and Subaleurone of 2-rowed Barley. Jpn J Crop Sci, 53: 64-70

Hrmova M, Harvey A J, Wang J, Shirley N J, Jones G P, Stone B A, Hoj P B, Fincher G B (1996) Barley Beta-D-glucan Exohydrolases with Beta-D-glucosidase Activity - Purification, Characterization, and Determination of Primary Structure from a CDNA Clone. J Biol Chem, 271: 5277-5286

Hrmova M, Fincher G B (1998) Barley Beta-D-glucan Exohydrolases. Substrate Specificity and Kinetic Properties. Carbohydr Res, 305: 209-221

Hrmova M, De Gori R, Smith B J, Fairweather J K, Driguez H, Varghese J N, Fincher, G B (2002) Structural Basis for Broad Substrate Specificity in Higher Plant Beta-D-glucan Glucohydrolases. Plant Cell, 14: 1033-1052

Hrmova M, Burton R A, Biely P, Lahnstein J, Fincher G B (2006) Hydrolysis of (1,4)-β-D-Mannans in Barley (*Hordeum Vulgare*): Purification and Characterization of a (1,4)-β-D-Mannan Endohydrolase and Transcriptional Activity of the Corresponding Gene. Biochem J, 399: 77-90

Hrmova M, Farkas V, Lahnstein J, Fincher G B (2007) A Barley Xyloglucan Xyloglucosyl Transferase Covalently Links Xyloglucan, Cellulosic Substrates, and (1,3;1,4)-beta-D-glucans. J Biol Chem, 282: 12951-12962

Imberty A, Monier C, Bettler E, Morera S, Freemont P, Sippl M, Flockner H, Ruger W, Breton C (1999) Fold Recognition Study of Alpha 3-galactosyltransferase and Molecular Modeling of the Nucleotide Sugar-binding Domain. Glycobiology, 9: 713-722

Jones R L, Jacobsen J V (1991) Regulation of Synthesis and Transport of Secreted Proteins in Cereal Aleurone. International Review of Cytology-a Survey of Cell Biology, 126: 49-88

Juge N, Payan F, Williamson G (2004) XIP-I, a Xylanase Inhibitor Protein from Wheat: a Novel Protein Function. Biochim Biophys Acta - Proteins and Proteomics, 1696: 203-211

Keegstra K, Raikhel N (2001) Plant Glycosyltransferases. Curr Opin Plant Biol, 4: 219-224

Kilian B, Ozkan H, Kohl J, Von Haeseler A, Barale F, Deusch O, Brandolini A, Yucel C, Martin W, Salamini F (2006) Haplotype Structure at Seven Barley Genes: Relevance to Gene Pool Bottlenecks, Phylogeny of Ear Type and Site of Barley Domestication. Mol Genet Genomics, 276: 230-241

Kim H S, Park K G, Baek S B, Son Y G, Lee C W, Kim J C, Kim J G, Nam J H (2004) Inheritance of (1,3)(1,4)-beta-D-glucan Content in Barley (*Hordeum Vulgare* L.) Proceedings of the 9th International Barley Genetics Symposium, Brno, Czech Republic.

Kleene R, Berger E G (1993) The Molecular and Cell Biology of Glycosyl Transferases. Biochim Biophys Acta, 1154: 283-325

Kroon-Batenburg L M, Kroon J (1997) The Crystal and Molecular Structures of Cellulose I and II. Glycoconjugate J, 14: 677-90

Kuntz R J, Bamforth C W (2007) Time Course for the Development of Enzymes in Barley. J Inst Brew, 113: 196-205

Kuo A L, Cappelluti S, Cervantes M, Rodriguez M, Bush D S (1996) Okadaic Acid, a Protein Phosphatase Inhibitor, Blocks Calcium Changes, Gene Expression, and Cell Death Induced by Gibberellin in Wheat Aleurone Cells. Plant Cell, 8: 259-269

Leah R, Kigel J, Svendsen I, Mundy J (1995) Biochemical and Molecular Characteriztion of a Barley Seed Beta-glucosidase. J Biol Chem, 270: 15789-15797

Lee R C, Burton R A, Hrmova M, Fincher G B (2001) Barley Arabinoxylan Arabinofuranohydrolases: Purification, Characterization and Determination of Primary Structures from CDNA Clones. Biochem J, 356: 181-189
Lee R C , Hrmova M, Burton R A, Lahnstein J, Fincher G B (2003) Bifunctional Family 3 Glycoside Hydrolases from Barley with α-L-Arabinofuranosidase and β-D-Xylosidase Activity: Characterization, Primary Structures and COOH-terminal Processing. J Biol Chem, 278: 5377-5387
Li C D, Tarr A, Lance R C M, Harasymow S, Uhlmann J, Westcot S, Young K J, Grime C, Cakir M, Broughton S, Appels R (2003) A Major QTL Controlling Seed Dormancy and Pre-harvest Sprouting/grain α-amylase in Two-rowed Barley (*Hordeum Vulgare* L.). Aust J Agric Res, 54: 1303-1314
Li Y, Lu H, Gu G X, Shi Z P, Mao Z G (2005) Studies on Water-extractable Arabinoxylans during Malting and Brewing. Food Chem, 93: 33-38
Liepman A H, Wilkerson C G, Keegstra K (2005) Expression of Cellulose Synthase-like (Csl) Genes in Insect Cells Reveals that CslA Family Members Encode Mannan Synthases. Proc Natl Acad Sci USA, 102: 2221-2226.
Litts J C, Simmons C R, Karrer E E, Huang N and Rodriguez R L (1990). The Isolation and Characterization of a Barley 1,3-1,4-beta-glucanase Gene. Eur J Biochem, 194: 831-838
Loi L, Barton P A, Fincher G B (1987) Survival of Barley (1-3, 1-4)-β-glucan Isoenzymes during Kilning and Mashing. J Cereal Sci, 5: 45-50
MacGregor A W, Fincher G B (1993) Carbohydrates of the Barley Grain. In: Barley: Chemistry and Technology. In: MacGregor A W, Bhatty R S (eds.) American Association of Cereal Chemists Monograph: 73-130
Macgregor A W, Dushnicky L G, Schroeder S W, Ballance G M (1994) Changes in Barley Endosperms during Early of Germination. J Inst Brew, 100: 85-90
MacGregor A W (1996) Malting and Brewing Science: Challenges and Opportunities. J Instit Brew, 102: 97-102
McCallum C M, Comai L, Greene E A, Henikoff S (2000) Targeting Induced Local Lesions in Genomes (TILLING) for Plant Functional Genomics. Plant Physiol, 123: 439-442
McCann M C, Roberts K (1991) Architecture of the Primary Cell Wall. In: Lloyd C W (ed.) The Cytoskeletal Basis of Plant Growth and Form. Academic Press, London, 109-129
McFadden G I, Ahluwalia B, Clarke A E, Fincher G B (1988) Expression Sites and Developmental Regulation of Genes Encoding (1-3,1-4)-beta-glucanases in Germinated Barley. Planta, 173: 500-508
Mitchell R A C, Dupree P, Shewry P R (2007) A Novel Bioinformatics Approach Identifies Candidate Genes for the Synthesis and Feruloylation of Arabinoxylan. Plant Physiol, 144: 43-53
Molina-Cano J L, Conde J (1982) Genetic and Environmentl Variation of Gum Content in Barley. J Inst Brew, 88: 30-33
Molina-Cano J L, Romera E, Aikasalo R, Perez-Vendrell A M, Larsen J, Rubio A (2002) A Reappraisal of the Differences between Scandinavian and Spanish Barleys: Effect of Beta-glucan Content and Degradation on Malt Extract Yield in the Cv. Scarlett. J Inst Brew, 108: 221-226
Molina-Cano J L, Moralejo M, Elia M, Munoz P, Russell J R, Perez-Vendrell A M, Ciudad F, Swanston J S (2007) QTL Analysis of a Cross between European

and North American Malting Barleys Reveals a Putative Candidate Gene for Beta-glucan Content on Chromosome 1H. Mol Breed, 19: 275-284

Moore J, Bamforth C W, Kroon P A, Bartolome B, Williamson G (1996) Ferulic Acid Esterase Catalyses the Solubilization of Beta-glucans and Pentosans from the Starchy Endosperm Cell Walls of Barley. Biotechnol Lett, 18: 1423-1426

Morgan A G, Gill A A, Smith D B (1983) Some Barley Grain and Green Malt Properties and Their Influence on Malt Hot-water Extract I. β-Glucan, β-Glucan Solubilase and Endo-β-Glucanase. J Inst Brew, 89: 283-291

Morrall P, Briggs D E (1978) Changes in Cell-wall Polysaccharides of Germinating Grains. Phytochemistry, 17: 1495-1502

Mundy J, Brandt A, Fincher G B (1985) Messenger RNAs from the Scutellum and Aleurone of Germinating Barley Encode $(1\rightarrow3,1\rightarrow4)$-$\beta$-D-glucanase, α-amylase and Carboxypeptidase. Plant Physiol, 79: 867-871

Mundy J, Brandt A, Fincher G B (1985) Messenger-rnas from the Scutellum and Aleurone of Germinating Barley Encode $(1\rightarrow3,1\rightarrow4)$-beta-D-glucanase, Alpha-amylase and Carboxypeptidase. Plant Physiol, 79: 867-871

Narzi L (1993) β-Glucan and Filterability. Brauwelt Int, 5: 435-442

Nischwitz R, Cole N W, MacLeod L (1999) Malting for Brewhouse Performance. J Inst Brew, 105: 219-227

Olsen O A (2004) Nuclear Endosperm Development in Cereals and Arabidopsis Thaliana. Plant Cell, 16: S214-S227

Pagny S, Bouissonnie F, Sarkar M, Follet-Gueye M L, Driouich A, Schachter H, Faye, L, Gomord V (2003) Structural Requirements for Arabidopsis Beta 1,2-xylosyltransferase Activity and Targeting to the Golgi. Plant J, 33: 189-203

Palmer G H, Sattler R (1996) Different Rates of Development of Alpha-amylase in Distal Endosperm Ends of Germinated (malted) Chariot and Tipper Barley Varieties. J Inst Brew, 102: 11-17

Pena M J, Zhong R, Zhou G K, Richardson E A, O'Neill M A, Darvill A G, York W S, Ye Z H (2007) Arabidopsis Irregular Xylem8 and Irregular Xylem9: Implications for the Complexity of Glucuronoxylan Biosynthesis. Plant Cell, 19: 549-63

Perrin R, Wilkerson C, Keegstra K (2001) Golgi Enzymes that Synthesize Plant Cell Wall Polysaccharides: Finding and Evaluating Candidates in the Genomic Era. Plant Mol Biol, 47: 115-130

Persson S, Caffall K H, Freshour G, Hilley M T, Bauer S, Poindexter P, Hahn M G, Mohnen D, Somerville C (2007) The Arabidopsis Irregular Xylem8 Mutant Is Deficient in Glucuronoxylan and Homogalacturonan, which Are Essential for Secondary Cell Wall Integrity. Plant Cell, 19: 237-255

Potokina E, Caspers M, Prasad M, Kota R, Zhang H, Sreenivasulu N, Wang M, Graner A (2004) Functional Association between Malting Quality Trait Components and CDNA Array Based Expression Patterns in Barley (*Hordeum Vulgare* L.). Mol Breed, 14: 153-170

Prentice N, Faber S (1981) Beta-D-glucan in Developing and Germinating Barley Kernels. Cereal Chem, 58: 77-79

Richmond T A, Somerville C R (2001) Integrative Approaches to Determining Csl Function. Plant Mol Biol, 47: 131-143

Richmond T, Somerville C (2000) The Cellulose Synthase Superfamily. Plant Physiol, 124: 495-498

Rosen M L, Edman M, Sjostrom M, Wieslander A (2004) Recognition of Fold and Sugarlinkage for Glycosytransferases by Multivariate Sequence Analysis. J Biol Chem, 279: 38683-38692

Ross J, Li Y, Lim E, Bowles D J (2001) Higher Plant Glycosyltransferases. Genome Biol, 2: 3004

Saulnier L, Sado P E, Branlard G, Charmet G, Guillon F (2007) Wheat Arabinoxylans: Exploiting Variation in Amount and Composition to Develop Enhanced Varieties. J Cereal Sci, 46: 261-281

Saxena I, Brown R (2000) Cellulose Synthases and Related Enzymes. Curr Opin Plant Biol, 3: 523-531.

Scheible W R, Pauly M (2004) Glycosyltransferases and Cell Wall Biosynthesis: Novel Players and Insights. Curr Opin Plant Biol, 7: 285-295

Shibuya N, Misaki A (1978) Structure of Hemicelluloses Isolated from Rice Endosperm Cell Wall: Mode of Linkages and Sequences in Xyloglucan, β-glucan and Arabinoxylan. Agric Biol Chem, 42: 2267-2274

Shibuya N, Nikane R (1984) Pectic Polysaccharides of Rice Endosperm Cell Walls. Phytochemistry, 23: 1425-1429

Simos G, Panagiotidis C A, Skoumbas A, Choli D, Ouzounis C, Georgatsos J G (1994) Barley β-glucosidase: Expression during Seed Germination and Maturation and Partial Amino Acids Sequences. Biochim Biophys Acta-General Subjects, 1199: 52-58

Simpson D J, Fincher G B, Huang A H C, Cameron-Mills V (2003) Structure and Function of Cereal and Related Higher Plant (1$\rightarrow$4)-β-xylan Endohydrolases. J Cereal Sci, 37: 111-127

Slade A J, Fuerstenberg S I, Loeffler D, Steine M N, Facciotti D (2005) A Reverse Genetic, Nontransgenic Approach to Wheat Crop Improvement by TILLING. Nature Biotechnol, 23: 75-81

Slade A, Hoj P, Morrice N, Fincher G B (1989) Purification and Characterization of Three (1$\rightarrow$4)-β-D-Xylan Endohydrolases from Germinating Barley. Eur J Biochem, 185: 533-539

Slakeski N, Fincher, G B (1992) Developmental Regulation of (1$\rightarrow$3,1$\rightarrow$4)-β-glucanase Gene Expression in Barley. Tissue-specific Expression of Individual Isoenzymes. Plant Physiol, 99: 1226-1231

Slakeski N, Baulcombe D C, Devos K M, Ahluwalia B, Doan D N P, Fincher G B (1990) Structure and Tissue-specifc Regulation of Genes Encoding Barley (1-3, 1-4)-beta-glucan Endohydrolases. Mol Gen Genet, 224: 437-449

Staudte R G, Woodward J R, Fincher G B, Stone B A (1983) Water-soluble (1$\rightarrow$3), (1$\rightarrow$4)- β -glucans from Barley (Hordeum Vulgare) Endosperm. III. Distribution of Cellotriosyl and Cellotetraosyl Residues. Carbohydr Polymers, 3: 299-312

Sterling J, Quigley H F, Orellana A, Mohnen D (2001) The Catalytic Site of the Pectin Biosynthetic Enzyme Alpha-1,4-galacturonosyltransferase Is Located in the Lumen of the Golgi. Plant Physiol, 127: 360-371

Stewart D C, Hawthorne D, Evans D E (1998) Cold Sterile Filtration: A Small Scale Filtration Test and Investigation of Membrane Plugging. J Inst Brew, 104: 321-326

Stewart R J, Varghese J N, Garrett T P J, Hoj P B, Fincher G B (2001) Mutant Barley (1,3;1,4)-beta-glucan Endohydrolases with Enhanced Thermostability. Protein Eng, 14: 245-253

Stuart I M, Loi L, Fincher G B (1986) Development of (1-3,1-4)-beta-D-glucan Endohydrolase Isoenzymes in Isolated Scutella and Aleurone Layers of Barley (*hordeum-vulgare*). Plant Physiol, 80. 310-314

Stuart I, Loi L, Fincher G B (1988) Varietal and Environmental Variations in (1-3)(1-4)-β-glucan Levels and (1-3)(1-4)-β-Glucanase Potential in Barley: Relationships to Malting Quality. J Cereal Sci, 7: 61-71

Sugimoto N, Takeda G, Nagato Y, Yamaguchi J (1998) Temporal and Spatial Expression of the Alpha-amylase Gene during Seed Germination in Rice and Barley. Plant Cell Physiol, 39: 323-333

Sungurtas J, Swanston J S, Davies H V, McDougall G J (2004) Xylan-degrading Enzymes and Arabinoxylan Solubilisation in Barley Cultivars of Differing Malting Quality. J Cereal Sci, 39: 273-281

Szyjanowicz P M J, McKinnon I, Taylor N G, Gardiner J, Jarvis M C, Turner S R (2004) The Irregular Xylem2 Mutant Is an Allele of Korrigan that Affects the Secondary Cell Wall of Arabidopsis Thaliana. Plant J, 37: 730-740

Taiz L, Honigman W A (1976) Production of Cell Wall Hydrilysing Enzymes by Barley Aleurone Layers in Response to Gibberellic Acid. Plant Physiol, 58: 380-386

Taiz L, Jones R L (1970) Gibberellic Acid, Beta-1,3-glucanase and Cell Wals of Barley Aleurone Layers. Planta, 92: 73

Taiz L, Jones R L (1973) Plasmodesmata and an Associated Cell Wall Component in Barley Aleurone Tissue. Am J Bot, 60: 67-75

Trethewey J A K, Campbell L M, Harris P J (2005) (1$\rightarrow$3),(1$\rightarrow$4)-β-D-Glucans in the Cell Walls of the Poales (*sensu Lato*): An Immunogold Labeling Study Using a Monoclonal Antibody. Am J Bot, 92: 1669-1683

Van Campenhout S, Pollet A, Bourgois T M, Rombouts S, Beaugrand J, Gebruers K, De Backer E, Courtin C M, Delcour J A, Volckaert G (2007) Unprocessed Barley Aleurone Endo-beta-1,4-xylanase X-I Is an Active Enzyme. Biochem Biophys Res Commu, 356: 799-804

Varghese J N, Garrett T P J, Colman P M, Chen L, Hoj P B, Fincher G B (1994) Three-dimensional Structures of Two Plant Beta-glucan Endohydrolases with Distinct Substrate Specificities. Proc Natl Acad Sci USA, 91: 2785-2789

Varghese J N, Hrmova M, Fincher G B (1999) Three-dimensional Structure of a Barley Beta-D-glucan Exohydrolase, a Family 3 Glycosyl Hydrolase. Structure, 7: 179-190

Vergara C, Carpita N (2001) Beta-D-glycan Synthases and the CesA Gene Family: Lessons to Be Learned from the Mixed-linkage (1,3),(1,4)-beta-D-glucan Synthase. Plant Mol Biol, 47: 145-160

Viëtor R J, Voragen A G J, Angelino S (1993) Composition of Non-starch Polysaccharides in Wort and Spent Grain from Brewing Trials with Malt from a Good Malting Quality Barley and a Feed Barley. J Inst Brew, 99: 243-248

Viëtor R J, Hoffmann R A, Angelino S A, Voragen A G, Kamerling J P, Vliegenthart J F (1994) Structures of Small Oligomers Liberated from Barley Arabinoxylans by Endoxylanase from *Aspergillus Awamori*. Carbohydr Res, 17: 245-255

Wallwork (1997) Investigation of the Physiolological Basis of Malting Quality of Grain Developing under High Temperature Conditions. Department of Plant Science, Adelaide, University of Adelaide, Thesis for Doctor of Philosophy.

Wang J M, Zhang G P, Chen J X, Wu F B (2004) The Changes of Beta-glucan Content and Beta-glucanase Activity in Barley Before and after Malting and Their Relationships to Malt Qualities. Food Chem, 86: 223-228

Waterhouse P M (2006) Defense and Counterdefense in the Plant World. Nature Genet, 38: 138-139

White J, Pacey-Miller T, Crawford A, Cordeiro G, Barbary D, Bundock P, Henry R (2006) Abundant Transcripts of Malting Barley Identified by Serial Analysis of Gene Expression (SAGE). Plant Biotechnol J, 4: 289-301

Wilson S M, Burton R A, Doblin M S, Stone B A, Newbigin E, Fincher G B, Bacic A (2006) Temporal and Spatial Appearance of Wall Polysaccharides during Cellularization of Barley (*Hordeum Vulgare*) Endosperm. Planta, 224: 655-667

Wood P J, Weisz J, Blackwell B A (1994) Structural Studies of the (1→3),(1→4)-β-D-glucans by 13C-nuclear Magnetic Resonance Spectroscopy and by Rapid Analysis of Cellulose-like Regions Using High Performance Anion-exchange Chromatography of Oligosaccharides Released by Lichenase. Cereal Chem, 71: 301-307

Woodward J R, Fincher G B (1982) Purification and Chemical Properties of Two 1,3:1,4-β-glucan Endohydrolases from Germinating Barley. Eur J Biochem, 121: 663-669

Woodward J R, Phillips D R, Fincher G B (1983) Water-soluble (1→3),(1→4)-β-D-glucans from Barley (Hordeum Vulgare) Endosperm. I. Physicochemical Properties. Carbohydr Polym, 3: 143-156

8

Barley Genomics and Malting Quality Improvement

A. Druka and R. Waugh
SCRI. Invergowrie Dundee, DD2 5DA Scotland, UK
Arnis.Druka@scri.ac.uk, Robbie.Waugh@scri.ac.uk

8.1 Introduction

Genomics can be defined as 'the generation of information about living things by a systematic approach that can be performed on an industrial scale' (Brent 2000). It can be subdivided into structural and functional genomics, with the former being genome mapping and sequencing, and the latter all of the other genomics; trascriptomics, proteomics, metabolomics, etc. In principle, genomics is not very different from conventional genetics performed on a large scale. However, this increase in scale, when accompanied by an increase in the resolution of the data, makes it possible to uncover novel links and relationships between components of a biological system.

The cellular, biochemical and physiological aspects of barley development that are directly associated with malting quality have been studied extensively. This has resulted in a detailed picture of the processes taking place within a narrow developmental window - usually the first few days of post-germinative development. This picture however seldom takes account of the potential impact of associated traits that may condition the variation observed during germination. The power of genetic analysis is that it can unify data collected on different, perhaps component traits by identifying the chromosomal loci that influence them. Genomics technologies and resources available at the time of study enable genetic analysis. They largely determine how straightforwardly, rapidly and cost-efficiently the genes underlying these chromosomal loci can be isolated. At a basic level, knowledge of the position of genes, frequently based on the outputs from large scale experiments that were not necessarily conducted to address a specific biological question, can provide a direct route to linking phenotype, genotype and gene. This is particularly relevant when viewed in the context of conserved synteny with the model rice and *Brachypodium* genome sequences as they, not mapped barley genes, may provide candidate genes for the trait under investigation. For simple traits, identifying candidate genes is contingent upon a number of factors that include marker density, recombination frequency and chromosomal rearrangements between the barley and model genomes. The situation is more complex when considering quantitative traits such as malting quality.

In this chapter we describe the application of two genomics platforms that have relatively recently been developed for barley and that have already made a significant impact on phenotypic trait analysis. Both are excellent examples of international collaboration in barley research. Using case studies we will speculate as to how these technologies can potentially be applied to address malting quality traits. We have focused this discussion on a single mapping population - the reference Steptoe x Morex (St/Mx) population that was originally developed to examine a range of traits that included malting quality.

8.2 Platform Technologies for Barley

To conduct genetic analysis of any trait, but particularly quantitative traits, a genetic linkage map is a basic necessity. Initially, a genetic map allows a trait phenotype to be associated with a chromosomal region. When applied at increasing resolution (i.e., in larger populations) linkage mapping can facilitate the identification of the specific gene in the region that controls the trait. Traditionally barley linkage maps were constructed using RFLP or SSR markers (Kleinhofs *et al.*, 1993; Ramsay *et al.*, 2000; Varshney *et al.*, 2007). These were seldom gene-based (with the exception of cDNA probes and EST-SSRs) but, by sharing markers across different labs, some consistency in trait analysis in terms of the underlying genetic maps has been achieved. Unfortunately, both of these marker types suffer from the fact that they are implemented using serial assays (i.e., they are low throughput) that generally depend upon the availability of a specialized lab. Furthermore, success is often determined by the experience and skill of the person running the genotyping assays. Similar problems are encountered when applying other non-sequence based approaches such as AFLP (Vos *et al.*, 2005).

The generation of a high-throughput, representative and unified genotyping platform became a high priority a few years ago in the barley genetics community because it was widely recognized that such a resource would enable the investigation of novel approaches for trait analysis such as association mapping. Based on the outputs of community EST sequencing programs two highly parallel gene-based technologies have been developed that have been used for genetic analysis. The first is a dedicated platform for genotyping single nucleotide polymorphisms (SNPs) and is based on Beadarray technology commercialized by Illumina (Fan *et al.*, 2003). The second was not designed for genotyping at all – but by combining innovative ways of interrogating microarray-based mRNA profiling data from segregating populations with QTL mapping (see section 2.3) it has been possible to map thousands of genes. In this chapter we will describe the technical features of these genotyping and mRNA profiling platforms.

8.2.1 The Beadarray by Illumina

The BeadArray is a genotyping system commercialized by Illumina and exploits an assay known as 'GoldenGate' (Fig. 8.1) (Fan *et al.*, 2003). In this assay 1536 different SNP loci are interrogated by a series of extension and ligation reactions, followed by PCR. The assay employs sets of locus specific oligodeoxynucleotides designed to interrogate each individual SNP. One of these is locus specific (Locus Specific Oligo, LSO) but non-discriminatory, and the other two are designed so that their 3'- base is complementary to one or other alternative allele sequences (Allele Specific Oligo's, ASO). After a complex series of reactions, the yield of a PCR product that contains one or other of the discriminating ASOs can be quantified and the SNP genotype at the specific locus ascertained (Fig. 8.1(A)).

The technology relies on the latest developments in laser fibre optics as this provides the possibility to test >1500 loci in a single DNA sample in a single tube. Discrimination between alleles is achieved by indirectly incorporating Cy3 or Cy5 dyes into the allele-specific primers (Fig. 8.1(B)). The assay has three steps:

Step 1. Generation of allele specific DNA templates: Test DNA is activated by the addition of streptavidin and then three primers per locus added and allowed to anneal to the template. Two ASO's and an LSO are needed for each SNP. The ASO's contain i) a sequence domain that is homologous to the locus specific sequence flanking the SNP to be assayed and ii) different universal primer domains at their 5' ends (U1 or U2). Each pair of ASO's terminate at their 3' end in the alternative bases at the SNP to be assayed. The LSO contains three domains. The first (5' domain) is complementary to the locus specific sequence located 5-20 nt downstream of the polymorphic SNP. It binds to the same strand as the ASO. The second domain is an addressing domain (Illumicode address) used in the third step of the assay and the third domain another universal primer domain (U3). The initial activation step allows the DNA to be bound to paramagnetic beads that facilitates the various washing stages carried out during the assay. After hybridisation of the three oligos to the activated DNA, a highly specific polymerase is added that only extends the ASO's with a matching terminal nucleotide. This polymerase has no exonucleolytic activity and no displacement activity and simply falls off the template when it bumps into the LSO. A ligase is then added to repair the gap between the extended ASO and the LSO yielding a single stranded allele-specific DNA template with universal

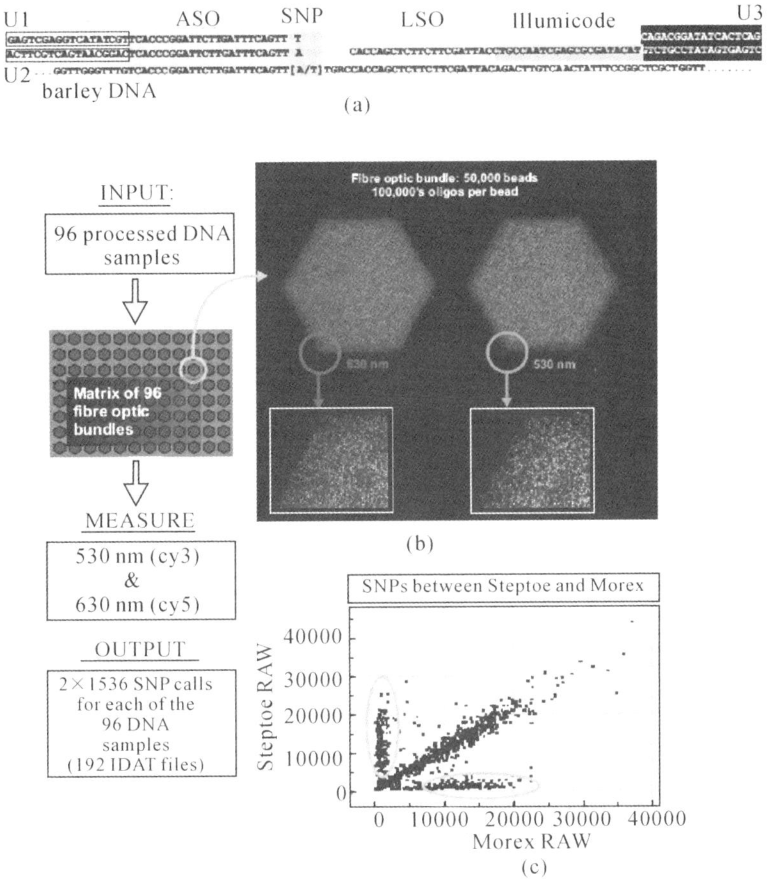

Fig. 8.1. Principle behind Illumina's Beadarray technology. (A) Allele specific template preparation and differential PCR. Multi-domain ASO and LSO primers anneal to the same strand of DNA and are extended and ligated to form a template for allele specific amplification (see text). The Figure shows the primer sequences and DNA template sequence (partial) for a single SNP. Highlighted in yellow is region containing SNP, blue are regions of specificity that relate to the target DNA sequence. The grey sequence shows the position of the Illumicode address and the sequence in the LSO homologous to the U3 primer and U3 primer is shown as white text in a black background. Boxed regions show the U1 and U2 sequences within the ASO. (B) Steps involved in processing the DNA samples, signal detection and de-convolution. (C) Output from BeadStudio analysis software showing cy3 signals. Signal values were generated from 1536 SNPs assayed on Steptoe and Morex DNA and plotted against each other. The signals that deviate from the diagonal are polymorphic between two cultivars. The map positions of the majority of these SNPs have been determined by genotyping experimental bi-parental populations, thus enabling instant selection of markers for the chromosomal region of interest primer sequences on each end. Extensive washing and cleanup is conducted to remove random interactions and reagents and leave clean single strand allele specific template that can be released from the activated DNA by melting. All 1536 SNP templates are generated at the same time in this phase.

Step 2. Amplification and labelling of the allele specific template: The 1536 templates from step 1 terminate in either U1 or U2 and U3 sequences. To generate sufficient material for detection, primers homologous to the universal primer domains are added and the mixture subjected to PCR to amplify the template. Each of the ASO specific universal primers (U1, U2) is labelled with either cy3 or cy5 to differentiate the alleles based on the emission spectrum (or "colour") of the products. The products from each SNP locus are generally artificially visualised as either green (AA) red (aa) or red and green (Aa).

Step 3. Detection of labelled template from within a complex mixture: The complex mix of products from step 2 is hybridised to a beadarray. The bead array is constructed from an array of fibre optic bundles in 96 well plate format (Fig. 8.1(B)) called a Sentrix Array Matrix (SAM). Each bundle contains c. 50K separate screened optical fibres. The end of each fibre is acid etched to create a pocket, and a suspension of 1536 different beads coated with oligo's complementary to the Illumicode addresses incorporated in each LSO is allowed to self assemble across the fibre bundle such that a single bead becomes lodged in the end of each fibre. This process results in a randomly ordered array of capture Illumicode beads. The manufacturer performs QC to verify that beads have been deposited, to map the individual Illumicode beads to each fibre (by a series of hybridisation reactions) and to verify that each Illumicode is represented a minimum of 5 times within each of the 96 arrays in a SAM. On average there is c. 30X redundancy within a given array. After hybridisation to the SAM, each array is interrogated to align each illumicode address with the fluorescence assigned to the alternative bases at each SNP locus.

The 1536-plex GoldenGate assay therefore requires a unique pool of oligonucleotides corresponding to the allele and locus specific probes (ASO1, ASO2 and LSO) for each of the 1536 loci that will be interrogated. To develop an assay, sequence information from at least two genotypes exhibiting SNPs is required.

While the assay appears complex, it is simplified enormously by the provision of dedicated reagent kits and protocols by Illumina. As expected, the assay is also conducted using instrumentation developed by Illumina. This can be purchased and run in house, accessed directly thorough Illumina or genotyping services can be subcontracted to one of several genome centres that operate the instrument. Once the array has been developed and used, data are exported as two IDAT files for each DNA sample representing the signals of 1536 SNPs measured after excitation by different laser wave lengths that differentiate Cy3 and Cy5 signals. These are merged using software also developed by Illumina called BeadStudio to obtain quantitative measurements and associated quality scores for each of the 1536 SNPs. These measurements then can be transformed into allele calls and, depending on the context of the experiment, used either to find relationships between samples or markers (Bohuon *et al.*, 1998) (Fig. 8.1(C)).

8.2.2 The GeneChip by Affymetrix

Affymetrix has for some time been one of the leading companies in microarray development and production. There are over 12,179 publications found in the database (as of March 2008) using or describing results from using Affymetrix' GeneChip mRNA profiling technology. The basic building blocks of Affymetrix' GeneChip microarrays are 25-base long oligodeoxynucleotides (Fig. 8.2(A)) that are synthesized at specific locations on a coated quartz surface by photolithography. Each 25-mer is called a feature. Over a million features per microarray are usually available for probe synthesis, allowing multiple, typically twenty-two, probes per gene (Fig. 8.2(B)). Eleven of these are perfectly complimentary to the mRNA sequence of a representative exemplar gene. They are called PM (perfect match) probes. The remaining eleven called MM (mismatch) probes have a non-complementary middle base designed to detect and eliminate any false or contaminating signal within the gene expression measurement. The initial assumption was that the MM probe would hybridize to non-specific sequences (background) as effectively as the PM probe and allow spurious signals to be quantified and subtracted. For Affymetrix GeneChips, the labelling process and signal generation start with synthesis and labelling of cRNA with biotin, followed by a 2-step signal enhancement procedure involving cDNA, cRNA and fluorescently labelled streptavidin-biotin based immunochemical detection. The procedures have been described many times before and are illustrated very well on the affymetrix web portal.(http://www.affymetrix.com/corporate/outreach/educator.affx).

The design features of the Affymetrix' expression array GeneChip technology can also be exploited for genotyping. Hybridization efficiency to the 25-base long oligonucletide is affected by small nucleotide sequence differences (SNPs and indels) between the probe and its target (Fig. 8.3(A)). As a result, for mRNA profiling using tissues from two genetically distinct individuals such differences can cause spurious results that may reflect mRNA binding rather than real mRNA abundance differences (Fig. 8.3(B)). Such differences have been termed single-feature polymorphisms (SFPs) (Borevitz, 2003; Rostoks *et al.*, 2005) and these can be exploited as genetic markers when followed in segregating populations. In reality, it is often found that individuals differ both in gene expression and in one or more feature polymorphisms. Thus it is possible to explore both general expression effects and specific SFP polymorphisms using the same data set. The former could be the result of either a genetic polymorphism in a regulatory sequence that affects transcription, while the second is most likely, though not exclusively, due to polymorphism in the sequence of the gene that affects hybridization success (Luo *et al.*, 2007;

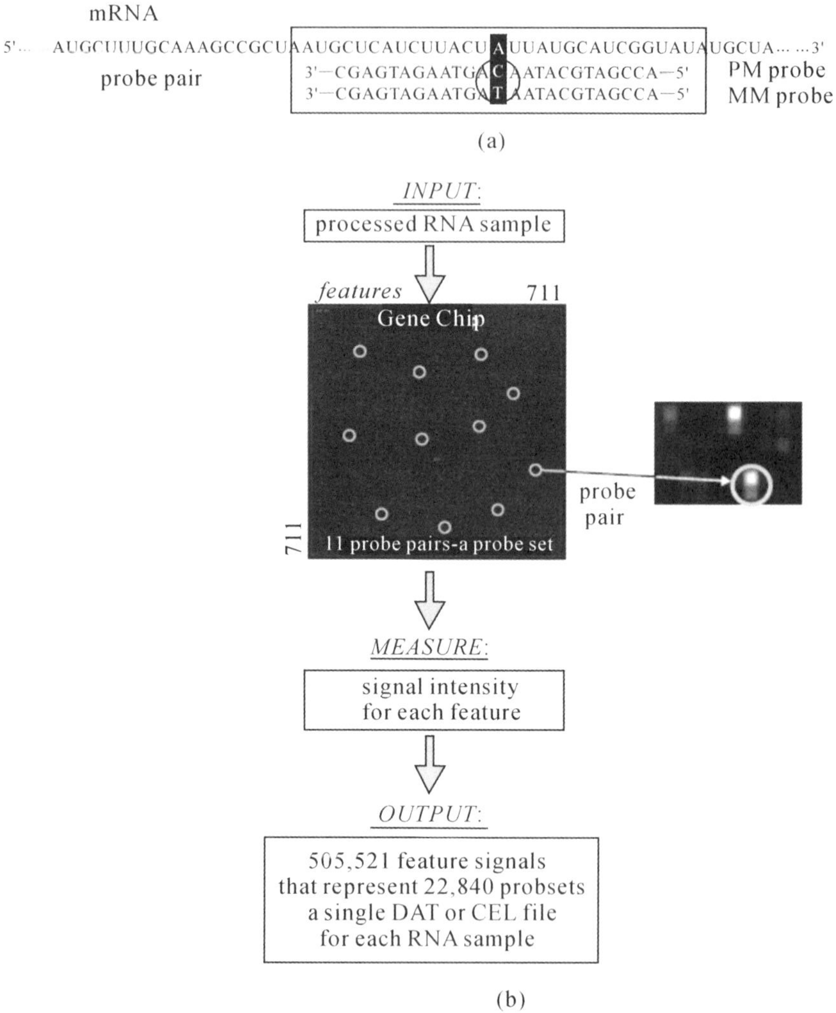

Fig. 8.2. Design principle and use of the GeneChip by Affymetrix. (A) the perfect match (PM) and mismatch (MM) probes are shown aligned against an exemplar DNA sequence. (B) the steps involved in GeneChip analysis highlighting the differential hybridisation signal between PM and MM probes (Inset)

Rostoks *et al.*, 2005; West *et al.*, 2006). SFP or expression differences can be exploited as genetic markers alone or as genetic markers and component traits respectively in genetic analyses.

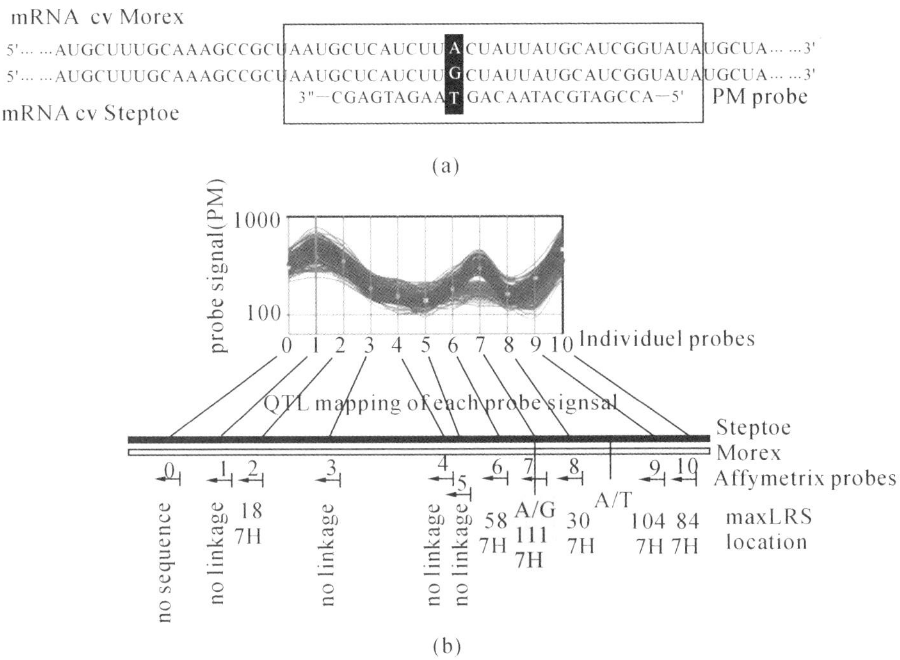

Fig. 8.3. Principle of using GeneChip's for genotyping. (A) The ideal situation for the detection of SNPs or indels: Polymorphisms between the sequence of the target cRNAs and the probes are located close to the middle. Such polymorphisms lower the hybridisation affinity between the target and the probe and this is observed as a reduced mRNA abundance signal value for a given probe. (B) In the example shown only one of the 11 probes are affected. The affected probe has been called a Single Feature Polymorphism (SFP) and can be used as a marker for genotyping. The figure shows a plot of the raw signal values reported by the PM probes across 150 SM lines. Coloring is based on the signal separation of the SFP probe #7 (highlighted in yellow), which contains a SNP in Steptoe. Therefore all the lines that have Steptoe allele at that locus would yield lower probe signal. A smoothed line graph is chosen only for enhanced visualization of the group separation

8.3 Linking Genotype and Phenotype via Component Traits

Expression differences, assayed as variation in mRNA or transcript abundance (often referred to as gene expression level) can be analyzed as a trait using the same approaches as any other conventional trait. The major difference from a classical trait is that microarrays allow mRNA abundance values for tens of thousands of genes to be assayed in a single experiment. Each individual dataset can then be considered a component trait that can be correlated with traditional phenotypic data. There is of course a profound difference between interrogating mRNA abundance and classical phenotypic trait QTL. For the latter, the underlying genes are either unknown, or a hypothesis involving specific genes that are usually inferred from other experiments is being tested. For mRNA abundance, the gene is known by the definition. Therefore, the first question when addressing transcript abundance QTL (also called expression QTL or eQTL) is whether the location of the structural gene coincides with its transcript abundance QTL. If it does, this generally means that the factors regulating mRNA abundance are in the proximity of the gene (so-called cis-regulation) (Fig. 8.4(A)). However, if the eQTL is at a different location from that of its structural gene then the eQTL locus is likely to encode a factor that controls the expression of the gene in *trans*- (i.e. *trans*-regulation) (Fig. 8.4(B)).

Parallel quantification of mRNA abundance of thousands of genes in a segregating population is the basis of the systems genetics approach coined "genetic genomics" (Jansen and Nap, 2001). Genetical genomics aims to decompose phenotypic variation into a series of individual components by simultaneously analyzing "trait" and "molecular" phenotypes across genetically defined populations. The approach was originally tested by Damerval *et al.* in 1994 who applied protein profiling to an F2 population of maize (Damerval *et al.*, 1994). More recently, genetic genomics has been applied to a range of species using mRNA abundance phenotypes (Brem *et al.*, 2002; Schadt *et al.*, 2003; Chesler *et al.*, 2003). In mice, such analyses have been used to understand how regulatory networks controlling transcription relate to phenotypic traits at the genome-wide level (Schadt *et al.*, 2005; Chesler *et al.*, 2005). Genetical genomics experiments in barley have recently been reported (Potokina *et al.*, 2007) and demonstrate that the control of gene expression is complex but provides valuable insights into the relationships between gene expression and phenotype. For species with unsequenced genomes, the robust detection of allelic differences in cis-regulation has considerable added value because it allows transcript abundance to be used as a molecular marker, which can accurately predict a gene's chromosomal position.

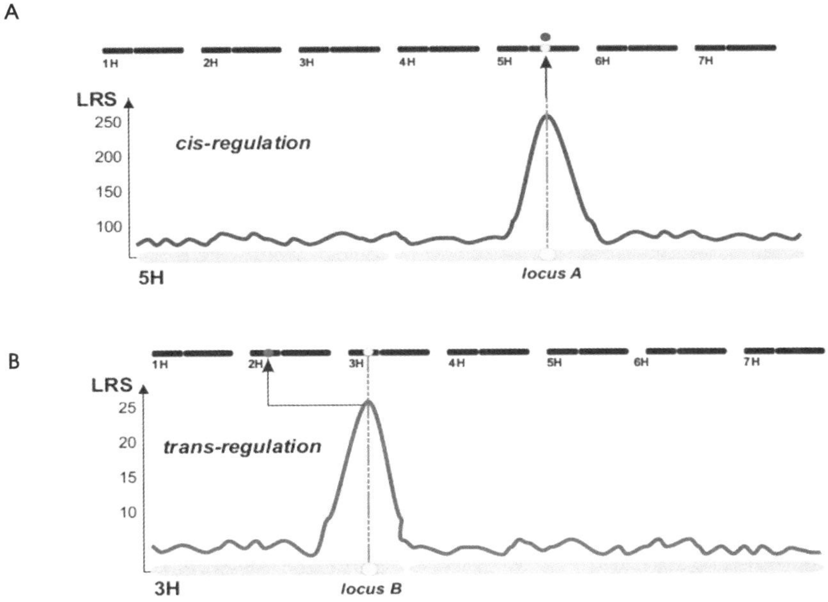

Fig. 8.4. Detection of cis- and trans-regulation using eQTL mapping. (A) Cis-regulation is the most likely cause of mRNA abundance variation if an eQTL (blue line graph) coincides with the position of its structural gene (yellow circle) on chromosome 5H. (B) trans-regulation is implicated if the eQTL (red line graph) is located at a different location (yellow circle on chromosome 3H) from its structural gene (red circle on chromosome 2H). Black bars along the top of each graph represent the 7 barley chromosomes. Line graphs show a QTL scan across a single chromosome. Note that the LRS value for the cis-regulated gene is significantly higher than that of the *trans*-regulated gene

8.4 Identifying Genes Underpinning Malting Quality

Balancing good malting quality with high yield is one of the basic problems in barley breeding. High yield, often dependant on the application of nitrogenous fertilizers, is associated with high protein and β-glucan content, which are not desired for good malt. To identify loci controlling malting and yield-related traits in barley, the St/Mx population was made in the early 1990s. Steptoe is a high-yielding cultivar and Morex has superior malting quality traits. One of the first barley genetic linkage maps was constructed based on this population (Kleinhofs *et al.*, 1993). Since then, many phenotypes–including biochemical, developmental and disease resistance traits have been mapped in this population. More recently, we have collected data on the mRNA abundance of over 16,000 barley genes that we have mapped as eQTL (Potokina *et al.*, 2007). We collected this dataset from germinating embryos (‘whole plants’). This has allowed the original genetic map to be both extended and improved by integrating over 1500 gene-based SFP markers (Luo *et al.*, 2007). However it has been even further improved by the integration of over 2000 gene-based SNP markers using the Illumina Beadarray technology (Close, pers. comm.).

These new resources allow the many independent experiments where phenotypic information has been collected to be combined with mRNA abundance eQTL and re-analyzed to investigate functional genetic relationships.

In the following sections we present some preliminary analyses of these genomics datasets in relation to malting quality QTL in the St/Mx population. We are necessarily speculative, with the discussion simply designed to illustrate the potential power of these emerging genomic technologies and to stimulate further exploration of some of our observations.

8.4.1 Malting QTLs

In the previous sections we have described how the genetic locations of thousands of genes have been placed on the St/Mx map. Here we ask how that information can, at least potentially, be related to traditional malting quality phenotypic trait data. Specifically, we want to explore if is there an overlap between the location of a gene that has been implicated in malting quality, its mRNA abundance QTL(s) and phenotype-derived malting quality QTL. In order to address this question we assembled a list of genes that have previously been associated with various aspects of malting quality. The list was comprised of 41 'malting' genes that had associated eQTL and/or SNP mapping information available to us. We extracted all the information and jointly analysed the genotypic (SNP), eQTL and phenotypic trait QTL mapping data. We then asked:

(1) Is *cis*- or *trans*- regulation implicated in determining mRNA abundance of these "malting" genes? This prediction can be made for all genes that have both eQTL and SNP mapping information.
(2) Do any of these genes or their eQTL underlies major malting quality QTL? This can be assessed by the co-location of phenotypic QTL and eQTL or gene location.

Twenty-four of the genes on our list were mapped as SNPs using the Illumina platform and all but one reported a significant eQTL (i.e., had a Likelihood Ratio Statistic (LRS) of >11.5 which we determined was the cut-off significance threshold based on 1000 permutations). Forty of the 41 genes revealed significant eQTL (LRS>11.5) and 25 of these had an LRS>30. From previous analyses, we know that a large proportion of the eQTL showing an LRS between 11.5 and 30 are likely to be trans-regulated, while eQTL with LRS >30 are most likely regulated in cis- (Fig. 8.4(C) and Fig. 8.5, also Druka *et al.*, 2008). The dataset contained 11 likely cis-combinations where a gene had an eQTL LRS>30 and was also mapped as a SNP to the same genetic locus (blue connecting lines on Fig. 8.5). Most of the remaining eQTL

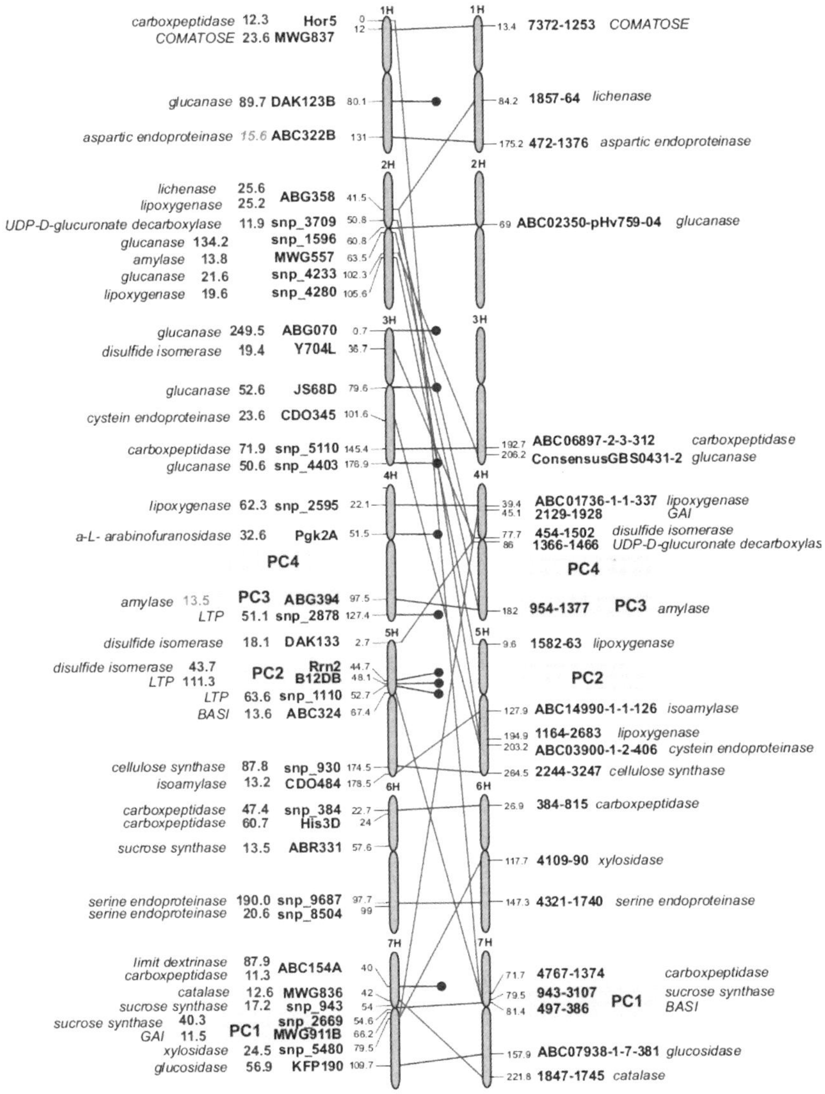

Fig. 8.5. Barley genetic maps showing alignment of eQTL loci (map on the left) and SNP-loci (map on the right) corresponding to barley genes associated with malting quality traits. Malting trait QTL locations are shown in yellow and are extracted from the whole genome scan shown in Fig.8.6 and as described in the text. For eQTLs, the LRS values are shown. They are coloured blue for cis-eQTLs and crimson for *trans*-eQTLs which is extended to the connecting lines. Blue 'lollypops' are loci that have only been mapped as eQTLs, their high LRS values indicating that they are most likely regulated in *cis*- (i.e., the positions shown are the likely location of their structural genes)

to SNP connections were low-LRS eQTLs suggesting that these genes were largely regulated in trans type. Ten high-LRS probe sets (shown as lollypops) identified unique eQTL loci that are highly likely to be cis-regulated based on their LRS values. Two loci (ABC322B (aspartic endoproteinase) and ABG394 (amylase)) had low LRS (shown in grey), but were mapped as SNPs to the same positions and were therefore regarded as cis-eQTLs. Low LRS can sometimes be the result of low overall expression in the tissues used for eQTL mapping.

We then aligned this molecular data with malting quality trait loci QTL. For this analysis, we combined data for a number of malting quality traits and analysed the dataset by Principle Component Analysis (PCA). The outcome revealed that PC1 to PC4 accounted for slightly over 40% of the total phenotypic variation (see Fig .8.6(A)). We analysed the data underlying the first four individual PC's as a QTL and plotted their position on the St/Mx genetic map alongside the malting quality gene eQTL and SNP data. We observed that three out of four PC QTLs contained malting quality associated genes (Fig. 8.5; Fig. 8.6). PC1 on chromosome 7H co-locates with a sucrose synthase cis-eQTL and its corresponding SNP. Two other malting quality genes, BASI (α-amylase/subtilisin inhibitor) and carboxypeptidase also reside within the PC1 interval. However, they exhibit eQTL that map to different chromosomal loci, neither of which appear to be associated with malting quality in this population. In contrast, GAI (Gibberelic Acid Insensitive) that maps on chromosome 4H and xylosidase on 6H have significant eQTLs that coincide with PC1. Thus, from this limited dataset, sucrose synthase and/or a locus that controls the expression of GAI and xylosidase are possible candidates for the PC1 QTL locus.

The other QTL can be analysed using a similar approach. PC2 is associated with relatively strong eQTL (LRS 43.7 and 111.3) for disulfide isomerase and LTP (Lipid Transfer Protein 1) but there are no corresponding mapped SNPs available for these genes. The only gene from our malting quality gene list underlying the PC3 QTL was β-amylase. β-amylase had a weak eQTL, but it mapped to the same locus on the long arm of chromosome 4H as its corresponding SNP. As we have mentioned previously, the low LRS observed for β-amylase may be explained by its low level of expression in tissues used for eQTL mapping. PC4 appears to map in a region where none of the genes from our list has any strong association.

8.4.2 Starch Synthesis and Degradation

The malting traits analysed primarily addressed a single biological process - starch degradation during germination. However, the starch conversion rate is determined not only by the activity of the hydrolytic starch degrading enzymes and properties of the starch granule itself. It can also be affected by grain shape, the amount of starch per cell and/or number of cells in the grain that accumulate starch. The genes involved in starch biosynthesis are

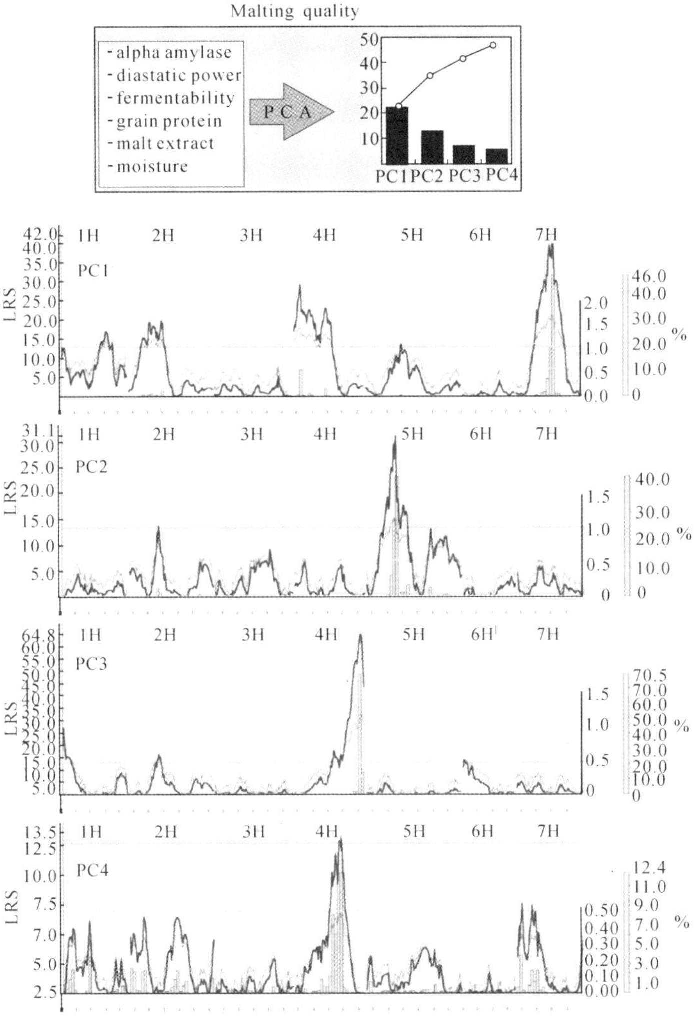

Fig. 8.6. Principal Components Analysis (PCA) of data from several malting traits including fermentability, grain protein, malt extract, moisture, diastatic power and -amylase activity. (B) Trait values were collected previously on the St/Mx population. The individual principal component (PC) scores were then used for QTL mapping. This treatment of the data identified a major QTL on chromosome 7H and several minor QTLs on chromosomes 1H, 2H, 4H and 5H. Altogether these loci account for just over 40% of overall variability, emphasising that malting quality is a highly complex trait. (B) Interval mapping of the malting quality PC1-4 scores. The blue line graph shows the LRS profile, and the green or red line graphs the allelic effects (green - Morex, red - Steptoe). Yellow bars represent the confidence intervals supported by 1,000 bootstrap tests and red and grey horizontal lines the upper and lower significance (LRS) thresholds determined by 1,000 permutation tests thus as important as those involved in starch degradation. To our knowledge there are no data available on the amount and structure of the starch in the St/Mx population. Starch biosynthetic genes are expressed during grain development, therefore our germinating tissue-based eQTL dataset may not seem adequate to address starch biosynthesis. Nevertheless, putative associations can be found between malting quality and 'starch genes' in our eQTL dataset that are supported by further independent data.

We performed correlation analysis between PC1 and our mRNA abundance dataset. This identified starch synthase IIa (*SsIIa*) as a possible candidate gene for this QTL (Druka *et al.*, 2008a). *SsIIa* was not one of the genes that we had identified on our list of malting quality genes. As a result, based on this observation we interrogated additional publicly available genomics resources in search of support for the hypothesis that *SsIIa* was a strong candidate for PC1. We examined an atlas of gene expression in barley (Druka *et al.*, 2006), which records levels of transcript abundance in a wide range of barley tissues, and observed that *SsIIa* mRNA accumulates at high level in developing endosperm. We then surveyed the location of classical grain mutants that we had recently mapped using the Illumina genotyping platform to see if any of them mapped in this region (see Fig. 8.7(A)-(C)). We found that two shrunken endosperm mutations *sex6* and *seg8* coincide with the PC1 QTL. These mutants have normal carpel development, but endosperm development is severely disrupted compared to the wild type (Fig. 8.7(A)). This produces an abnormal grain shape that is the result of a decrease in the starch content and unusual amylopectin / amylose ratio. Importantly, *sex6* has recently been shown to be a mutation in *SsIIa* (Morell *et al.*, 2003).

Given these observations it is possible to establish the following trail of inference that links *SsIIa* to malting quality: A malting quality QTL in St/Mx maps to 7H. The location of this QTL is highly correlated with an eQTL for *SsIIa* which is differentially expressed between St and Mx (which have contrasting quality), is highly expressed in the developing endosperm and is centrally involved in the starch biosynthetic pathway. Because the eQTL has a high LRS value (LRS>45) it is likely that the gene maps to the same position as its eQTL (i.e., we can infer it is cis-regulated). Two extreme starch biosynthesis mutants, *sex6*, a mutation in *SsIIa*, and *seg8* were mapped to the same location. The eQTL analysis shows that natural alleles–as observed in St and Mx—exhibit a spectrum of less severe genetic diversity. This could be elaborated as either differences in protein activity or, as we observed, patterns / levels of gene expression. It is therefore feasible that the variation in *SsIIa* expression that we observed in St/Mx could putatively lead to variation in starch content and/or composition which influences malting quality. The resulting hypothesis is that *SsIIa* underpins the observed malting quality PC1 QTL in the St/Mx population. This hypothesis needs sound experimental testing.

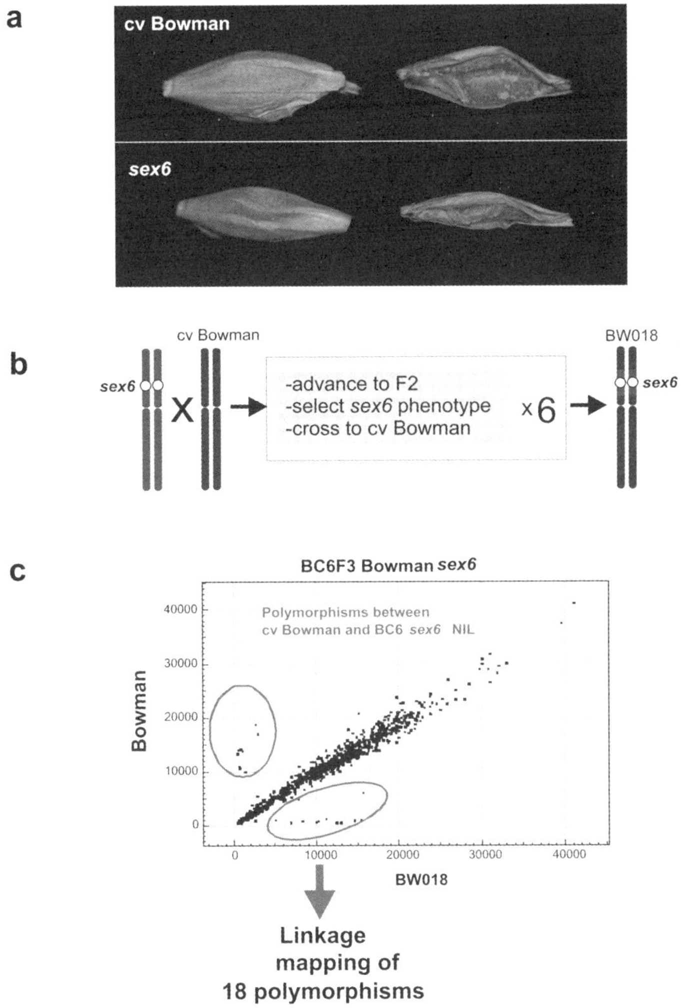

Fig. 8.7. Application of the Beadarray technology to map near isogenic lines. (A) As an example, two *shrunken endosperm* mutants were (B) backcrossed to the cv Bowman to create nearly isogenic lines (NILs). These consist of the mutant allele on a small introgressed segment in an otherwise Bowman genetic background. (C) genotyping Bowman and the Bowman *sex6* NIL (BW018) reveals polymorphisms (i.e., the spots off the diagonal) only in the introgressed segment. As the vast majority of the SNPs on the Illumina platform have already been mapped a look-up table can immediately assign sex6 to the region containing the PC1 QTL on chromosome 7H

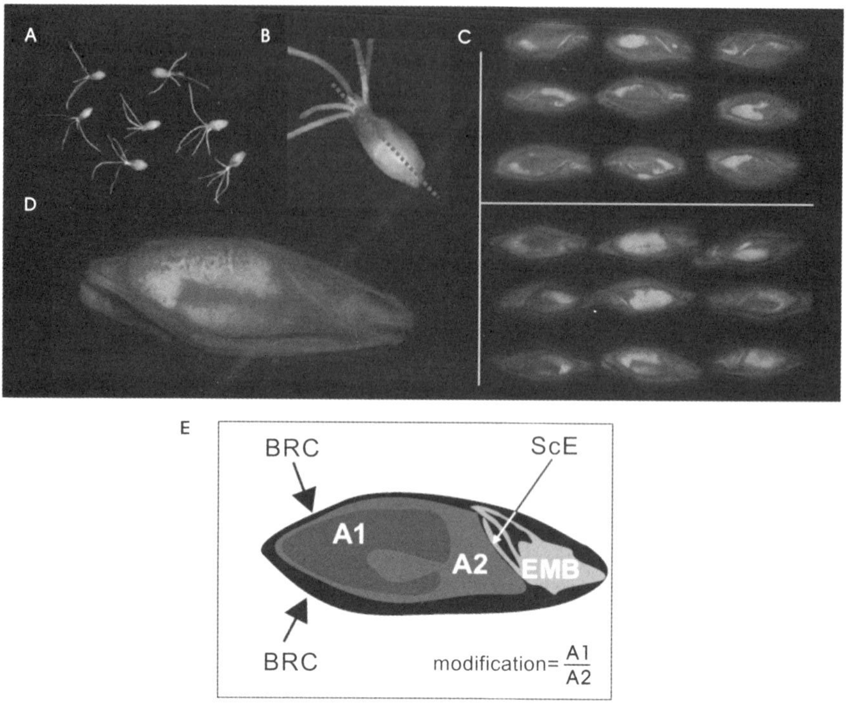

Fig. 8.8. *In situ* method to measure endosperm modification during barley germination. (A) Four days after imbibition, germinating barley grains were (B) cut in half (along the red dotted line), (C) stained with calcuflor and photographed under UV light (C, D). (E) Schematic of cut barley grain. The perimeter and areas of the calcuflour stained and non-stained parts of the endosperm were measured using ImageJ software. The ratios between stained / non stained (A1 and A2) and the intensity measurements were used for QTL mapping. ScE - Scutellum, EMB - Embryo, BRC - Bracts

8.4.3 Endosperm Cell Wall Degradation: Developmental Variation Adds Complexity

While starch synthesizing and degrading enzymes are prime targets for having a role in determining malting quality, many other genes may be equally important. However, they may or may not be linked to starch synthesis and conversion. Upon rehydration of dried and ripened barley grain, hydrolytic enzymes are released by the aleurone and scutellar epithelial cells. These enzymes modify the endosperm cell walls and allow starch degrading enzymes to access the starch granules. Products of starch degradation provide energy for germination, growth and development. Malting, the process of controlled germination and thus cell wall modification generates access to the starch and sugar substrates that support fermentation by yeast during commercial brewing and distilling. Achieving uniform and sufficient–but not excessive–cell wall modification is crucial to good malting quality.

In practice, cell wall modification is measured as the (1,3;1,4)-β-D-glucan content (or simply β-glucan content) in either germinating grains or malt extract. The (1,3;1,4)-β-D-glucans are grass-family specific non-cellulosic cell wall polysaccharides. After malting, low (1,3;1,4)- β-D-glucan content generally reflects a high level of cell wall degradation and is characteristic of good malt for brewing and distilling. β-glucan content is frequently determined in malt extract by flow injection analysis using calcofluor staining of soluble, high molecular weight β-glucan (Jorgensen, 1988). However, temporal and spatial characteristics of endosperm cell wall degradation during germination have also been extensively studied using a calcuflor based technique that was developed for measuring (1,3;1,4)-β-D-glucan content *in situ*. Calcuflor binds to hexapyranose polymers having a β-configuration ((1,3;1,4)-β-D-glucans). It fluoresces at a short wavelength, and more fluorescence translates as more intact β-glucan and less endosperm cell wall modification. Recently, this technique was applied to the St/Mx recombinant DH population (Fig. 8.8) (Druka 2008b). β-glucan content was quantified as a ratio of the calcuflor stained area to the whole endosperm area and was mapped and revealed a major QTL on chromosome 2H. Once again, using the eQTL dataset described above, correlation analysis highlighted a lichenase (1,3-1,4-β-D-Glucanase) *trans*-eQTL that coincides with the major β-glucan content locus. As a SNP in this lichenase maps to 1H (Fig. 8.5), the 2H locus may harbour a *trans*-acting factor that controls lichenase mRNA abundance and, potentially β-glucan content within endosperm.

(1,3; 1,4)-β-D-glucan content has been mapped previously in barley to two loci on chromosome 2HS; a major QTL peak coinciding with marker ABG019 and a minor QTL with ABG703 (Han *et al.*, 1995) (Fig. 8.8(A)). A member of the rice cellulose synthase-like protein family, *CslF*, located at the orthologous genomic position was recently shown to be a (1,3;1,4)- β-D-glucan synthase by analyzing gain-of-function transgenic Arabidopsis plants (Burton *et al.*, 2006). By combining β-glucan content QTL mapping data and synteny with rice, it was suggested that β-glucan content in St/Mx population is determined by *CslF* family members. Cloning and mapping of barley *CslF* gene family members did identify a cluster of three genes on chromosome 2H (Burton *et al.*, 2008), but this locus is proximal to the endosperm cell wall degradation locus that we determined by *in situ* analysis. Another glucanase gene has been mapped about 20 cM proximal to the lichenase eQTL (Fig. 8.5). It has a strong *cis*-eQTL and possibly underlies one of the previously reported β-glucan content QTLs on chromosome 2H (Han *et al.*, 1995).

8.5 Future Developments

The genotyping and mRNA profiling technologies, and experimental approaches we describe here, have been only recently adopted in barley and are still being evaluated and explored. Currently, two large barley association mapping projects, AGOUEB and BarleyCAP in the UK and USA respectively - that we have not discussed here - rely on the Illumina genotyping platform. Preliminary results from these association mapping studies are encouraging (Rostoks *et al.*, 2006) and several other parallel projects have already been initiated. In contrast, use of Affymetrix GeneChips for mRNA profiling in barley is currently restricted by their relatively high price. However, alternative platforms (e.g. Agilent arrays) have recently been developed and successfully tested.

So what comes next? The discussion above suggests that a significant number of leads could emerge by further exploration of the datasets that we have described, and a considerable amount of experimentation aimed at validating some of these putative associations is required. Furthermore, high throughput protein and metabolite profiling on mapping populations in the same way as mRNA abundance is described above, has not yet been reported in the barley literature but their application may enhance phenotypic analysis and bring further breakthroughs. Advanced high throughput phenotyping is also on the horizon and the level of detail derived from quantifying specific physiological and cellular states may further tease apart the complexities of traits such as malting quality. As highly parallel systems for data collection continue to increase, linking genetics with the developmental, physiological, morphological and environmental components of biological complexity will be a significant challenge for years to come.

To conclude, what we have tried to do in this chapter is outline how the new genomics data and approaches may enhance our understanding of traditional phenotypic traits, especially when considered alongside extant information and genetic materials. Simply, we have presented a list of associations between a collection of malting genes through their genetic location, the location of the sequences that "control" their expression, and traditionally measured malting trait QTLs. Mapping malting genes has been done before, but the eQTL analysis adds a new and powerful dimension to these investigations. We have presented a number of hypotheses, especially about candidate genes for certain traits. However we stress that the noted associations are entirely speculative at this stage and to determine how relevant they are will require further experimental testing and analysis.

Acknowledgements

We would like to thank Catherine Jamar and Pat Hayes for sharing a list of "Malt Genes", some of which we have included in this chapter. AD and RW are funded by the Scottish Government Rural and Environment Research and Analysis Directorate (RERAD).

References

Bohuon E J, Ramsay L D, Craft J A, Arthur A E, Marshall D F, Lydiate D J, Kearsey M J (1998) The Association of Flowering Time Quantitative Trait Loci with Duplicated Regions and Candidate Loci in Brassica Oleracea. Genetics, 150: 393-401

Brem R B, Yvert G, Clinton R, Kruglyak L (2002) Genetic Dissection of Transcriptional Regulation in Budding Yeast. Science, 296: 752-755

Brent R (2000) Genomic Biology. Cell, 100: 169-183

Borevitz J O, Liang D, Plouffe D, Chang H S, Zhu T, Weigel D, Berry C C, Winzeler E, Chory J (2003) Large-scale Identification of Single-feature Polymorphisms in Complex Genomes. Genome Res, 13: 513-523

Burton R A, Wilson S M, Hrmova M, Harvey A J, Shirley N J, Medhurst A, Stone B A, Newbigin E J, Bacic A, Fincher G B (2006) Cellulose Synthase-like *CslF* Genes Mediate the Synthesis of Cell Wall (1,3;1,4)-beta-D-glucans. Science, 311: 1940-1942

Burton R A, Jobling S A, Harvey A J, Shirley N J, Mather D E, Bacic A, Fincher G B (2008) The Genetics and Transcriptional Profiles of the Cellulose Synthase-Like *HvCslF* Gene Family in Barley (*Hordeum Vulgare* L). Plant Physiol, 146: 1821-1833

Chesler E J, Lu L, Shou S, Qu Y, Gu J, Wang J, Hsu H C, Mountz J D, Baldwin N E, Langston M A, Threadgill D W, Manly K F, Williams R W (2005) Complex Trait Analysis of Gene Expression Uncovers Polygenic and Pleiotropic Networks that Modulate Nervous System Function. Nat Genet, 37: 233-242

Chesler E J, Wang J, Lu L, Qu Y, Manly K F, Williams R W (2003) Genetic Correlates of Gene Expression in Recombinant Inbred Strains: a Relational Model System to Explore Neurobehavioral Phenotypes. Neuroinformatics, 1: 343-357

Damerval C, Maurice A, Josse J M, De-Vienne D (1994) Quantitative Trait Loci Underlying Gene Product Variation: a Novel Perspective for Analyzing Regulation of Genome Expression. Genetics, 137: 289-301

Druka A, Muehlbauer G, Druka I, Caldo R, Baumann U, Rostoks N, Schreiber A, Wise R, Close T, Kleinhofs A, Graner A, Schulman A, Langridge P, Sato K, Hayes P, McNicol J, Marshall D, Waugh R (2006) An Atlas of Gene Expression from Seed to Seed through Barley Development. Funct Integr Genomics, 6: 202-211

Druka A, Druka I, Centeno A G, Li H, Sun Z, Thomas W T B, Bonar N, Steffenson B J, Ullrich S E, Kleinhofs A, Wise R P, Close T J, Potokina E, Luo Z, Wagner C, Schweizer G F, Marshall D F, Kearsey M J, Williams R W, Waugh R (2008a). Towards Systems Genetic Analyses in Barley: Integration of Phenotypic, Expression and Genotype Data into GeneNetwork, Theor Appl Genet, 18: 9-73

Druka A, Potokina E, Luo Z, Bonar N, Druka I, Zhang L, Marshall D F, Steffenson B J, Close T J, Wise R P, Kleinhofs A, Williams R W, Kearsey M J, Waugh R (2008b) Exploiting Regulatory Variation to Identify Genes Underlying Loci of

Quantitative Resistance to the Wheat Stem Rust Pathogen Puccinia Graminis F. Sp. Tritici in Barley. Theor Appl Genet, 117: 261-272

Fan J B, Oliphant A, Shen R, Kermani B G, Garcia F, Gunderson K L, Hansen M, Steemers F, Butler S L, Deloukas P, Galver L, Hunt S, McBride C, Bibikova M, Rubano T, Chen J, Wickham E, Doucet D, Chang W, Campbell D, Zhang B, Kruglyak S, Bentley D, Haas J, Rigault P, Zhou L, Stuelpnagel J, Chee M S (2003) Highly Parallel SNP Genotyping. Cold Spring Harb Symp Quant Biol, 68: 69-78

Han F, Ullrich S E Chirat S, Menteur S, Jestin L, Sarrafi A, Hayes P M, Jones B L, Blake T K,Wesenberg D M, Kleinhofs A, Kilian A (1995) Mapping of Beta-glucan Content and Beta Glucanase Activity Loci in Barley Grain and Malt. Theor Appl Genet, 91: 921-927

Jansen R C and Nap J P (2001) Genetical Genomics: the Added Value from Segregation. Trends Genet, 17: 388-391.

Jorgensen K G (1988) Quantification of High Molecular Weight Glucan Using Calcofiuor Complex Formation and Flow Injection. Carlsberg Res Commun, 53: 277

Kleinhofs A, Kilian A, Saghai Maroof M A, Biyashev R M, Hayes P, Chen F Q, Lapitan N, Fenwick A, Blake T K, Kanazin V, Ananiev E, Dahleen L, Kudrna D, Bollinger J, Knapp S J, Liu B, Sorrells M, Heun M, Franckowiak J D, Hoffman D, Skadsen R, Steffenson B J (1993) A Molecular, Isozyme and Morphological Map of the Barley (*Hordeum Vulgare* L.) Genome. Theor Appl Genet, 86: 705-712

Luo Z W, Potokina E, Druka A, Wise R, Waugh R, Kearsey M (2007) SFP Genotyping from Affymetrix Arrays Is Robust but Largely Detects Cis-acting Expression Regulators. Genetics, 176: 789-800

Morell M K, Kosar-Hashemi B, Cmiel M, Samuel M S, Chandler P, Rahman S, Buleon A, Batey I L, Li Z (2003) Barley Sex6 Mutants Lack Starch Synthase IIa Activity and Contain a Starch with Novel Properties. Plant J, 34: 173-185

Potokina E, Druka A, Luo Z W, Wise R, Waugh R, Kearsey M (2008) EQTL Analysis of 16,000 Barley Genes Reveals a Complex Pattern of Genome Wide Transcriptional Regulation. Plant J, 53: 90-101

Ramsay L, Macaulay M, Degli Ivanissevich S, MacLean K, Cardle L, Fuller J, Edwards K J, Tuvesson S, Morgante M, Massari A, Maestri E, Marmiroli N, Sjakste T, Ganal M, Powell W, Waugh R (2000). A Simple Sequence Repeat-based Linkage Map of Barley. Genetics, 156: 1997-2005

Rostoks N, Borevitz J O, Hedley P E, Russell J, Mudie S, Morris J, Cardle L, Marshall D F, Waugh R (2005) Single-feature Polymorphism Discovery in the Barley Transcriptome. Genome Biol, 6, R54

Rostoks N, Ramsay L, MacKenzie K, Cardle L, Bhat P R, Roose M L, Svensson J T, Stein N, Varshney R K, Marshall D F, Graner A, Close T J, Waugh R (2006) Recent History of Artificial Outcrossing Facilitates Whole-genome Association Mapping in Elite Inbred Crop Varieties. Proc Natl Acad Sci USA, 103: 18656-18661

Schadt E E, Monks S A, Drake T A, Lusis A J, Che N, Colinayo V, Ruff T G, Milligan S B, Lamb J R, Cavet G, Linsley P S, Mao M, Stoughton R B, Friend S H (2003) Genetics of Gene Expression Surveyed in Maize, Mouse and Man. Nature, 422: 297-302

Schadt E E, Lamb J, Yang X, Zhu J, Edwards S, Guhathakurta D, Sieberts S K, Monks S, Reitman M, Zhang C, Lum P Y, Leonardson A, Thieringer R, Metzger J M, Yang L, Castle J, Zhu H, Kash S F, Drake T A, Sachs A, Lusis A J (2005) An Integrative Genomics Approach to Infer Causal Associations between Gene Expression and Disease. Nat Genet, 37: 710-717

Varshney R K, Marcel T C, Ramsay L, Russell J, Roder M S, Stein N, Waugh R, Langridge P, Niks R E, Graner A (2007) A High Density Barley Microsatellite Consensus Map with 775 SSR Loci. Theor Appl Genet, 114: 1091-1103

Vos P, Hogers R, Bleeker M, Reijans M, De Lee T V, Hornes M, Frijters A, Pot J, Peleman J, Kuiper M, Zabeau M (1995) AFLP a New Technique for DNA Fingerprinting. Nucl Acids Res, 23: 4407-4414

9

Genetic Improvement of Malting Quality through Conventional Breeding and Marker-assisted Selection

C. D. Li [1,2], M. Cakir[2] and R. Lance [1,2]
[1]Department of Agriculture & Food, Government of Western Australia, 3 Baron-Hay Court, South Perth, Western Australia 6151
[2]Western Australian Agricultural Biotechnology Centre, Murdoch University, Murdoch, WA6150
Cli@agric.wa.gov.au

A significant high value use of barley is for malting to produce malt as a raw material for the brewing of beer and fermentation and distillation for the production of whisky. Thus, malting quality is the primary objective of various barley breeding programs around the world as there is a premium for malting quality in many markets. This chapter summarizes progress on genetic improvement of malting quality through conventional breeding and marker-assisted selection.

9.1 Conventional Approaches to Breeding for Malting Quality

Conventional approaches to the improvement of malting quality have been successful where there is an understanding of the components of genetic variation for the malting quality traits sought to be improved. This fundamental understanding needs to be from the basis of the physiological processes, the biochemistry of the pathways involved and a detailed understanding of the genetic control of these processes and their interaction with the environment. Recent reviews (Fox *et al.*, 2003; Swanston and Ellis, 2002) have described in detail the state of knowledge of the processes involved in malting quality and the known genes involved. The list as such is far from being complete.

9.1.1 Definition of Malting Quality

It is important to have a fundamental understanding of the end-use for the malt in relation to the brewing processes. As an example, in relation to diastatic power (DP) (Table 9.1), a simple consideration could be given to whether the brewer uses an "all malt" process or if adjuncts are used. If adjuncts are used then it needs to be understood whether sugar or cereal adjuncts are used. Secondly, consideration has to be given to the mashing process employed by the brewer; if it is an infusion mash or a decoction mash.

Table 9.1. Relationship between brewing processes and level of diastatic power required for malt

	All malt	Sugar adjunct	Starch adjunct
Mashing Process	None	Sugar	Cereal
Infusion	Mod DP	Mod DP	High DP
Decoction	Low DP	Low-Mod DP	Mod DP

Before barley is considered for malting it must be acceptable for grain receival standards. These consist of standards for purity, freedom from contamination of "foreign matter" and assessments of basic grain quality parameters. The later often include but not exclusively: grain type (2 or 6 row), grain husk (bright, thin, adhering), thousand kernel weight, kernel assortment (< 2.2mm, 2.2–2.5 mm, 2.5–2.8 mm, >2.8 mm) or % plumpness (%> 2.5 mm), % screenings (% < 2.2mm), hectolitre weight, kernel brightness and % protein.

Malting quality traits are measured either from the dried malt or various "wet chemistry" assays and derived parameters. These may include: malt extract (fine and/or coarse grind), malt protein, soluble protein (from the ME), Kolbach index (Sol N/Malt N), free amino nitrogen (FAN), diastatic power, apparent attenuation limit, and wort viscosity. Other traits may include: alpha-amylase, beta glucanase, wort beta-glucan.

The malting quality methods of analysis have been used by the industry with the major methods being those of the Institute of Brewing, the European Brewing Convention and the American Society of Brewing Chemists. For earlier generation material, typically from breeding trials, less grain is available and so small scale methods and either destructive (flour) or non-destructive near infra-red (NIR) procedures have been developed to estimate some of the major traits. Small scale methods require smaller sample sizes and can often be processed in large batch sizes and as such there is a compromise between sample numbers and precision or reliability of the assay.

Most of these traits are phenotypically expressed and as such can be envisaged to have components of genetic control, environmental control and the interaction between the genotype and the environment as the following equation:

$$V_P = V_G + V_E + V_{GxE}$$

In order to obtain estimates for these components, adequate experimental designs need to be employed with a realistic set of genotypes (test lines and controls) in an appropriate set of environments. Typically, these are from advanced breeding and these trials are repeated over 2 or more years.

Barley Breeding Australia (BBA) is the new joint collaboration between State Governments and the Grains Research and Development Corporation. In addition to improving malting quality *per-se* BBA seeks to increase grain yield and agronomic performance, enhance resistance to multiple diseases and select for appropriate levels of tolerance to abiotic stresses.

Australian breeding programs seek to have release varieties to satisfy both domestic (sugar adjunct) and export (starch adjuncts) markets. The malting quality of new varieties is assessed by the Malting and Brewing Industry Barley Technical Committee (MBIBTC) and traits are weighted as in Table 9.2. For comparison, the European Brewing Convention Barley and Malt Committee has a similar Barley Rating System (Molina-Cano, 1987). Even though the weighting system is the same for Australian Domestic and Export malting barley there are higher points given for higher diastatic power for export barleys. Traditionally Australian barleys were low in diastatic power whereas European barleys were mostly adequate, which is why the Australian system places more emphasis on diastatic power than the EBC system.

Table 9.2. Australian and EBC malting barley rating systems

Character	MBIBTC Weighting	EBC Weight (wj)
Extract	3.5	0.45
Diastatic Power	1.5	0.05
Kohlbach Index	2.5	0.10
Viscosity	2.0	0.25
DMS Precursor	0.5	
Apparent Attenuation		0.15
Total	10.0	1.00

9.1.2 Breeding Methods for Malting Quality

Breeding methods for malting quality improvement are typical of any self pollinated crop breeding program. These may include: Pedigree, F_2 or F_3 progeny, Mutation, Backcross, Topcross, Intercross, Bulk - population, Doubled haploids (anther or microspore culture; *Hordeum bulbosum*), Single Seed Descent, and Male Sterile Facilitated Recurrent Selection (MSFRS). In the Barley Breeding Australia - West program, the major methods employed are the F_2 Progeny and Doubled Haploid methods (Table 9.3). Increasingly more complex crosses are being made involving combinations of topcrosses and intercrosses where the desired outcome is combining high yield with good agronomic performance, enhanced malting quality, multiple disease resistances and superior tolerance to abiotic stresses. Marker assisted selection (MAS)(see later sections) is often utilised in both the conventional and DH programs to "enhance the genetic potential" of the lines prior to taking them to the field.

Table 9.3. Barley Breeding Australia - West F_2 Progeny Breeding Method

Year	Year (DH)	Season	Stage	Generation	Traits Assessed
0	0	W	G/H	P1 x P2	Ag Y Q D Ab
0	0	S	G/H	F1	
1	1	W	Stage 1 SP	F_2	Ag
2		W	S1, (1-0)	F_2—3	Ag Y Q D
3		W	Stage 1-1	F_2—4	Ag Y Q D
4		W	Stage 1-2	F_2—5	Ag Y D
4		W	Stage 2 SP	F5	Ag
5	2	W	S2, (2-0)	F5—6	Ag Q D
6	3	W	Stage 2-1	F5—7	Ag Y Q D Ab
7	4	W	Stage 2-2	F5—8	Ag Y Q D Ab
8	5	W	Stage 3-1	F5—9	Ag Y Q D Ab
9	6	W	Stage 4-1	F5—10	Ag Y Q D Ab
10	7	W	Stage 4-2	F5—11	Ag Y Q D Ab
11	8	W	Release	F5—12	

Traits Assessed : Ag = Agronomic, Y = Yield, Q = Quality, D = Disease, Ab = Abiotic, W =Winter, S=Spring; F_2—5 means $F_{2|5}$ derived F5 families

The F_2 progeny method is a robust breeding method but it can be criticised for taking a long time. The introduction of a doubled haploid program based on either anther culture or microspore culture has reduced the time to release by 3 years. Testing for malting quality begins early in the breeding process for both the conventional and DH program. Whole grain near infra red (NIR) methods between harvest and planting have further improved efficiencies within the BBA programs. The combined technologies of marker assisted selection with the F_2 progeny and DH methods (see later) provides even further improvements in genetic efficiencies.

The complication of selection for malting quality is that many traits are correlated with one another (Rutger *et al.*, 1967; Arends *et al.*, 1995; Molina-Cano *et al.*, 1997). The relationship between malting quality traits is illustrated from malts derived from the Australian malting variety Schooner grown in South Australia (Fig. 9.1). The data serve to illustrate a number of points; firstly there is a typical negative relationship between % Extract and % Protein and a positive relationship between Diastase and % Protein. As a consequence there is a strong negative relationship between % Extract and Diastase. Protein content is a key "driver" of malting quality. The second point is that improvements in both % Extract and Diastase were required in new varieties to satisfy the changing requirements of the export malting markets. To a large extent this has been achieved with the newer malting varieties; Gairdner, Baudin, Hamelin, Flagship, Buloke and Vlamingh released in the past ten years in Australia.

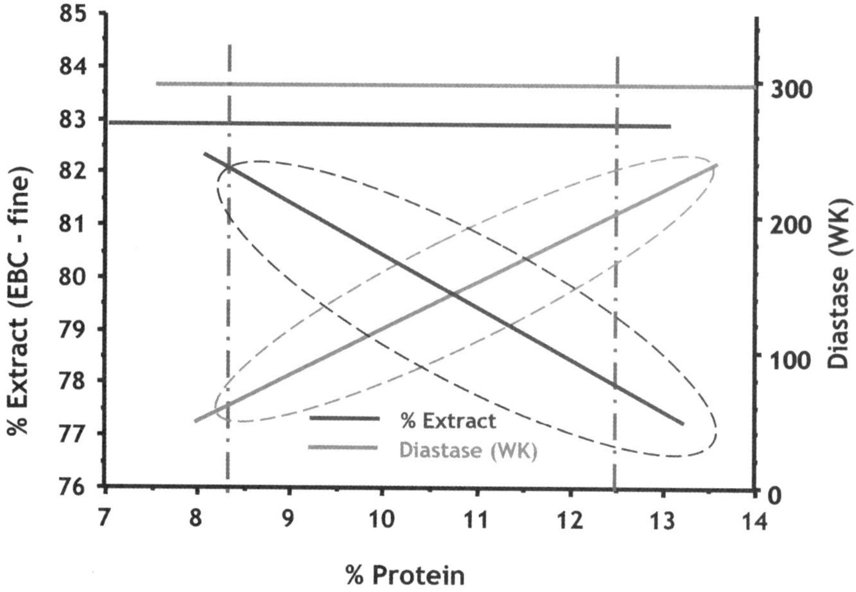

Fig. 9.1. Relationship between % Extract and % Protein content and Diastase (WK) and % Protein content of malt from Schooner barley grown in South Australia

9.1.3 Genetic Advance under Selection

The values in Table 9.4 were used to construct the values for the "parental" population derived from a cross between P1 (=80.0) and P2 (=82.0) (mid parent value = 81.0) and the selected progeny. The selection coefficient in this example is k = 0.20. The goal was to achieve the progeny with a mean of 81.5.

Several things should be noted in this example; The rate of genetic gain under selection; Gs = (81.5∼81.0) = 0.5 %. The genetic variation in the progeny (σ_A = 0.30) is reduced compared to the "parental" (σ_A = 0.60) population.

Historically other malting quality traits have been studied and in most cases heritability estimations in the broad sense (b.s.) are reasonably high,

Table 9.4. Parameters used for malt extract genetic gain under selection example

	Parent	Progeny
Mean	81.0	81.5
σ_A	0.60	0.30
σ_E	0.60	0.60
σ_P	0.85	0.67
σ_A^2	0.36	0.09
σ_E^2	0.36	0.36
σ_P^2	0.72	0.45
h^2	0.50	0.20

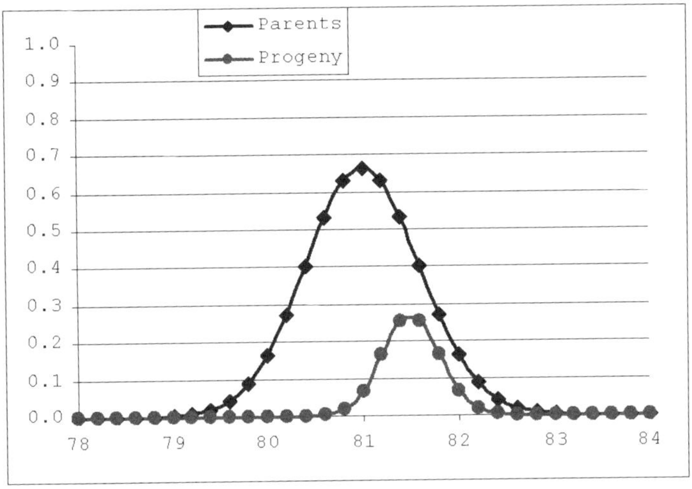

Fig. 9.2. Genetic advance under selection for malt extract

indicating that, even though these traits would be considered to be under quantitative control, perhaps only a few genes would be involved (Table 9.5).

The major difficulties arise when there is a need to select for multiple traits simultaneously. We can consider estimations for population size with the following equation derived from Mather:

$$N = population\ size = log(1-b)/log(1-a)$$
$$Where: a = fraction\ of\ desired\ genotype;$$
$$b = acceptable\ probability\ for\ outcome.$$

In most scenarios a breeder would like to achieve his or her objectives with a high degree of certainty so a probability of success of 0.99 is not unreasonable. As the number of traits increases and the heritability decreases then the fraction of the desired genotype decreases. If it was considered that a trait was controlled by a single recessive gene with an h^2 of 1.00, then for a 99% chance an F_2 population size of 17 would be required. Similarly if the trait(s)

Table 9.5. Heritability estimations for malting quality related characters

Author	Year	Quality Trait	Heritability h^2 (b.s.)
Hsi and Lambert	1954	Diastatic Power	0.84
Rutger *et al.*,	1966	Diastatic Power	0.86
"	"	alpha-Amylase	0.78
"	"	beta-Amylase	0.89
"	"	% Extract	0.69
"	"	Malt N	0.77
"	"	Wort N	0.31
Rasmusson and Glass	1967	Diastatic Power	0.56
"	"	% Extract	0.22-0.66
Foster *et al.*,	1967	Diastatic Power	0.07-0.49
"	"	Diastatic Power	0.40-0.95
"	"	% Extract	0.71
"	"	Malt N	0.80
Baker *et al.*,	1968	beta-Amylase	0.94
"	"	beta-Amylase	0.97
Greenberg	1977	beta-Glucan	0.74
Berbigier *et al.*,	1981	alpha-Amylase	0.38
"	"	beta-Amylase	0.96
"	"	beta-Glucanase	0.17
Molina-Cano *et al.*,	1991	beta-Glucan	0.79
Eagles *et al.*,	1995	% Protein	0.20
"	"	Diastatic Power	0.44
"	"	% Extract	0.44
Molina-Cano *et al.*,	1997	Beta-Glucan	0.54
"	"	% Protein	0.26
"	"	% Extract	0.70
"	"	Kolbach Index	0.39
"	"	Apparent Attenuation Limit	0.29
Lu *et al.*,	1999	% Protein	0.12-0.73
"	"	% Extract	0.11-0.56
"	"	Diastatic Power	0.03-0.65
"	"	100-Grain Weight	0.27-0.55

were controlled by 5 genes then a population of 4,714 is required. For 10 genes it is 4,828,869 (Table 9.6).

9.1.4 Selection Strategies for Malting Quality

Independent culling involves selection against material by employing specified criteria. Several independent traits can be selected independently by this method.

Tandem selection recognises that traits are more often than not related to each other. In our example above % Extract, Diastatic Power and % Protein are interrelated. Two approaches could be used; a) % Extract and Diastatic

Table 9.6. Populations sizes required in the F_2 for selection of desired homozygous gene(s) at 2 levels of probability of success

No. Genes	Fract (a)	(b) 0.9	(b) 0.99
1	0.2500	9	17
2	0.0625	36	72
3	0.0156	147	293
4	0.0039	589	1,177
5	0.0010	2,357	4,714
10	0.0000	2,414,435	4,828,869

Power can be plotted against % Protein and selection could be practiced on the residual variation after % Protein is taken into account. b) either % Extract could be plotted against Diastatic Power or % Extract and Diastatic Power could both be adjusted for % Protein and could then be plotted against each other and the combined material with both low Extract and Diastatic Power could be culled.

Selection indices are considered to be the most efficient form of selection where there are many traits, many of which would be interrelated (Baker 1986). The form of the selection is as follows :

$$I = w_1x_1 + w_2x_2 + w_3x_3 \cdots + w_ix_i$$
$$where\ w_i\ is\ the\ weighting\ for\ trait\ x_i$$

Once the selection index is calculated for all individuals, an appropriate culling level can be decided and those with a low index culled. An advantage of index selection is that an individual that may by culled because it is poor in one trait but outstanding in other traits could be kept under index selection and be available for further crossing and selection.

In practice, the BBA-West program uses the SAS Institute Inc^{TM} program JMP® Statistical Discovery Software to visualise all the relevant trait data as both histograms and in 3-D graphical form. Controls are labeled. Using the Microsoft Excel function =NORMINV(probability, mean, standard deviation) decile levels for individual traits can be defined. For the 1st decile or bottom 10%, the probability would be P = 0.1. A form of selection index is then determined empirically by applying weightings to individual traits and a "decile score" assigned to each individual for each major group of traits such as grain quality, malting quality and agronomic performance. The advantage of this selection method is that selection can be practiced on a group of traits and selection pressures adjusted until appropriate population sizes are obtained. It should be noted that for the standardised decile information a score of 8 would indicate that the individual would be in the top 20% of the population. If the same individual had a score of 8 for agronomic performance (assuming independence) then it would be in the top 4% (P = 0.2 × 0.2 = 0.04). In JMP we use the "Row State" designations to assign a combination of markers and colours to individuals within selected groups. We create a second column which reads the colour and marker combination and then assigns a number such that changes to an individual colour/marker status can be immediately captured as a "Malting or Agronomy Score"(Table 9.7).

Table 9.7. The example of selection for malting quality and agronomic traits. When a more complex selection is practiced, selection pressure can be jointly applied

		Agronomy Score								
		1	2	3	4	5	6	7	8	9
Malting Score	1	1	1.5	2	2.5	3	3.5	4	4.5	5
	2	1.5	2	2.5	3	3.5	4	4.5	5	5.5
	3	2	2.5	3	3.5	4	4.5	5	5.5	6
	4	2.5	3	3.5	4	4.5	5	5.5	6	6.5
	5	3	3.5	4	4.5	5	5.5	6	6.5	7
	6	3.5	4	4.5	5	5.5	6	6.5	7	7.5
	7	4	4.5	5	5.5	6	6.5	7	7.5	8
	8	4.5	5	5.5	6	6.5	7	7.5	8	8.5
	9	5	5.5	6	6.5	7	7.5	8	8.5	9

By summarizing the data for individuals in an Excel spread sheet, selection formula can be constructed in the compiles "IF, AND, OR" form : =IF(OR(AND(MS>6,AS>5), AND(MS>7,AS>4), AND(MS>8,AS>3), AVERAGE(MS,AS)>6.5),1,0); MS represents the Malting Score and AS the Agronomy Score. A "1" indicate selection and "0" indicated to be culled. The advantage of this selection strategy is that if there is a correlation between the malting and agronomic performance then the best of both traits are selected and can then be used in further crossing. The formula can be changed to accommodate different levels of selection intensity. More complex formulae can be constructed to include for example grain quality and / or disease scores.

In summary, selection for malting quality improvements through conventional breeding methods in general has been quite successful. Clearly, new varieties around the world have resulted in improvements in malt extract, diastatic power, viscosity, Kolbach Index, apparent attenuation limit, to name a few traits. The future challenge is to combine these with increased yield, optimal phenological development and adaptation, multiple disease resistances and tolerances to abiotic stresses. Increases in the efficiency of the breeding methods will come from the integration and implementation of the biotechnologies such as doubled haploids, marker assisted selection and NIR spectroscopy.

9.2 Marker-assisted Selection for Malting Quality

Phenotypic selection for malting quality using laboratory scale malting of grain samples is expensive and the data on malt quality from one harvest is generally not available prior to selection for sowing the next season's yield trials. As most of malting quality is controlled by multiple genes, the results from micro malting analysis varied significantly with environments. Micro malting is also not feasible for early generation selection of a single plant. While selection using NIR calibrations of whole grain for some malting quality attributes, such as extract, has been successful, molecular markers for malting quality could allow more rapid and early generation selection for malt quality. It has been a major target in the molecular marker programs around the world to use molecular markers for selection of malting quality.

Successful implementation of marker-assisted selection in a breeding program consists of nine key steps (Barr *et al.*, 2000), which include; characterizing genetics of traits, choosing parents, constructing genetic population, multiplying seeds, constructing map, phenotyping mapping population, identifying genes/QTL for target traits, validating putative markers and implementing molecular markers in breeding programs. Although the attempt to use molecular markers (protein and isoenzyme) for selection of malting quality has a long history, it only became reality after the first generation of molecular linkage maps were available (Graner *et al.*, 1991; Kleinhofs *et al.*, 1993).

9.2.1 Marker and Mapping Technologies

Map construction is the most expensive and time-consuming step and it usually costs more than 40% of the total cost for MAS (Barr *et al.*, 2000). Developments of the marker technologies in the last few years have significantly reduced the cost of map construction. Since evelopment of the first generation of RFLP barley molecular maps (Graner *et al.*, 1991; Kleinhofs *et al.*, 1993), PCR-based molecular markers have become the dominant marker type for map construction and MAS (Varshney *et al.*, 2004).

9.2.1.1 Microsatellite or SSR markers

Among different types of molecular markers available for barley, microsatellite or simple sequence repeats (SSRs) have become the mostly commonly-used markers for genetic diversity study, map construction and MAS in breeding programs, as SSR markers are abundant in genomes, multi-allelic, highly informative, co-dominantly inherited. In addition, SSR markers are easily detectable and amenable to high throughput analysis. Several thousands of barley SSR markers have been developed from different research groups including SCRI (Bmac, Bmag, EBmac, EBmag, HVGeneName, scsssr), IPK (GBM, GBMS), WUR (GBM), Virginia Polytechnic Institute (HVM), and MPI for Plant Breeding (HVGeneName). These SSR markers have been mapped in various mapping populations. Ideally, all SSR markers should be mapped in the same mapping population. However, the limited polymorphism in the current mapping populations has not allowed all possible SSR markers to be mapped onto a single genetic map. Recently, a microsatellite consensus map of barley was constructed by joining six independent genetic maps based on the mapping populations "Igri/Franka", "Steptoe/Morex", "OWBRec /OWB-Dom", "Lina /Canada Park", "L94/ Vada" and "SusPtrit/Vada". The integrated map contained 775 unique microsatellite loci from 688 primer pairs. The marker density ranged from 93 on chromosome 6H to 132 on chromosome 2H with an average of 111 markers per chromosome. The genomic DNA-derived SSR marker loci had a higher polymorphism information content value (average 0.61) as compared to the EST/gene derived SSR loci (average 0.48). The consensus map spans 1,068 cM providing an average density of one SSR marker every 1.38 cM. Such a high-density consensus SSR map provides barley molecular breeding programs with a better choice regarding the quality of markers and a higher probability of polymorphic markers in an important chromosomal interval. This map also offers the possibilities of thorough alignment for the future physical map and implementation in haplotype diversity studies in barley (Varshney *et al.*, 2007).

Numerous studies have shown that PCR multiplexing and multi-pooling strategies can significantly reduce the cost of SSR genotyping and increase genotyping throughput (Narvel *et al.*, 2000; Tang *et al.*, 2003). Technical improvements to further improve PCR assay throughput and reduce genotyping costs have the potential to increase the efficiency and speed of marker assisted breeding. Recently, Hayden *et al.* (2008) developed multiplex-ready PCR for SSR genotyping using a fluorescence-based DNA fragment analyzer. The method allows SSR genotyping to be performed using a standardized protocol and simultaneously amplifies several SSRs in the same reaction. It combines the principles of the M13-tailed primer method (Oetting *et al.*, 1995) and two-step multiplex PCR amplification (Belgrader *et al.*, 1996) to facilitate highly paralleled, fluorescence-based SSR genotyping. The method enables flexible fluorescence labeling of SSRs, generates a relatively constant amount of PCR product for each marker, and has a high amenability to multiplex PCR. A database of more than 2,800 barley and wheat SSRs, and a suite of bioinformatic tools were developed to support the deployment of multiplex-ready PCR for various genetic applications, and are accessible at http://www.genica.net.au. Multiplex-ready PCR makes it possible to implement marker assisted breeding strategies to select for multiple chromosomal regions in a single cross, and even attempt to manipulate entire genomes by simultaneously using a large number of markers to select and combine the most favorable regions of chromosomes. This method is widely used for barley map construction and marker assisted selection in Australia.

9.2.1.2 Diversity array technology (DArT)

Diversity arrays technology (DArT) is a hybridization-based molecular marker method deployed on a microarray platform. The technology was developed and commercialized by Triticarte P/L (http://www.diversityarrays.com). DArT is genome sequence-independent and allows high throughput screening of hundreds of molecular markers simultaneously. It is especially suitable for generating genome-wide markers for genetic linkage mapping and identifying markers closely linked to genes or QTLs for MAS applications. Barley was the first species for which DArT markers were made available. Approximately 2.3 million data points for 4,000 lines have been generated for barley breeders and researchers (Wenzl *et al.*, 2006). More than 2000 DArT markers have been mapped in various barley populations and integrated into a consensus map with RFLP, SSR and STS markers (Wenzl *et al.*, 2006). The consensus map comprised 2,935 loci (2,085 DArT, 850 other loci) and spanned 1,161 cM. It contained a total of 1,629 'bins' (unique loci), with an average inter-bin distance of 0.7 1.0 cM (median = 0.3 cM). More than 98% of the map could be covered with a single DArT assay. The arrangement of loci was very similar to, and almost as optimal as, the arrangement of loci in component maps built for individual populations. The locus order of a synthetic map derived from merging the component maps without considering the segregation data was only slightly inferior. The distribution of loci along chromosomes indicated centromeric suppression of recombination in all chromosomes except 5H. DArT markers appeared to have a moderate tendency toward hypomethylated, gene-rich regions in distal chromosome areas. On average, 14 9 DArT loci were identified within 5 cM on either side of SSR, RFLP or STS loci previously identified as linked to agricultural traits. This consensus map provides a framework for transferring genetic information between different marker systems and for deploying DArT markers in molecular breeding schemes. DArT can also be used for quantitative bulked segregant analysis (BSA) (Wenzl *et al.*, 2007). DArT-BSA identifies genetic loci that influence phenotypic characters in barley with at least 5 cM accuracy and should prove useful as a generic tool for high-throughput, quantitative BSA in plants, irrespective of their ploidy level. DArT provides a rapid and cheap approach to develop molecular markers for MAS in barley.

9.2.1.3 Single nucleotide polymorphisms (SNPs)

Single Nucleotide Polymorphisms (SNPs) are the most abundant form of DNA polymorphism and now predominate applications in modern genetic analysis. Over 17 000 candidate SNPs have been identified from mining 350 000 public barley expressed sequences in barley (Appleby *et al.*, 2008). However, we should be cautious about this number as the greatest challenge of *in silico* SNP discovery is the differentiation between true polymorphisms and sequence error. In another study, more than 2,000 genome-wide barley single nucleotide polymorphisms (SNPs) were developed by resequencing unigene fragments from eight diverse accessions (Rostoks *et al.*, 2005). The average genome-wide SNP frequency observed in 877 unigenes was 1 SNP per 200 bp DNA sequence. However, SNP frequency was highly variable with the least number of SNP and SNP haplotypes observed within European cultivated germplasm reflecting effects of breeding history on genetic diversity. More than 300 SNP loci were mapped genetically in three experimental mapping populations which allowed the construction of an integrated SNP map incorporating a large number of RFLP, AFLP and SSR markers (1,237 loci in total). Recently, a total of 12,615 SNPs were identified in 3509 EST-derived contigs using 36 pairwise comparisons between nine barley genotypes (Bhat *et al.*, 2007). A linear relationship was observed between the number of SNPs and the product of the number of ESTs for each pair of genotypes. An additional 800 SNPs from genomic amplicon sequences were also used. An Illumina oligonucleotide pool assay (OPA) was designed for high throughput SNP genotyping, representing 1524 different barley unigenes with one SNP each. Details of the technique were discussed in chapter 8. The SNPs were mapped in three barley mapping populations Morex x Barke, Steptoe x Morex, and Oregon Wolfe Barley. In total, 1153 SNP loci were integrated into a consensus map with 1077 cM. Two Illumina Oligo Pool Assay (OPA) based SNP genotyping platforms were developed to use for genotyping barley SNP (http://barleycap.coafes.umn.edu/). The best 3072 SNPs were selected to construct Barley OPA1 and Barley OPA2 (BOPA1 and BOPA2). The high quality of data is reflected by the fact that the alignment of shared markers included no conflict in gene order, and that markers as a whole are highly syntenic with rice. The 3072 SNPs should provide an efficient tool in plant breeding programs for marker assisted selection, pedigree analysis and association genetics.

9.2.1.4 Expressed sequence tags (EST)

We are still far from having the whole barley genome sequenced as the barley genome size is 12 times larger than rice. Expressed sequence tags (EST) obtained through cDNA sequencing provide the link to gene information in barley. There is a large collection of expressed sequence tag (EST) presently comprising more than 400,000 entries in dbEST (http://www.ncbi.nlm.nih.gov/ dbEST/dbEST_summary.html). The available EST collection represented a large proportion of all barley genes (Zhang *et al.*, 2004), which can be exploited for the construction and anchoring of high-density genetic transcript and physical linkage maps in barley. Recently, Stein *et al.* (2007) constructed a consensus barley map comprising 1,032 expressed sequence tag (EST)-based markers (total 1,055 loci: 607 RFLP, 190 SSR, and 258 SNP), and 200 anchor markers from previously published data. The map covers 1118.3 cM with individual linkage groups ranging from 130 cM (chromosome 4H) to 199 cM (chromosome 3H), yielding an average marker interval distance of 0.9 cM. This barley transcript map is a valuable resource for targeted marker saturation and identification of candidate genes at important quality loci. The 475 EST-based markers showed a syntenic organization to known colinear linkage groups of the rice genome. It provides new anchor points for detailed studies in comparative grass genomics and will support future attempts towards the integration of genetic and physical mapping information.

9.2.1.5 Array-based technology and eQTL mapping

Development of array-based technology provides a new approach for identification of genes controlling malting quality. Potokina *et al.* (2004) used a cDNA array with 1400 ESTs for identification of genes involved in malting quality. Using this procedure between 17 and 30 candidate genes were identified for each of the six malting parameters analyzed. In addition to genes of unknown function, the list of candidates contains well-known malting-related genes. Five out of eight mapped candidate genes display linkage to known QTLs for malting quality traits. From the candidate genes serine carboxypeptidase I (Cxp1) was further used for expressed QTL mapping. The expressed QTL was mapped to the same region on chromosome 3H as was the structural gene and corresponds to a QTL for Diastatic Power (pQTL). They further identified 6 SNPs at the Cxp1 locus from 90 barley cultivars. Three SNPs display complete linkage disequilibrium and define 2 haplotypes. The Cxp1 expression level in a set of barley accessions showing haplotype I was significantly higher than that of accessions displaying haplotype II (Potokina *et al.*, 2006). The eQTL approach provides an efficient way to identify candidate genes for malting quality.

9.2.1.6 Association mapping

Traditional gene/QTL mapping was based on the few genotypes that can be sampled as parents of mapping populations. Association mapping provided an opportunity for analyzing large germplasm arrays. Association mapping was initially developed in human disease genetics. The concept was recently tested in barley (Cocram *et al.*, 2008). Rostoks *et al.* (2006) demonstrated that LD can occur at great distances (up to 60 cM) in barley. They ascribe this extensive LD to the "artificial outcrossing" accomplished by plant breeders using shared genetic resources. These controlled crosses are the crucial factor that has allowed for sufficient genetic recombination to remove spurious relationships between alleles at different loci but at the same time has preserved extensive nonrecombinant regions (haplotype). A large project is under way to assay more than 4,000 barley varieties using over 3000 mapped SNPs. Approximately 40 traits will be measured on the same germplasm (Hayes and Sczus, 2006). This very large data set will be stored in the Germinate database. Functional allelic variants will be found in a subset of candidate genes. This information will provide barley breeders with markers for identifying and manipulating haplotype blocks and for marker-assisted selection based on "perfect" markers. One of such examples was demonstrated in a recent study for a gene controlling diastatic power (Potokina *et al.*, 2006).

In conclusion, a large numbers of molecular markers and marker techniques have been developed in the last 20 years. SSR, SNP and DArT markers are becoming the major molecular markers for genetic mapping and marker assisted selection. All these markers can be adapted to high throughput and multiplex marker technology, which significantly reduce the cost for mapping the quality genes and made it possible to select multiple traits in a single PCR reaction. Recently, more than 1000 SSR and DArT markers were integrated into a map from a single cross (Hearnden *et al.*, 2007). This map provides valuable resources to identify molecular markers for marker assisted selection.

9.2.2 Mapping Quantitative Loci Controlling Malting Quality

Molecular technologies and QTL analysis provide opportunities for the detailed analysis of the genetic factors that influence complex traits such as malting quality. This analysis can lead to the identification of chromosome regions with significant gene effects. The first systematic quantitative trait loci (QTL) mapping in barley was reported by Hayes *et al.* (1993), in which 62 QTLs for 8 traits were mapped. Since then, great efforts have been made to map QTLs for malting quality by three major groups around the world (North American Barley Genome Mapping Project, Australia National Barley Molecular Marker Project and European group). A large number of genes and QTL related to malting quality have been mapped and reviewed by various authors (Kleinhofs and Han 2002; Fox *et al.*, 2003; Hayes *et al.*, 2003). Here we have compiled malting quality QTL mapping data published worldwide and present a summary map in relation to linked molecular markers from a consensus map (Wenzle *et al.*, 2006; Rajeev, 2007) and barley bin map (Kleinhofs and Graner 2001) (http://barleygenomics.wsu.edu/arnis/linkage_maps /maps-svg.html).

QTL regions are distributed across 7 chromosomes (Fig. 9.3). In an earlier study 156 QTLs for 19 malting traits in 9 barley populations were identified (Zale *et al.*, 2000; Hayes *et al.*, 2003). Further to these studies Fox *et al.* (2003) have reported many additional QTL regions from a variety of populations including Australian barley mapping populations. These studies have shown that QTLs for a given trait have been detected in the same regions of the chromosomes in multiple mapping populations. The detection of two or more traits in the same regions of the chromosomes was also prevalent, indicating possible pleitropic relationships among the traits. The possibility of the presence of gene clusters could also not be ruled out. In recent studies Gao *et al.* (2004) and Han *et al.* (2004) have undertaken fine mapping studies on chromosomes 4H and 7H, respectively. Both of these studies indicated the presence of gene clusters in those chromosomal regions.

The aim of marker-assisted selection (MAS) is to establish a sustained improvement of malting quality traits in a breeding program. The fact that an individual marker may be linked to a number of traits (i.e. Bmag0501 on 1H) may be useful for some MAS applications and may also have the disadvantage of bringing an undesired trait from the same source when a given parent carries the undesired allele in that linked locus.

QTL studies have also established the presence of a strong QTLxE interaction for malting quality traits. Ponazzo *et al.* (2007) reported that, from the five test environments, only one had identified a significant QTL for DP on chromosome 2HS. Likewise, HWE data have shown that not all of the environments expressed the QTLs.

Consensus maps are useful in comparing marker and QTL information from a number of independent maps that forms the consensus map. The availability of highly dense consensus maps and a bin map has allowed the direct comparison of malting QTL locations. Most significant markers identified for each trait were used for the approximate locations of the traits in the QTL summary bin map (Fig. 9.3). It was obvious that QTL locations for a number of traits were clustered in certain regions of the chromosomes such as the centromeric regions of chromosomes 1H, 2H and 7H and telomeric region of chromosome 5H.

The QTL summary map presented here is expected to be very useful to barley breeders in their crossing decisions, since the targeting of gene clusters may help to sustain malting quality in breeding populations(Table 9.8)

Table 9.8 QTLs mapped for barley malting quality

Traits	Populations	No. of QTLs	Chromosomes	References
Malt Extract	Steptoe/Morex	3	1HS, 2HS, 7HC	Hayes *et al.*, 1993
	Blenheim/E224	5	1H, 2H, 3H, 5H, 7H	Thomas *et al.*, 1996
	Dicktoo/Morex	2	2H, 5H	Oziel *et al.*, 1996
	Blenheim/Kym	6	1H, 2H, 3H, 4H, 5H, 6H	Bezant *et al.*, 1997
	Harrington/TR306	3	1HL, 5HS, 5HL	Mather *et al.*, 1997
	Harrington/Morex	3	1HS, 1HL, 2HS	Marquez-cedillo *et al.*, 2000
	Chebec/Harrinton	2	1H, 5HL	Collins *et al.*, 2001
	Galleon/Haruna Nijo	1	2H	Collins *et al.*, 2001
	Tallon/Kaputar	1	6H	Cakir *et al.*, 2003
	Amagi Nijo/WI2585	2	2H,5H	Pallotta *et al.*, 2001
	Tankard/Livet	3	1HC, 6HC, 7HL	Rajasekaran *et al.*, 2003

(to be continued)

Traits	Populations	No. of QTLs	Chromosomes	References
	VB9524/ND11231*12	2	2H, 7HL	Emebiri *et al.*, 2004
	Brenda/HS213	1	5HL	Li *et al.*, 2005
	Arapiles/Franklin	2	1H, 7HL	Panozzo *et al.*, 2007
	Alexis/Sloop	4	1H, 2H, 3H, 5H	Panozzo *et al.*, 2007
	BCD47/Baronesse	3	3H, 5H	Castro *et al.*, 2007
	Steptoe/Morex	10	1HC, 1HL, 3HS, 3HL, 4HS, 4HL, 5HS, 5HL, 6HS, 6HL	Rae *et al.*, 2007
	Scarlett/ISR42-8	4	1HS, 1HC, 3HL, 7HS	Van Korff, *et al.*, 2008
Diastatic Power	Steptoe/Morex	4	1HS, 2HS, 4HL, 7HC	Hayes *et al.*, 1993
	Blenheim/E224	3	1H, 3H, 5H	Thomas *et al.*, 1996
	Dicktoo/Morex	3	4H, 5H, 7H	Oziel *et al.*, 1996
	Harrington/TR306	2	5HL, 6HC	Mather *et al.*, 1997
	Harrington/Morex	2	2HL, 7HC	Marquez-cedillo *et al.*, 2000
	Chebec/Harrinton	1	4H, 5HL	Collins *et al.*, 2001
	Tallon/Kaputar	1	1H, 2H, 5H	Cakir *et al.*, 2003
	Amagi Nijo/WI12585	1	4HL	Pallotta *et al.*, 2003
	VB9524/ND11231*12	4	2H, 4HS, 4HL, 5H	Emebiri *et al.*, 2004
	Arapiles/Franklin	3	1H, 2H, 6H	Panozzo *et al.*, 2007
	Alexis/Sloop	3	2H, 3HL, 5HL	Panozzo *et al.*, 2007
	BCD47/Baronesse	3	1H, 2H, 5H	Castro *et al.*, 2007
α-amylase	Steptoe/Morex	5	1HL, 2HS, 4HC, 5HC, 7HC	Hayes *et al.*, 1993

(to be continued)

Traits	Populations	No. of QTLs	Chromosomes	References
	Dicktoo/Morex	2	5H, 7H	Oziel *et al.*, 1996
	Harrington/TR306	2	5HL, 6HC	Mather *et al.*, 1997
	Harrington/Morex	2	4HS, 5HL	Marquez-cedillo *et al.*, 2000
	Tallon/Kaputar	2	2H, 7H	Cakir *et al.*, 2003
	VB9524/ND11231*12	5	2H, 3H, 4H, 5H, 6H	Emebiri *et al.*, 2004
	Arapiles/Franklin	3	1H, 2H, 6H	Panozzo *et al.*, 2007
	Alexis/Sloop	5	1H, 2H, 3H, 5HL, 6HS	Panozzo *et al.*, 2007
	BCD47/Baronesse	3	4H, 5H, 6H	Castro *et al.*, 2007
	Scarlett/ISR42-8	8	1HS, 2HL, 4HS, 4HL, 5HS, 5HL, 6H	Van Korff, *et al.*, 2008
Malt protein	BCD47/Baronesse	1	2H	Castro *et al.*, 2007
Grain protein	Steptoe/Morex	6	2HS,3H,4HL,5HC	Hayes *et al.*, 1993
	Blenheim/E224	5	1H,2H,3H,5H,6H	Thomas *et al.*, 1996
	Dicktoo/Morex	5	1HL, 4HS, 5HS, 6HS, 7H	Oziel *et al.*, 1996
	Harrington/TR306	3	4HL, 5HC, 7HC	Mather *et al.*, 1997
	Harrington/Morex	5	2HS, 2HL, 4HS, 5HC, 7HC	Marquez-cedillo *et al.*, 2000
	Tallon/Kaputar	2	2H, 5H	Cakir *et al.*, 2003
	VB9524/ND11231*12	4	4H, 5HS, 5HL, 7HS	Emebiri *et al.*, 2004
	Brenda/HS213	2	2HS, 7HL	Li *et al.*, 2005
	BCD47/Baronesse	3	2H, 4H	Castro *et al.*, 2007
	Scarlett/ISR42-8	8	1HS, 1HC, 1HL, 2HS, 4HS, 4HC, 4HL, 6HS	Van Korff, *et al.*, 2008

(to be continued)

Traits	Populations	No. of QTLs	Chromosomes	References
Viscosity	Blenheim/E224	4	2H, 3H, 4H, 5H	Thomas *et al.*, 1996
	Harrington/TR306	3	5H, 6H, 7H	Mather *et al.*, 1997
	VB9524/ND11231*12	4	2HS, 2HL, 3H, 4H	Emebiri *et al.*, 2004
	Arapiles/Franklin	1	1H	Panozzo *et al.*, 2007
	Alexis/Sloop	3	1H, 2H, 3HL	Panozzo *et al.*, 2007
	Scarlett/ISR42-8	7	1HS, 1HL, 2HS, 3HC, 3HL, 5H, 6HL	Van Korff, *et al.*, 2008
β-Glucanase	VB9524/ND11231*12	4	1H, 3H, 5H, 7H	Emebiri *et al.*, 2004
	Arapiles/Franklin	2	1H, 4H	Panozzo *et al.*, 2007
	Alexis/Sloop	1	1H	Panozzo *et al.*, 2007
	BCD47/Baronesse	3	4H, 5H, 6H	Castro *et al.*, 2007
FAAN	VB9524/ND11231*12	4	2H, 3H, 4H, 5H	Emebiri *et al.*, 2004
	Arapiles/Franklin	1	1H	Panozzo *et al.*, 2007
	Alexis/Sloop	5	1H, 2H, 3H, 4H, 5H	Panozzo *et al.*, 2007
	BCD47/Baronesse	2	5H, 6H	Castro *et al.*, 2007
Soluble protein	Blenheim/E224	2	2H, 5H	Thomas *et al.*, 1996
	Harrington/TR306	2	5H, 7H	Mather *et al.*, 1997
	Harrington/Morex	3	1HL, 4HS, 5HL	Marquez-cedillo *et al.*, 2000
Milling energy	Blenheim/E224	4	2H, 3H, 5H, 7H	Thomas *et al.*, 1996

(to be continued)

Traits	Populations	No. of QTLs	Chromosomes	References
	Steptoe/Morex	7	1HL, 3HS, 4HL, 5HS, 5HC, 5HL, 7HS	Rae *et al.*, 2007
	Derkado/B83-12/21/5	2	3HS, 7HL	Rae *et al.*, 2007
Ferment-ability	Derkado/B83-12/21/5	4	4H, 5H, 7H	Meyer *et al.*, 2001
	Scarlett/ISR42-8	3	4HS, 4HL, 6HS	Van Korff, *et al.*, 2008
Kernal plumpness	Harrington/TR306	2	4HL, 5HC, 7HC	Mather *et al.*, 1997
	Harrington/Morex	6	1HS, 1HL, 2HL, 4HS, 5HL, 7HC	Marquez-cedillo *et al.*, 2001
	Chebec/Harrinton	4	1HS, 2HS, 5HS, 5HL	Barr *et al.*, 2003b
	Blenheim/E224-3	2	1HS, 3HL	Thomas *et al.*, 1995
	Blenheim/Kym	1	6HL	Bezant *et al.*, 1997
Test weight	Blenheim/E224-3	2	3HS, 4HL, 5HL,	Thomas *et al.*, 1995
	Blenheim/Kym	8	2HS, 3HS, 3HL, 5HL, 6HS, 6HL, 7HS, 7HL	Bezant *et al.*, 1997
	Harrington/Morex	5	2HL, 4HS, 4HC, 6HS, 6HL	Marquez-cedillo *et al.*, 2001
	Alexis/Sloop	3	2HS, 3HL, 7HL	Barr *et al.*, 2003a
	Chebec/Harrinton	2	1HS, 2HC	Barr *et al.*, 2003b
	Brenda/HS213	2	2HL, 6HS	Li *et al.*, 2005
WBG	Dicktoo/Morex	1	5H	Oziel *et al.*, 1996
Malt β-glucan	Steptoe/Morex	4	1H, 3H, 4H, 7H	Hayes *et al.*, 1993
Grain β-glucan	Steptoe/Morex	2	1H, 2H	Hayes *et al.*, 1993
Starch granule traits	Steptoe/Morex	3	2H, 4H, 5H	Hayes *et al.*, 1993

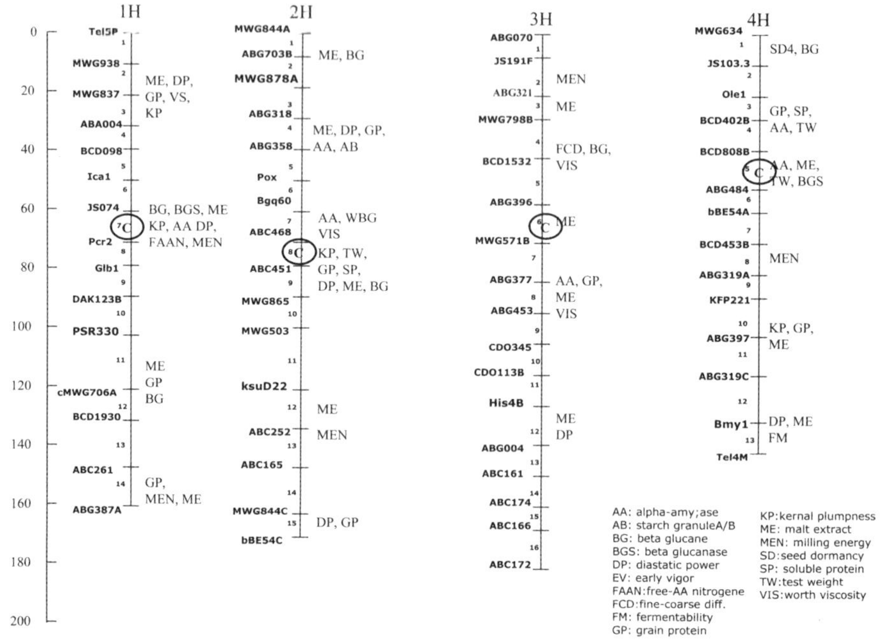

(a)

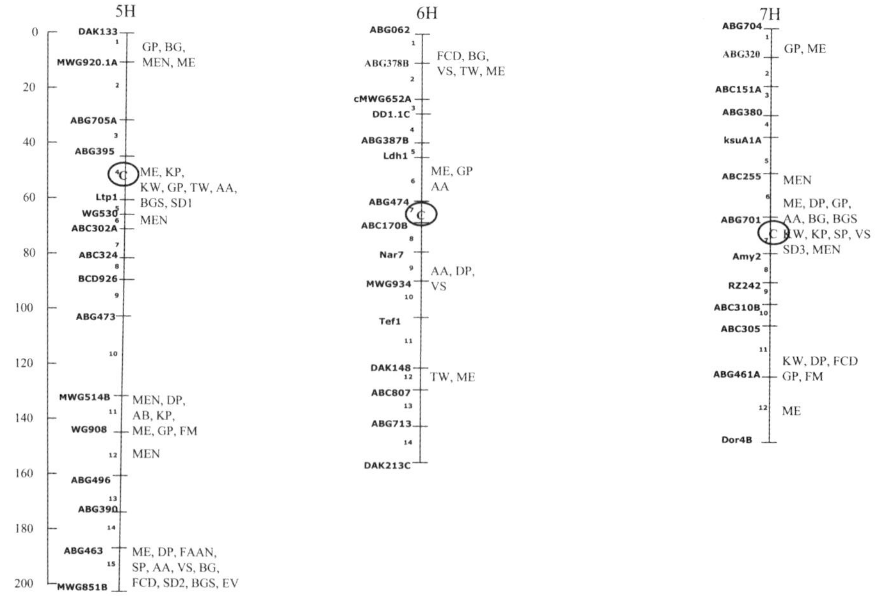

(b)

Fig. 9.3 Summary of malting QTL location in relation to the barley bin map (a) Summary of malting QTL locations in relation to the barley bin map (b) Summary of malting QTL bin locations in relation to the barley bin map, updated April 30

9.3 Marker Assisted Selection for Major Genes Controlling Malting Quality

One of the most exciting applications of MAS in backcrossing is the potential to transfer multiple traits. It is difficult to conceive and manage practical backcrossing strategies using phenotypic selection to simultaneously transfer two traits, let alone more. In contrast, the application of MAS is only limited by handling an appropriate BC×F1 population to ensure recovery of the required number of individuals heterozygous for all target loci.

One of the successful examples for improvement of malting quality using MAS is beta-amylase. Beta-amylase is the most important enzyme to determine the diastatic power and fermentability in Australian barley. Details of the biochemistry and genetics have been discussed in Chapter 4. The *Bmy1* locus on chromosome 4H controls enzyme activity, free/bound enzyme ratio and thermostability. Up to seven different isozyme types were identified among the barley germplasm. A suit of simple PCR-based molecular markers have been developed for selection of different alleles of beta-amylase for different quality requirements. The markers were developed in the multiplex-ready format, but can also be detected on agarose and non-denaturing polyacrylamide gels. This is achieved using standard methods to electrophoretically separate the multiplex-ready PCR product. The markers can be used individually for MAS when the allelic constitution of the germplasm is known. However, greater allele discrimination is achieved when the markers are used in combination to generate haplotypes (Hayden *et al.*, 2007). Barley varieties with different beta-amylase alleles were characterized in various studies (Evans *et al.*, 1997; Polakova *et al.*, 2003). Depending on different brewing requirements, beta-amylase alleles can be selectively transferred into commercial varieties using MAS. In case of the requirement of high diastatic power and high fermentability, the Sd2-H and Sd3 alleles should be selected. The four alleles could be identified unambiguously by codominantly genotyping two of the cSNPs using a duplex single nucleotide primer extension (SNuPE) assay. Two genotyping primers with their three ends directly flanking the selected SNPs were annealed to the amplified target sequences and extended by single dideoxynucleotides complementary to the polymorphic nucleotides (Paris *et al.*, 2002).

The Sd3 type beta-amylase was identified in *Hordeum spontaneum* (Eglinton *et al.*, 1998). Molecular markers and doubled haploid technique were used to facilitate transferring the Sd3 gene into Gairdner and Fitzgerald barley. Four backcrosses were completed with the recurrent parent. On average, the backcrossing lines contain 92.5% of the elite genetic background. Two DH populations (Gairdner/AB75(Sd3) and Fitzgerald/AB75) were tested using molecular markers to distinguish the DH lines with different isoenzyme types. Quality analysis showed that the Sd3 gene has significantly enhanced the diastatic power by increasing the beta-amylase enzyme activity and thermostability. There was little affect on malt extract and wort viscosity but a slight increase in wort soluble nitrogen content was observed. The malt beta-amylase activity and thermostability have been increased by an average 30% and 77% respectively by introducing the Sd3 gene (Li *et al.*, 2004).

Epiheterodendrin (EPH) is a cyanogenic glucoside produced during germination of barley. Under certain conditions, a breakdown product released during mashing and fermentation can react with ethanol in copper stills to form ethyl carbamate (a carcinogen) during whisky production (Cook *et al.*, 1990). The presence or absence of EPH can be assayed phenotypically but does pose a health and safety risk (Thomas, 2003). The industry has utilized cultivars which have inherited an allele at the *eph* locus that appears to block the formation of epiheterodendrin. The EPH gene has been mapped on barley chromosome 1H with a linkage distance of 6 cm to the SSR marker Bmac213. The non-EPH barley varieties included Aramir, Athos, Corniche, Maris Mink, Derkado, Decanter and Delibes, which may be derived from the barley variety Emir which contains both alleles of the EPH gene (Swanston *et al.*, 1999). The *eph* gene is also linked with the powdery mildew resistance gene Mla12. More close linked SSR markers have been identified, thus can be used in MAS schemes to select the non-EPH producers and powdery mildew resistance.

Formation of haze in beer is a serious quality problem. Haze active proteins isolated from beer have been found to be derived primarily from barley (*Hordeum vulgare* L.) storage proteins (hordeins), composed of fragments of several different molecular weights (MWs), and are relatively rich in proline (Asano *et al.*, 1982). Genetic analysis of the hordeins has shown that they are encoded by families of genes at a single, linked locus on chromosome 1H (Shewry, 1993). The N-terminal sequences of the hordeins contain repeats and motifs rich in glutamine and proline. During the colloidal stabilisation of beer, silica is used to remove proteins rich in proline that have the ability to interact with polyphenols to form haze in bright beer. In addition to the hordeins, a silica eluate (SE) protein fraction of MW 12,000 was identified (Evans *et al.* 2003). Interestingly, it was found that some varieties contained the MW 12,000 band (SE+ve) while in other varieties it was absent (SE-ve). Pilot brewing trials have shown that the use of SE-ve malt for brewing, improved haze stability in the resultant beer (Evans *et al.* 2003; Robinson *et al.*, 2004). A survey of 219 international barley varieties, including a comprehensive selection of current and past malting varieties, identified 181 varieties as SE+ve, and 38 varieties as SE-ve (Robinson *et al.*, 2007). The SE-ve barley varieties include Annabel, Arra, Barke, Beka, Bowman, Brenda, conquest, Copeland, defra, Dicktoo, Drummond, Excel, fatran, Guardian, Haruna Nijo, landlord, legacy, Moravian III, Olli, Onslow, Osiris, Pasadena, ploygena, Pomo, Prefect, Robust, Saana, TR306, Trebon, Unicorn, Vantmore, Vigis and Viskosa. The polymorphism of the SE protein was mapped to the short arm of chromosome 3H (Robinson *et al.*, 2007). Thus, molecular markers can be used to transfer the SE-ve gene into commercial varieties. Recently, the barley SE protein has been identified as a barley trypsin inhibitor of the chloroform/methanol type (Robinson *et al.*, 2007b). This may result in development of gene-specific diagnostic molecular markers for MAS.

Haze may also be derived from proanthocyanidins (PA) originating from barley and hops. Intensive efforts have been made to breed PA-free malting barley varieties. Yield performance and disease resistance of the PA-free barley lines have been improved obviously in recent years and PA-free varieties are now comparable with the traditional malting barley varieties on the market. Extensive malting and brewing trials on a production scale, especially with the PA-free variety Caminant, have shown that PA-free barley can be malted and brewed to meet the breweries quality specifications. The haze stability of unstabilized PA-free beers has been excellent and it is possible to mix PA-free and traditional malt in the proportions 1:1 and still obtain a beer with satisfactory haze stability without technical stabilization (Jende-Strid, 1997). The PA-free gene was mapped to chromosome 3HL (Garvin and Miller-Garvin 1998).

Foam stability is another beer quality. Despite the availability of various foam enhancement options including propylene glycol alginate (PGA) or widgets, the use of malt with enhanced levels of foam promoting proteins is desirable as this option is relatively less expensive and is increasingly being perceived by the consumer as being more natural. Key malt derived foam-positive proteins including protein Z4, Z7, LTP1 and some members of the hordein storage protein family interact principally with hop acids to stabilise foam (Evan *et al.*, 2001). These genes have been sequenced in barley, thus gene-specific molecular markers can be used for MAS.

Lipoxygenase (EC.1.13.11.12) (Lox) catalyzes hydroperoxidation of polyunsaturated fatty acids with 1,4-*cis-cis*-pentadiene structure. There are two isozymes of Lox in barley. Lox-1 and Lox-2 have been identified and well characterized (Yabuuchi, 1976; Baxter, 1982; Doderer *et al.*, 1992; Yang *et al.*, 1993). The Lox-2 activity was detected only in germinating barley, while Lox-1 was present in silenced grain. Lox-1 catalyzes the formation of 9-hydroperoxide (9-HPOD) which forms beer-deteriorating substances in further reaction in brewing processes. Sapporo's Biological Resources Development Laboratories (BRDL) has identified no Lox-1 mutants by screening 1,500 genetic resources of Okayama University. Immunological analysis showed that these lines lacked the authentic Lox-1 protein. Genetic analysis of the F_2 population revealed that this trait was governed by a single recessive gene located at the LoxA locus on chromosome 4H (Hirota *et al.*, 2005). LOX-1 deficiency is caused by a single nucleotide mutation at the splicing donor site of its fifth intron, which creates the polymorphism at the AfaI site and provides the AfaI-CAPS marker for direct selection of the Lox-null gene (Nahiko *et al.*, 2006). New barley breeding lines of lox-less were developed through marker assisted selection in three years, in contrast to the usual 10 years needed to develop the desired breed. Beer brewed using Lox-less barley showed that the occurrence of substances that cause an aged flavor due to oxidation was reduced by 75% compared to beer containing conventional barley malt. In addition, the use of Lox-less barley malt led to a significant improvement in the beer foam (http://www.sapporobeer.jp/english /kenkyu/bio/loxres.html). Barley varieties also showed different thermostability of Lox. It seems that the thermostability is controlled by a structural gene (Hirota *et al.*, 2006).

9.4 Marker-assisted Selection for QTLs Controlling Malting Quality

In contrast to a large number of QTLs mapped in the literature, only a small number of QTLs were used in the breeding programs (Thomas, 2003; Tuvesson *et al.*, 2007). Most research of the QTL mapping for malting quality was based on a cross with two extreme parents for malting quality, e.g. malting variety x feed variety. This is not consistent with the general practice for malting barley breeding. Thus, the QTLs for malting quality are required to be validated in a breeding program before they can be used for MAS. The second challenge is that the exact location and effect of individual QTLs is highly unreliable in the initial QTL mapping studies. A survey of some early QTL studies showed that the confidence limits attached to QTLs were often more than 20 cM (Kearsey and Farquhar, 1998). Such intervals often overlap with those of other important characters and frequently the orientation of desirable alleles is in repulsion. This then makes MAS less practical if selection for improvement in one key character is going to adversely affect others. Validation of QTL effects in independent samples has shown that QTL effects and the proportions of phenotypic and genotypic variance accounted for by the detected QTL is over-estimated by mapping studies (Melchinger *et al.*, 1998). Validation of results by further genotyping and phenotyping of more material from the same cross consumes many resources and the process of cross validation within the mapping data appears to yield more realistic results than the original data (Utz *et al.*, 2000). The cross validation method is recommended to identify markers linked to QTLs with the greatest effect on the target trait. Studies in maize suggest that large populations of up to 1000 are required for reliable identification of QTLs (Beavis 1998).

Collins *et al.* (2003) reported validation of 10 QTLs controlling malt extract, diastatic power, alpha-amylase and beta-amylase. These QTLs were mapped from the Canadian malting barley variety Harrington, the European variety Alexis and Japanese variety Haruna Nijo (Barr *et al.*, 2003; Karakousis *et al.*, 2003). One QTL on chromosome 2H associated with RFLP marker Xpsr108 consistently increased malt extract. This QTL is believed to be linked with the thin husk from the Japanese malting barley variety. The Canadian variety Harrington contributed two positive QTLs for malt extract on chromosomes 2H and 5H. The 2H QTL was located in the similar position of the high malt extract detected from the Japanese variety. The QTL at the terminal region of 5H is the most significant for increasing malt extract. Two QTLs from Alexis contribute to increase malt extract on chromosomes 2H and

5H, but these two QTLs are located on different chromosome positions compared to the QTLs from Harrington or Haruna Nijo. One QTL from Alexis on chromosome 1H, associated with SSR marker Ebmac501, increased alpha-amylase, beta-amylase and diastatic power. The other QTLs showed positive effects but not statistically significant. Using similar genetic materials, Coventry *et al.* (2003) reported improvement of diastatic powder and beta-amylase and alpha-amylase activity using MAS. Based on these results, a breeding strategy was designed to convert feed variety into malting types (Vasos *et al.*, 2004). In detail, a feed variety Keel was used as the recurrent parent to produce backcross lines containing the key quality genes from Alexis, Haruna Nijo and AC Metcalfe. The BC1 and BC2 generations for each introgression stream were screened using molecular markers for a range of malt quality loci. Populations derived from intercrosses between the introgression streams were used to pyramid the malt quality genes from the three international varieties into the Keel genetic background. Selected individuals were screened as $BC1F_2$ single plants to identify individuals homozygous for the target loci. Elite lines were identified and promoted for further agronomic evaluation in yield plots. The preliminary results look promising for improvement of malting quality using MAS.

Although QTLs for malting quality are distributed to all chromosomes, the QTLs for a wide range of traits are clustered in several chromosomal regions. This is the case especially for chromosomes 5H and 7H. This may be due to multi-locus clusters, pleiotrophic effects or reduced recombination in regions of genome with increased marker density (Hayes *et al.*, 1996). In term of breeding strategies, the net effect is that multiple phenotypes will be inherited as a unit. The availability of markers that define such key regions should be of great utility in classifying germplasm and designing breeding strategies. For malting quality, this has both positive and negative effects. For example, the QTLs for high levels of enzyme activity, high DP and high malting extract were inherited as one unit, which will increase the breeding efficiency for improvement of malting quality. On the other hand, the QTLs for high grain nitrogen were consistent with most QTLs for high levels of enzyme activity. Two putative adjacent overlapping QTLs were mapped to the centromeric region of chromosome 7H, each of which has effects on malt-extract percentage, alpha-amylase activity, diastatic power, and malt beta-glucan content. Fine mapping could dissect this region into two QTLs each for malt extract and for alpha-amylase, and two to three for diastatic power were identified in certain environments and resolved into 1 to 8-cM genome fragments. There was a significant QTL environment interaction for diastatic power (Han *et al.*, 1997). The telomere region of chromosome 5HL controls malt extract, malt yield, free amino acid nitrogen, activity of alpha-amylase and beta-glucanase, early vigor, seed dormancy and pre-harvest sprouting tolerance (Li *et al.*, 2004). Although the possibility of multiple gene clusters for different malting quality cannot be excluded, it is more likely that this is due to the pleiotrophic effect of the same gene(s). Comparative genetic analysis of the barley and rice maps has established that the terminal region of the barley chromosome 5HL is syntenic to rice chromosome 3L near the telomere end. The rice BAC sequences covering this region of chromosome 3L were thus used to blastn the barley EST database. Homologous barley EST sequences were used to design primers to analyze the genes in this chromosome region. Polymorphisms between different varieties of the genes in the region were identified. Six new PCR based molecular markers were developed and mapped in three doubled haploid barley populations (Stirling x Harrington, Baudin x AC Metcalf and Chebec x Harrington). According to the linkage maps, the closest marker was recognized and showed cosegregation with multiple malting quality (Zhang *et al.*, 2007). These gene specific markers provide valuable tools for characterization of barley germplasm with various quality profiles and MAS for malting quality. It will lead to ultimately isolate and identify the major gene(s) controlling multiple quality traits in barley chromosome 5HL.

Breeding for malting quality has concentrated on a restricted gene pool with high malting quality donors and avoids the use of exotic germplasm, as malting quality is often viewed as one of the most complex and sensitive traits. We reported the introduction of high thermostable beta-amylase from wild barley for improvement of diastatic power and fermentability (Li *et al.*, 2004). Recently, AB-QTL approach was used to map and introduce QTLs for the improvement of malting quality from Hordeum spontaneum (von Korff *et al.*, 2008). In total, 18 favourable exotic alleles were identified from the European malting barley variety Scarlett. The exotic allele on chromosome 4HL improved alpha-amylase activity by 16.3%, fermentability by 0.8% and reduced raw protein by 2.4%. The allele on chromosome 6HL increased alpha-amylase by 16%, fermentability by 1.3%, friability by 7.3% and reduced viscosity by 2.9%. However, just like most of the other QTLs for malting quality, these QTLs are required to be further validated in different genetic backgrounds. These results demonstrated that wild barley may be an under-utilized gene pool for the improvement of malting quality. MAS provides an effective tool to transfer these exotic alleles.

9.5 Summary

In the past decade, high density maps of SSR, DArT and SNP markers have been developed in barley. These molecular markers can be easily adapted to a high throughput molecular system. Simple and rapid DNA extraction methods as well as high-throughput MAS systems make marker assisted selection more efficient and cost-effective for practical breeding programs. MAS has evolved from selection of single gene/traits toward multiple traits/loci and whole genome. More and more functional genes related to barley malting quality have been identified, which provide functional molecular markers for MAS. QTLs for malting quality have been mapped in large numbers of mapping populations, which provided a basis for the understanding of the genetic controls of malting quality. However, most of the QTLs required further validation and fine-mapping before they can be used in the breeding program for MAS. It is essential to understand epistatic interactions for malting quality. A large numbers of barley ESTs are available for development of functional molecular markers. Significant progress has been made to construct a barley physical map and toward sequencing of the barley genome. New developments in functional genomics, expression genetics and transcriptomics providing novel tools to identify functional genes and metabolic pathways controlling malting quality (Varshney *et al.*, 2005). EcoTILLING, association mapping and advanced backcrosss QTL analysis will facilitate exploitation of natural variation for malting quality in germplasm collections. These developments will allow breeders to design a genotype *in silico* and to practice whole genome selection for improvement of malting quality (Fig. 9.4).

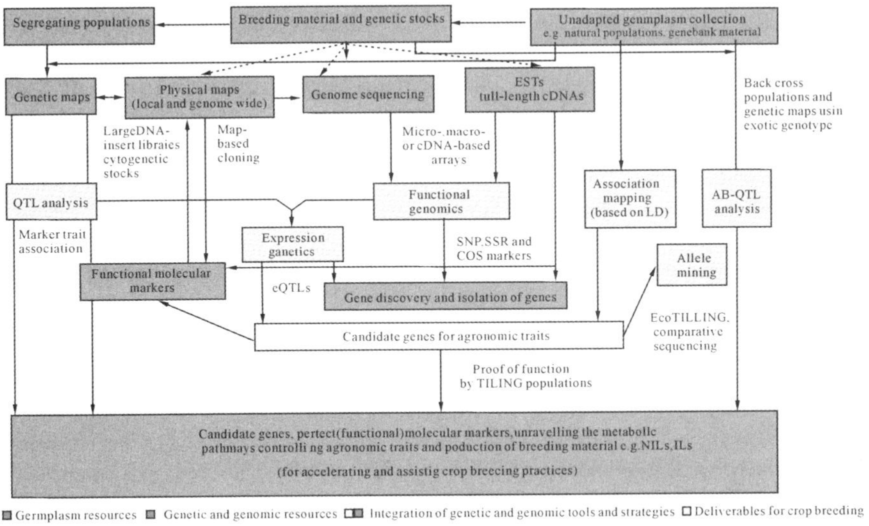

Fig. 9.4 An integrated view of exploitation of genomic resources for crop improvement via different genetic and genomic strategies (Varshney *et al.*, 2005). Abbreviations: AB-QTL, advanced backcross QTL; COS, conserved orthologous set; DHs, doubled haploids; eQTLs, expression QTLs; ESTs, expressed sequence tags; ILs, introgression lines; LD, linkage disequilibrium; NILs, near isogenic lines; QTL, quantitative trait locus; RILs, recombinant inbred lines; SNP, single nucleotide polymorphism; SSR, simple sequence repeat or microsatellite; TILLING, targeted induced local lesions in genome

References

Arends A M, Fox G P, Henry R J, Marschke R J, Symons M H (1995) Genetic and Environmental Variation in the Diastatic Power of Australian Barley. J Cereal Sci, 21: 63-70

Appleby N, Duran C, Imelfort M, Wood D L A, Edwards D, Batley J (2008) Discovery and Annotation of Barley SNP and SSR Markers with Autosnpdb. Plant & Animal Genomes XVI Conference, San Diego, CA, January: 12-16

Asano K, Shinagawa K, Hashimoto N (1982) Characterization of Haze-forming Proteins of Beer and Their Roles in Chill Haze Formation. J Am Soc Brew Chem, 40: 147-154

Baker R J (1986) Selection Indices in Plant Breeding. Florida: CRC Press Boca Raton.

Baker R J, Bendelow V M, Buchannon K W (1968) Early Generation Inheritance of Malting Quality Characters in a Barley Cross. Crop Sci, 8: 446-448

Belgrader P, Marino M M, Lubin M, Baraby F (1996) A Multiplex PCR-ligase Detection Reaction Assay for Human Identity Testing. Genome Sci Technol, 1: 77-87

Barr A R, Karakousis A, Lance R C M, Logue S J, Manning S, Chalmers K J, Kretschmer J M, Collins H M, Roumeliotis S, Coventry S, Moody D B, Read B J, Poulsen D, Li C D, Platz G J, Inkerman A, Panozzo J F, Cullis B R, Smith A B, Lim P, Langridge P, Bord W J R (2003) Mapping and QTL Analysis of the Barley Population Chebec X Harrington. Aust J Agric Res, 54: 1125-1130

Roose M, Close T J (2007) High-throughput SNP Genotyping and Mapping in Barley Provide Ample Markers for Plant Breeders. Plant & Animal Genomes XV Conference, San Diego, CA

Bos I, Caligari P (1995) Selection Methods in Plant Breeding. Chapman & Hall, London, Glasgow, Weinheim, New York, Tokyo, Melbourne, Madras, 347

Cockram J, White J, Leigh F J, Lea V J, Chiapparino E, Laurie D A, Mackay I J, Powell W, O'Sullivan D M (2008) Association Mapping of Partitioning Loci in Barley. BMC Genetics, 9: 16

Barr A R, Jefferies S P, Warner P, Moody D B, Chalmers K J, Langridge P (2001) Marker Assisted Selection in Theory and Practice. 8th International Barley Genetics Symposium: 167-178

Cook R, McCaig N, McMillan J M B, Lumsden W B (1990) Ethyl Carbamate Formation in Grain-based Spirits Part III. The Primary Source. J Inst Brew, 96: 233-244

Coventry S J, Collins H M, Barr A R, Jefferies S P, Chalmers K J, Logue S J, Langridge P (2003) Using of Putative QTLs and Structural Genes in Marker Assisted Selection for Diastatic Power in Malting Barley (*Hordeum Vulgare*). Aust J Agric Res, 54: 1241-1250

Eagles H A, Bedgood A G, Panozzo J F, Martin P J (1995) Cultivar and Environmental Effects on Malting Quality in Barley. Aust J Agric Res, 46: 831-844

Evans D E, Macleod L C, Eglinton J K, Gibson C E, Zhang X, Wallace W, Skerritt J H, Lance R C M (1997) Measurement of Beta-amylase in Malting

Barley (Hordeum Vulgare L.). I. Development of a Quantitative ELISA for Beta-amylase. J Cereal Sci, 26: 229-239

Evans D E, Robinson L H, Sheehan M C, Tolhurst R L, Hill A, Skerritt J S, Barr A R (2003) Application of Immunological Methods to Differentiate between Foam-positive and Haze Active Proteins Originating from Malt. J Am Soc Brew Chem, 61: 55-62

Foster A E, Peterson G A, Banasik O J (1967) Heritability of Factors Affecting Malting Quality of Barley. Crop Sci, 7: 611-613

Fox G, Panozzo J F, Li C D, Lance R C M, Inkerman A, Henry R J (2003) Molecular Basis of Barley Quality. Aust J Agric Res, 54: 1081-1101

Garvin D F, Miller-Garvin J E (1998) Identification of Molecular Markers Linked to Ant28-484, a Mutation that Eliminates Proanthocyanidin Production in Barley Seeds. Crop Sci, 38: 1250-1255

Graner A, Jahoor A, Schondelmaier J, Siedler H, Pillen K, Fischbeck G, Wenzel G, Herrmann R G (1991) Construction of an RFLP Map of Barley. Theor Appl Genet, 83: 250-256

Goblirsch C A, Horsley R D, Schwarz P B (1996) A Strategy to Breed Low-protein Barley with Acceptable Kernel Color and Diastatic Power. Crop Sci, 36: 41-44

Greenberg D C (1977) A Diallel Cross Analysis of Gum Content in Barley (*Hordeum Vulgare*). Theor Appl Genet, 50: 41-46

Karakousis A, Barr A R, Kretschmer J M, Manning S, Logue S J, Roumeliotis S, Collins H M, Chalmers K J, Lance R C M, Langridge P, Li C D, (2003) Mapping and QTL Analysis of the Barley Population Galleon X Haruna Nijo. Aust J Agric Res, 54: 1131-1135

Kleinhofs A, Kilian A, Saghai-Maroof M A, Biyashev R M, Hayes P, Chen F Q, Lapitan N, Fenwick A, Blake T K, Kanazin V, Ananiev E, Dahleen L, Kudrna D, Bollinger J, Knapp S J, Liu B, Sorrells M, Heun M, Franckowiak J D, Hoffman D, Skadsen R, Steffenson B J (1993) A Molecular, Isozyme and Morphological Map of Barley (*Hordeum Vulgare*) Genome. Theor Appl Genet, 86: 705-712

Han F, Ullrich S E, Kleinhofs A, Jones B L, Hayes P M, Wesenberg D M (1997) Fine Structure Mapping of the Barley Chromosome-1 Centromere Region Containing Malting-quality QTLs. Theor Appl Genet, 95: 903-910

Hayden M J, Nguyen T M, Waterman A, McMichael G L, Chalmers K J (2008) Application of Multiplex-ready PCR for Fluorescence-based SSR Genotyping in Barley and Wheat. Mol Breeding, 21: 271-281

Hayes P M, Liu B H, Knapp S J, Chen F, Jones B, Blake T K, Franckowiak G, Rasmusson D, Sorrells M, Ullrich S E, Wesenberg D, Kleinhofs A (1993) Quantitative Trait Locus Effects and Environmental Interaction in a Sample of North American Barley Germplasm. Theor Appl Genet, 87: 392-401

Hayes P M, Briceno G, Cerono J (1996) Using QTL in Barley. Proc V IOC and VII IBGS, Saskatoon: 182-187

Hayes P M, Castro A, Marquez-Cedillo L, Corey A, Henson C, Jones B L, Kling J, Mather D, Matus I, Rossi C, Sato K (2003) Genetic Diversity for Quantitatively Inherited Agronomic and Malting Quality Traits. In: Von Bothmer R, Van Hintum Th, Knupffer H, Sato K (eds). Diversity in Barley (*Hordeum Vulgare*), Elsevier Science B V, Amsterdam, The Netherlands: 201-226

Hayes P, Szucs P (2006) Disequilibrium and Association in Barley: Thinking outside the glass. Proc Natl Acad Sci USA, 103: 18385-18386

Hayter A M, Riggs T J (1978) The Inheritance of Diastatic Power and Alpha-amylase Contents in Spring Barley. Theor Appl Genet, 52: 251-256

Hearn P R, Eckermann P J, McMichael G L, Hayden M J, Eglinton J K, Chalmers K J (2007) A Genetic Map of 1000 SSR and DarT Markers in a Wide Barley Cross. Theor Appl Genet, 115: 383-391

Hirota N, Kaneko T, Kuroda H, Kaneda H, Takashio M, Ito K, Takeda K (2005) Characterization of Lipoxygenase-1 Null Mutants in Barley. Theor Appl Genet, 111: 1580-1584

Hirota N, Kaneko T, Ito K, Takeda K (2006) Mapping a Factor Controlling the Thermostability of Seed Lipoxygenase-1 in Barley. Plant Breeding, 125: 231-235

His C H, Lambert J W (1954) Inter-and Intra-annual Relationships of Some Agronomic and Malting Quality Characters of Barley. Agron J, 46: 470-474

Jende-Strid B (1997) Proanthocyanidin-free Barley - A Solution of the Beer Haze Problem. Proceedings of the European Brewery Convention Congress, Maastricht: 101-106

Kleinhofs A, Han F (2002) Molecular Mapping of the Barley Genome. In: Slafer G A, Molian-Cano J L, Savin R, Araus J L, Romagosa I (Eds.) Barley Science: Recent Advances from Molecular Biology to Agronomy of Yield and Quality. Food Products Press, New York: 31-63

Lu M Q, O'Brien L, Stuart I M (1999) Environmental and Genetic Variation for Grain Yield and Barley Malting Quality Attributes. Aust J Agric Res, 50: 1425-1434

Molina-Cano J L(1987) The EEC Barley and Malt Committee Index for the Evaluation of Malting Quality in Barley and Its Use in Breeding. Plant Breeding, 98: 249-256

Molina-Cano J L, Francesch M, Perez-Vendrell A M, Ramo T, Voltas J, Brufau J(1997) Genetic and Environmental Variation in Malting and Feed Quality of Barley. J Cereal Sci, 25: 37-47

Molina-Cano J L, Rugio A, Royo C, Ramo T, Perez-Vendrell A (1991) Environmental Effects on Mendelian Quality Genes in Barley. In: Munk L (ed.) Barley Genetics VI, Proceedings of the 6th International Barley Genetics Symposium, Helsingborg. 1: 513-515

Naohiko H, Hisao K, Kiyoshi T, Takafumi K, Hirotaka K, Ikuya Y, Masachika T, Kazutoshi I, Kazuyoshi T (2006) Development of Novel Barley with Improved Beer Foam and Flavor Stability : The Impact of Lipoxygenase-1-less Barley in the Brewing Industry . Technical Quarterly - Master Brewers Association Am, 43: 131-135

Narvel J M, Chu W C, Fehr W R, Cregan P B, Shoemaker R C (2000) Development of Multiplex Sets of Simple Sequence Repeat DNA Markers Covering the Soybean Genome. Mol Breeding, 6:175-183

Oetting W S, Lee H K, Flanders D J, Wiesner G L, Sellers T A, King R A (1995) Linkage Analysis with Multiplexed Short Tandem Repeat Polymorphisms Using Infared Fluorescence and M13 Tailed Primers. Genomics, 30: 450-458

Paris M, Jones M G K, Eglinton J K (2002) Genotyping Single Nucleotide Polymorphisms for Selection of Barley -amylase Alleles. Plant Mol Biol Rep, 20: 149-159

Polakova K, Laurie D, Vaculova K, Ovesna J (2003) Characterization of Beta-Amylase Alleles in 79 Barley Varieties with Pyrosequencing. Plant Mol Biol Rep, 21: 439-447

Potokina E, Caspers M, Wang M, Prasad M, Kota R, Zhang H, Sreenivasulu N,Graner A (2004) Functional Association between Malting Quality Trait Components and CDNA Array Based Expression Patterns in Barley (*Hordeum Vulgare* L.). Mol Breeding, 14(2): 153-170

Potokina E, Prasad M, Malysheva L, R?der M S, Graner A (2006) Expression Genetics and Haplotype Analysis Reveal Cis Regulation of Serine Carboxypeptidase I (Cxp1), a Candidate Gene for Malting Quality in Barley (*Hordeum Vulgare* L.). Funct Integr Genomics, 6: 25-35

Rasmusson D C, Glass R L (1967) Estimates of Genetic and Environmental Variability in Barley. Crop Sci, 7: 185-188

Robinson L H, Healy P, Stewart D C, Eglinton J K, Ford C M, Evans D E (2007) The Identification of a Barley Haze Active Protein that Influences Beer Haze Stability: The Genetic Basis of a Barley Malt Haze Active Protein. J Cereal Sci, 45: 335-342

Robinson L H, Evans D E, Kaukovirta-Norja A, Vilpola A, Aldred P, Home S (2004) The Interaction between Malt Protein Quality and Brewing Conditions and Their Impact on Beer Colloidal Stability. Technical Quarterly Master Brewers Association Am, 41: 353-362

Robinson L H, Juttner J, Milligan A, Lahnstein J, Eglinton J K, Evans D E (2007) The Identification of a Barley Haze Active Protein that Influences Beer Haze Stability: Cloning and Characterisation of the Barley SE Protein as a Barley Trypsin Inhibitor of the Chloroform/methanol Type. J Cereal Sci, 45: 343-352

Rostoks N, Mudie S, Cardle L, Russell J, Ramsay L, Booth A, Svensson J T, Wanamaker S I, Walia H, Rodriguez E M, Hedley P E, Liu H, Morris J, Close T J, Marshall D F, Waugh R (2005) Genome-wide SNP Discovery and Linkage Analysis in Barley Based on Genes Responsive to Abiotic Stress. Mol Genet Genomics, 274: 515-527

Rostoks N, Ramsay L, MacKenzie K, Cardle L, Bhat P R, Roose M L, Svensson J T, Stein N, Varshney R K, Marshall D F, Graner A, Close T J, Waugh R (2006) Recent History of Artificial Outcrossing Facilitates Whole-genome Association Mapping in Elite Inbred Crop Varieties. Proc Natl Acad Sci USA, 103: 18656-18661

Rutger J N, Schaller C W, Dickson A D (1967) Variation and Covariation in Agronomic and Malting Quality Characters in Barley. II. Interrelationships of Characters. Crop Sci, 7: 325-326

Rutger J N, Schaller C W, Dickson A D, Williams J C (1966) Variation and Covariation in Agronomic and Malting Quality Characters in Barley I. Heritability Estimates. Crop Sci, 6: 231-234

Shewry P R (1993) Barley Seed Proteins. In: MacGregor A W, Bhatty R S (eds.) Barley: Chemistry and Technology, American Society of Cereal Chemists, Minnesota, USA: 131-197

Swanston J S, Thomas W T B, Powell W, Young G R, Lawrence P E, Ramsay L, Waugh R (1999) Using Molecular Markers to Determine Barleys Most Suitable for Malt Whisky Distilling. Mol Breeding, 5: 103-109

Stein N, Prasad M, Scholz U, Thiel T, Zhang H, Wolf M, Kota R, Varshney R K, Perovic D, Grosse I, Graner A (2007) A 1,000-loci Transcript Map of the Barley Genome: New Anchoring Points for Integrative Grass Genomics. Theor Appl Genet, 114: 823-839

Swanston J S, Ellis R P (2002) Genetics and Breeding of Malt Quality Attributes. In: Slafer G A, Molina-Cano J L, Savin R, Araus J L, Romagosa I (eds.) Barley Science, Recent Advances from Molecular Biology to Agronomy of Yield and Quality. Food Products Press: New York: 85-114

Tang S, Kishore V K, Knapp S J (2003) PCR-multiplexes for a Genome-wide Framework of Simple Sequence Repeat Marker Loci in Cultivated Sunflower. Theor Appl Genet, 107: 6-19

Thomas W T B (2003) Prospects for Molecular Breeding of Barley. Ann Appl Biol, 142: 1-12

Tuvesson S, Dayteg C, Hagberg P, Manninen O, Tanhuanpää P, Tenhola-Roininen T, Kiviharju E, Weyen J, Forster J, Schondelmaier J, Lafferty J, Marn M, Fleck A (2007) Molecular Markers and Doubled Haploids in European Plant Breeding Programmes. Euphytica, 158: 305-312

Varshney R K, Prasad M, Graner A (2004) Molecular Marker Maps of Barley: a Resource for Intra- and Interspecific Genomics. In: Lorz H, Wenzel G (eds.) Biotechnology in Agriculture and Forestry. Molecular Markers Systems, Springer, Heidelberg: 229-243

Varshney R K, Graner A, Sorrells M E (2005) Genomics-assisted Breeding for Crop Improvement. Trends Plant Sci, 10: 621-630

Varshney R K, Marcel T C, Ramsay L, Russell J, Roder M S, Stein N, Waugh R, Langridge P, Niks R E, Graner A (2007) A High Density Barley Microsatellite Consensus Map with 775 SSR Loci. Theor Appl Genet, 114: 1091-1103

Vassos E J, Barr A R, Eglinton J K (2004) Genetic Conversion of Feed Barley Varieties to Malting Types. Proceedings of the 9th International Barley Genetics Symposium. Czech, June: 20-26

Von Korff M, Wang H, Leon J, Pillen K (2008) AB-QTL Analysis in Spring Barley: III. Identification of Exotic Alleles for the Improvement of Malting Quality in Spring Barley (H. Vulgare Ssp. Spontaneum). Mol Breeding, 21: 81-93

Wenzl P, Raman H, Wang J, Zhou M, Huttner E, Kilian A (2007) A DArT Platform for Quantitative Bulked Segregant Analysis. BMC Genomics, 8: 196

Wenzl P, Li H, Carling J, Zhou M, Raman H, Paul E, Hearnden P, Maier C, Xia L, Caig V, Ovesna J, Cakir M, Poulsen D, Wang J, Raman R, Smith K P, Muehlbauer G J, Chalmers K J, Kleinhofs A, Huttner E, Kilian A (2006) A High-density Consensus Map of Barley Linking DArT Markers to SSR, RFLP and STS Loci and Phenotypic Traits. BMC Genomics, 7: 206

Zhang H, Sreenivasulu N, Weschke W, Stein N, Rudd S, Radchuk V, Potokina E, Scholz U, Schweizer P, Zierold U, Langridge P, Varshney R K, Wobus U, Graner A (2004) Large-scale Analysis of the Barley Transcriptome Based on Expressed Sequence Tags. Plant J, 40: 276-290

Index